The Sonoran Desert Tortoise

Arizona-Sonora Desert Museum Studies in Natural History

SERIES EDITORS

Gary Paul Nabhan
Richard C. Brusca
Thomas R. Van Devender
Mark A. Dimmitt

The Sonoran Desert Tortoise
Natural History, Biology, and Conservation

Edited by Thomas R. Van Devender

The University of Arizona Press and
The Arizona–Sonora Desert Museum | Tucson

The University of Arizona Press
©2002 The Arizona-Sonora Desert Museum

Library of Congress Cataloging-in-Publication Data
The Sonoran Desert tortoise : natural history, biology, and conservation /
edited by Thomas R. Van Devender.
p. cm.
Includes bibliographical references (p.) and index.
ISBN-13: 978-0-8165-2191-3 (cloth : alk. paper)—ISBN-10: 0-8165-2191-3
ISBN-13: 978-0-8165-2606-2 (pbk. : alk. paper)—ISBN-10: 0-8165-2606-0
1. Desert tortoise—Sonoran Desert. I. Van Devender, Thomas R.
QL666.C584 S65   2002
597.92–dc21
2001007091

Manufactured in the United States of America on acid-free, archival-quality
paper containing a minimum of 50% post-consumer waste and processed
chlorine free.

11 10 09 08 07 06   7 6 5 4 3 2

Contents

Preface

THOMAS R. VAN DEVENDER

Charles H. Lowe's 1964 *The Vertebrates of Arizona* was a comprehensive account of the diverse wildlife of Arizona. The first half of the book, printed separately as *Arizona's Natural Environment*, is still the best introduction to the natural history and ecology of the southwestern United States. In those days, the wildlife of Arizona was held in an almost mysterious, spiritual reverence by young biologists around the country who had read about burrowing tree frogs, ridge-nosed rattlesnakes, gray hawks, jaguars, and other New World tropical or Sierra Madrean animals.

The desert tortoise is special among Arizona wildlife because it is one of the few species that can legally be kept as a pet. By sharing our backyards with these living rocks for decades or longer, the general public has learned more about the behavior, seasonality, diets, health, and charming individual eccentricities of desert tortoises than any other Sonoran Desert reptile. This book is a comprehensive summary of the natural history, biology, and conservation of the desert tortoise in Arizona and Sonora by the biologists and veterinarians who have studied and cared for them. In consideration of the broad interest in desert tortoises, the writing style in this book was targeted at a sophisticated lay audience who has some experience with the animals, although the information in some chapters needed to be more technical.

The old Biological Sciences Building always brings back memories of my early graduate years at the University of Arizona. In 1968 I arrived from Texas clean shaven with fashionably long red hair, a deep concern about the Vietnam War and my draft number, and an intense desire to see Arizona mountain kingsnakes, Arizona coral snakes, and other southwestern reptiles and amphibians so elegantly pictured in the Houghton Mifflin field guides by Roger Conant and Robert Stebbins. I'll never forget the excitement of first hearing a banded rock rattlesnake rattling in a talus slide, seeing a Gila monster emerge from a packrat house in early morning light, finding a Sinaloan milksnake crossing the Alamos road on a rainy night, and watching a stone on a rocky slope transform into a desert tortoise. More importantly, these animals led me into wonderful landscapes and habitats, the Huachuca Mountains, Sycamore Canyon, and many other special places.

Graduate school worked its magic on me, broadening my world view and interests and leading me in unforeseen directions. A course in vertebrate paleontology by Everett Lindsay led to earth history and fossil reptiles and amphibians. Paul Martin enchanted me with tales of Paleoindians killing mammoths and giant tortoises, glaciers covering most of Canada and the top of the Rocky Mountains, enormous pluvial lakes in the modern Great Basin Desert, and much more. He taught me that the secrets of the past are often the key to understanding the environments of today. Under Paul, I completed a doctoral dissertation using fossil plant remains preserved in ancient packrat middens to reconstruct the Ice Age environments of the Sonoran Desert in southwestern Arizona and adjacent California and, in the process, became a botanist and paleoecologist.

The idea for this book came in August 1995 at Brent Martin's thesis defense for his master's degree in wildlife and fisheries science. His thesis was a decade-long study of the ecology of the desert tortoise in grassland in the Tortolita Mountains north of Tucson. As I sat in the crowded Biological Sciences classroom to celebrate Brent's accomplishment, I realized several things. Charles Lowe and his student Cecil Schwalbe were two of the professors on the graduate committee. Dr. Lowe, as we always called him, had been a huge presence, physically and intellectually, to most of the people in the room. From him, I learned about herpetology, biogeography, evolution and speciation, environmental gradients, the role of climate in limiting species ranges, vegetation ecology, and the power of really getting to know animals and plants in the field. I realized that most of the herpetologists in the room were his students—a rich legacy of more than forty years of collecting, research, teaching, and just tramping around observing natural history.

The second realization was that an immense amount had been learned about tortoises in Arizona in the previous fifteen years. Lowe had begun marking tortoises near his home in the Tucson Mountains in the 1950s. Schwalbe had been the herpetologist with the Arizona Game and Fish Department in Phoenix, where he helped establish permanent desert tortoise study plots. He had returned to the University of Arizona to become the tortoise biologist in the School of Renewable Natural Resources. Brent, Scott

Bailey, Peter Holm, Roy Murray, and Betsy Wirt were Lowe-Schwalbe students who had spent months on plots in Arizona cajoling tortoises into divulging their life-history secrets. Roy Murray (now Averill-Murray) would soon become the Arizona Game and Fish Department's tortoise biologist. Bob McCord and I knew about the fossil record of tortoises and the environmental history of North America for the last 50 million years. Skills learned in packrat midden analyses had preadapted me to study tortoise diets by identifying the seeds, fruits, and leaves in their fecal pellets. It was clear that the combined knowledge of the people assembled in that room, together with a few of our friends and colleagues, was enough to write a comprehensive book that would help highlight the uniqueness of the Sonoran tortoise and how different it is from its much-studied Mohave Desert cousin.

A symposium on the Sonoran desert tortoise was held at the Arizona-Sonora Desert Museum in January 1997. The eleven presentations at the symposium are the core chapters of this volume. Additional chapters on the fossil record, evolutionary environments, nutritional ecology, and growth were added to make the book a more comprehensive summary of the natural history, biology, and conservation of the Sonoran tortoise. Reviews of each chapter immensely improved the book. Most chapters were internally reviewed by authors of other chapters, especially by Roy Averill-Murray, Brent Martin, Jim Jarchow, Todd Esque, and Dave Morafka. Reviews by other scholars and colleagues not only refined the manuscripts but also encouraged us to reach the highest academic standards and the broadest syntheses possible. They were Linda Allison, Laurie Averill-Murray, Carlos Castillo, Terry Johnson, Chris Klug, Earl McCoy, Steve McLaughlin, Phil Medica, Cristina Meléndez, Becky Moser, Henry Mushinsky, Ana Lilia Reina G., Amadeo Rea, Martin Whiting, and two anonymous reviewers for the University of Arizona Press.

The Arizona-Sonora Desert Museum has long been interested in desert tortoise conservation, husbandry, and natural-history interpretation. Howard Lawler began the museum's Tortoise Adoption Program in the early 1980s and collaborated with me on five studies of diet and nutrition of tortoises and chuckwallas funded by the museum's Roy Chapman Andrews Fund and the Southwest Parks and Monuments Association in the early 1990s. Howard, Cecil Schwalbe, Mary Erickson, Barbara Yates, and Carol

Cochran helped organize the symposium. The museum's support of tortoise conservation and of this book is greatly appreciated. Tom Wootten of T & E, Inc., provided financial support.

This volume is dedicated to Dr. Charles H. Lowe Jr., in appreciation of his life's work and the profound influences that he had on our lives. His contributions to the herpetology, natural history, ecology, and biogeography of the greater Southwest lie beneath the lines, enriching the book throughout.

The Sonoran Desert Tortoise

CHAPTER 1

Natural History of the Sonoran Tortoise in Arizona

Life in a Rock Pile

THOMAS R. VAN DEVENDER

The desert tortoise *(Gopherus agassizii)* is one of the most recognizable and, to many, the most endearing animals of the deserts of the southwestern United States (fig. 1.1). The desert tortoise's large geographic distribution ranges from northern Sinaloa, northward through Sonora and Arizona
to California, Nevada, and Utah (fig. 1.2). Although most of this range is in
the Sonoran Desert or more tropical communities from Arizona southward,
tortoises in the Mohave Desert of California were initially studied in much
greater detail than those in other biomes. This was largely an artifact of the
concentrations of people, universities, and biologists on the West Coast and
the charismatic leadership of some of these researchers. After the Mohave
population of the desert tortoise was listed as threatened by the U.S. Fish and
Wildlife Service in 1990, the research spotlight broadened to include desert
tortoises in the Sonoran Desert in Arizona (Johnson et al. 1990). Sonoran
desert tortoises are not federally listed but are protected statewide by Arizona law (Howland and Rorabaugh, ch. 14 of this volume) and are listed as
amenazada ("threatened") in Mexico (Secretaría de Desarrollo Social 1994).

In this chapter I describe the habitat of the desert tortoise including
vegetation and climate of the Sonoran Desert in Arizona, in part by portray-
ing the life of a model female tortoise in an ideal Arizona Upland habitat. I
discuss the probable factors limiting tortoise distribution and abundance at
lower and higher elevations in Arizona as well as the habitat transitions to
thornscrub in the south and the Mohave Desert to the northwest. Finally, I
summarize the ecological and behavioral differences between Sonoran and
Mohave tortoises.

Figure 1.1. Sonoran desert tortoise. (Photograph by Cecil Schwalbe)

Arizona Upland

It is a clear, pleasant day in March that reaches 23 °C. Unlike many years, the rains since autumn have been generous. About 200 mm have fallen since October, twice as much as the normal rainfall for this period. Wild-flowers, primarily species of Mohave Desert origin, thrive, painting the landscape blue with lupines *(Lupinus sparsiflorus)* and orange-gold with Mexican poppies *(Eschscholtzia mexicana)*, but mostly the landscape is green from a myriad of different annuals, perennial herbs, and grasses. Some of these have not been seen for years, passing the dry years as seeds or bulbs in the soil. For awhile, the foothills paloverde *(Parkinsonia [Cercidium] microphylla)* — saguaro *(Carnegiea gigantea)* desertscrub community will have a respite from the foresummer heat and aridity that will surely follow.

This vegetation is typical of the Arizona Upland, the cooler, wetter northeastern subdivision of the Sonoran Desert (fig. 1.3). In contrast to the spring annuals, most of the perennial plants have New World tropical origins with adaptations to the heat and aridity that evolved in the dry season

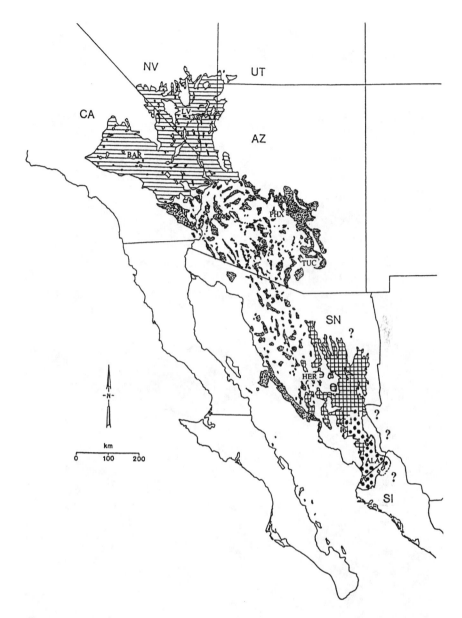

Figure 1.2. The range of the desert tortoise *(Gopherus agassizii)*. Habitats: Mohave desertscrub (horizontal lines), Sonoran desertscrub (stippling), foothills thornscrub (crosshatching), and tropical deciduous forest/coastal thornscrub (closed circles). The map is modified from Germano et al. (1994) and includes additional occupied area in Sonora and Sinaloa based on recent field observations, as well as possible range extensions (denoted by question marks) in Sonora and Chihuahua based on reports of local residents and the existence of potential habitat. ALA = Alamos; BAR = Barstow; HER = Hermosillo; LV = Las Vegas; and TUC = Tucson.

Figure 1.3. Prime Sonoran desert tortoise habitat northeast of Phoenix, Maricopa County, Arizona. The vegetation is a foothills paloverde–saguaro desertscrub in the Arizona Upland subdivision of the Sonoran Desert. Top: Four Peaks tortoise study plot, foothills of the Mazatzal Mountains. Saguaro *(Carnegiea gigantea)*, teddybear cholla *(Opuntia bigelovii)*, creosotebush *(Larrea divaricata)*, and ocotillo *(Fouquieria splendens)* are visible. (Photograph by Thomas Van Devender) Bottom: Chris Klug tracking tortoises with a radio-transmitter at Sugarloaf Mountain. (Photograph by Roy Averill-Murray)

in ancient tropical deciduous forests or thornscrub long before the Sonoran Desert existed (Van Devender, ch. 2 of this volume). The plants were pre-adapted to live in the drier and more extreme (i.e., hotter and colder) temperatures of the younger deserts.

The Arizona Upland is mostly a landscape of boulders and ridges broken in places by canyons and washes, the scars of periodic high-energy rains. The vegetation is as diverse, both in structure and species composition, as the physical landscape (Rondeau et al. 1996). Paloverdes, saguaros, and ironwoods *(Olneya tesota)* are the most visible plants, although many smaller shrubs and cacti are present. Some of them, such as brittlebush *(Encelia farinosa)*, triangleleaf bursage *(Ambrosia deltoidea)*, and California buckwheat *(Eriogonum fasciculatum)*, are often more numerous and cover more ground than the larger plants. On some steep slopes and upper *bajadas*, the saguaros are so numerous that they form forests. On the lower bajadas, several chollas and prickly pears *(Opuntia* spp.) form thickets, whereas large shrubs of mesquite *(Prosopis glandulosa, P. velutina)*, desert hackberry *(Celtis erhenbergiana [pallida])*, wolfberry *(Lycium andersonii, L. berlandieri)*, and jojoba *(Simmondsia chinensis)* are restricted to arroyos and washes.

Five Seasons of the Sonoran Desert

The unflagging heat and aridity envisioned by most nondesert dwellers are not found in most of the North American deserts—and especially not in the Sonoran Desert. In fact, the northeastern Sonoran Desert has five distinctive seasons that reflect much of the biological activity (Dimmitt 2000). The most predictable climate pattern is that of summer and winter precipitation separated by fall and spring droughts. The colder months from late November to February are familiar enough as winter. This period is about a month ahead of the astronomical winter, between the winter solstice (21 December) and the spring equinox (20 March). Winter days can be cool but are often warm and sunny. Nights are cold, although temperatures below 0°C can be expected usually fewer than ten to fifteen nights each year. At these latitudes, temperatures are often warm enough that reptiles including desert tortoises, Gila monsters *(Heloderma suspectum)*, lyre snakes *(Trimorphodon biscutatus)*, and blacktailed, diamondback, and tiger rattlesnakes *(Crotalus molossus, C. atrox,* and *C. tigris)* on rocky ridges may be active in the winter (Repp 1998). Gentle, widespread rains from Pacific frontal storms may occur one to three times a month. Once or twice a decade, El Niño circulation patterns in the eastern Pacific bring much more frequent storms and greater amounts of winter rain to the arid southwestern United States.

Spring, which lasts from late February through mid-April, is the time of wildflowers. Expected daytime temperatures rise above 20 °C and the winter rains may continue into March. The abundance and germination of spring annuals depend on the amount and timing of the winter-spring rains. Great displays of lupines and poppies are only found in years with above-normal rainfall that commences in October (Mark A. Dimmitt, pers. comm. 1999). The wildflowers usually peak in late March in Arizona Upland and earlier at lower elevations to the west. The multicolored cactus flowers of the chollas and the yellow flowers of prickly pear *(Opuntia engelmannii)* flower in April, followed by the yellow flowers of the region's most conspicuous trees, the foothills and blue *(Parkinsonia florida)* paloverdes. In tropical deciduous forest, from east-central Sonora southward, many trees flower in the spring to produce fruit and seed in time for the summer rains (Van Devender et al. 2000).

The arid foresummer, the third Arizona Upland season, begins when afternoon temperatures approach 35 °C and the last of the winter-spring rains have ended. Commencing in late April or early May and extending into July, this season is the hottest and driest time of the year. For weeks, skies are cloudless, temperatures reach 38 °C or higher, and relative humidity is in the single digits. May and June are frequently rainless. For many desert animals, it is a time for estivation, a time of reduced biological activity. In thornscrub and tropical deciduous forest in the southernmost portions of the desert tortoise's range in southern Sonora and northern Sinaloa, a single dry season begins in February and extends into July.

The summer rainy season from July through September could be called the monsoon summer, the Sonoran Desert's fourth season. The term *monsoon*, derived from an Arabic word meaning a seasonal wind, was used first to describe the winds that brought torrential summer rains to southeastern Asia. It was later applied to similar warm-season winds that bring tropical moisture and subsequent rainfall to northwestern Mexico and the southwestern United States. This season of higher humidity and rainfall begins with uncanny precision in late June in southern Sonora to early July in Arizona. In the Arizona Upland, as in tropical deciduous forests, the first summer rains initiate renewed, intense biological activity. Midday temperatures still average about 35°C, but late afternoon–early evening thunderstorms regularly wet the soil and cool the air. Plants leaf out and grow rapidly, and animals feed and breed. Amphibians, such as the red-spotted toad *(Bufo punctatus)*, Sonoran Desert toad *(B. alvarius)*, and Couch's spadefoot *(Scaphiopus couchi)*, emerge from underground retreats for the first time since the previous September to breed in temporary rain pools.

Autumn, the fifth season, begins in late September as summer rains diminish and afternoon temperatures fall below 35°C and extends until the first frosts, usually in late November. Although this season can be dry, periodic hurricanes originating off the West Coast of Mexico or early winter cyclonic storms can inundate the Sonoran Desert, sometimes leaving more rainfall in a day or two than some areas receive in most years. These storms can extend the lives of short-lived summer plants, but have questionable effect on prolonging summerlike activity or initiating fall inactivity cycles in desert reptiles.

Life in a Rock Pile

On this late March morning, a female desert tortoise emerges slowly from her burrow under a large granite boulder. Centuries before, she would have been called *komik'c-ed* ("shell with [living thing] inside") by the indigenous Tohono O'odham (Nabhan, ch. 15 of this volume), and later *tortuga de los cerros* ("hill tortoise") or "desert tortoise" by subsequent Spanish, Mexican, and Anglo-American immigrants. She has been hibernating in the shelter since late October. In hibernation her metabolic rate was very low, allowing her to pass long periods of inactivity using little energy (Peterson 1996). Curiously, this ability of tortoises to slow their metabolism to such low rates likely evolved as a survival mechanism in the dry season in tropical forests, but it serves equally well in the Sonoran Desert's temperate winters (Van Devender, ch. 2 of this volume). Now, she sits in the open, allowing the bright sunshine to heat her limbs and shell. Vessels carry the warm blood away from the surface of her body and stimulate her metabolism. Nearby, western diamondback rattlesnakes and coachwhips *(Masticophis flagellum)* bask in front of their dens.

It was not critical that she come out and feed in the spring, because she had plenty of high-energy fat stored in her body from the previous summer and fall (Jarchow et al., ch. 12 of this volume). In most years, Sonoran tortoises easily find enough food in the summer rainy season to fuel growth and store fat for the next year. Tortoises are very efficient herbivores who, like cows, have bacteria in their hindgut that allow them to digest the cellulose in plant walls (Jarchow et al., ch. 12 of this volume). Well-hydrated tortoises easily harvest the energy and nutrients in dried grasses and annuals (Nagy et al. 1998).

Throughout hibernation, tiny embryos have been developing inside the female's body using stored energy and water. Now she basks in the sun to warm up, helping the embryos develop. The eggs were fertilized with live sperm in her cloaca, stored from last summer's encounters with several eager males. Although the air temperatures are warm and the sun bright, some females and many of the males in the local population on the isolated mountain have not yet emerged from their winter quarters (Averill-Murray et al., ch. 7 of this volume). Considering that most of the males will remain in their shelters for several more weeks, or even until July (Martin 1995), sperm storage is a useful adaptation. She feeds on the protein-rich spring annuals and will not, as in drier years, need to wait for the more-dependable forage of the monsoon summer. For the remainder of April and May, she occasionally basks and moves around in a small portion of her home range, but otherwise spends days at a time sedentary in a few sheltersites. Once in April and again in May she is able to drink water from temporary rain pools.

It is now late June, the nesting season. By this time she has moved to a burrow underneath a large boulder. She excavates a hole in the soft soil near the entrance of this burrow and lays six eggs. The eggs are truly remarkable — they not only contain enough protein and water for the development of the baby tortoises over the next ninety days, but they also provide the hatchlings with energy for the next six months or more (Jarchow et al., ch. 12 of this volume). The hard shells protect the developing embryos from desiccation, effectively buffering them from the vagaries of climate. If they elude Gila monsters and other predators, the eggs will hatch in late September or early October. The baby tortoises will not interact with their mother and may not feed before hibernation. Hatchlings and juveniles have little bone in the shell and therefore provide rich sustenance for various predators. Only a few hatchlings will survive their first year.

The female, now thirty-six years old with a carapace length of 250 mm and a weight of 3 kg, has been laying eggs for the last fifteen years. Females begin to lay eggs at 220 mm carapace length and at about nineteen to twenty-one years of age (Germano et al., ch. 11 of this volume). Most of her lifetime output of eighty-five or so eggs and the subsequent hatchlings will not survive. Hopefully, at least two or three of her hatchlings will survive to replace her and a mate in the population. If young tortoises reach age 20, they can expect to live at least twenty years more (Averill-Murray et al., ch. 6 of this volume).

In Arizona, desert tortoise eggs are laid in June before the summer rains commence and hatch in late September and early October after the rainy season ends. This is in contrast to many other common Sonoran Desert reptiles, whose young appear in late August or early September in time for a month or more of active foraging before the fall inactive period begins. Many tropical trees have similar reproductive strategies, with flowers in the dry spring producing mature fruits and seeds that germinate at the beginning of the summer rains in July. For a desert animal derived from tropical ancestors, the desert tortoise's reproductive strategy is puzzling. It would appear to be an equally reasonable adaptive strategy for the desert tortoise to use the extra energy stored as fat in the summer to develop embryos and eggs in the summer and early fall, with eggs laid in early April soon after emergence from hibernation. Then the young would hatch at the beginning of the rainy season ready to partake in the flush of summer annuals and grasses. I postulate that the timing of egg laying by desert tortoise originally evolved to avoid the intense rainfall of the summer monsoon, and that the rich egg yolk in the hatchlings provided the nutrition needed to survive the spring drought.

It is now early July, and not a drop of rain has fallen since early May. With the beginning of the summer rainy season in July, the lives of female and male tortoises become more alike. With the first rains, the tortoises are able to drink and end their arid foresummer lethargy. Drinking free water is very important because it allows the tortoises to empty their bladders of concentrated urine and accumulated salts (Oftedal, ch. 9 of this volume). Water does not usually stand on the surface for long, so tortoises often emerge before the rains start, possibly in response to changes in barometric pressure or temperature. As soon as she drinks, the female tortoise begins to consume dried curly mesquite grass *(Hilaria belangeri)*. In ten days to two weeks, the newly germinated summer annuals and grasses will grow enough that she could shift to fresh plants. She is especially fond of spurges *(Euphorbia* spp.), several species of which thrive in the open areas between the boulders. Well-hydrated tortoises can eat fresh annuals without accumulating excessive potassium, which requires a lot of water to excrete, and can more readily extract the rich protein resources tied up in plant cellulose.

The monsoon summer in August and September is the most active time of the year for the Sonoran tortoise. After spending most of early and middle summer near her nesting burrow, she wanders through most of her home range, feeds frequently, and uses many other shelters. Some of these she shares with male tortoises, diamondback rattlesnakes, and even the potentially egg-predatory Gila monster (Averill-Murray et al., ch. 7 of this volume). Males may wander more extensively at this time. Over the summer, four amorous males come to visit. On the hottest nights, she leaves her stifling burrow for a cooler, exposed pallet under a shrub.

The summer rains end in September, and as midday temperatures noticeably cool in early October, the female tortoise switches her activity time from early morning and late afternoon to midday. Her body is ready for the following year—rehydrated with excess water stored in her bladder as well as sperm in her cloaca. She has fed well on dried and fresh grasses, especially curly mesquite and fluffgrass *(Erioneuron pulchellum)*. She also has eaten globe mallow *(Sphaeralcea ambigua)*, desert vine *(Janusia gracilis)*, and summer herbs, including spiderlings *(Boerhavia spp.)* and spurges. She has devoured the sweet fruits of the prickly pear, a summer dessert that stained her face purple (Van Devender et al., fig. 8.1 of this volume). The rich foods have been more than adequate to build up her energy-rich fat reserves.

Her daily movements bring her closer to the shelter that she used last winter. Finally, in late October, she enters her hibernaculum. Hibernation begins for some tortoises as early as late September, when afternoon temperatures are still 30°C or higher. Thus, both the inactive periods of the fall (hibernation) and late spring (estivation) reflect sedentary behavior during warm dry weather by some, but not all, tortoises.

During the summer, she had used many shelters, in part to buffer the extremes in temperature and humidity in her physical environment (Averill-Murray et al., ch. 7 of this volume). Some were relatively deep soil burrows under shrubs, others simply shallow, bowl-shaped pallets. Sometimes she used a packrat *(Neotoma albigula)* house under a cholla cactus or shrub, yet another intruder into the packrat's home. Curiously, nature appears to call a truce in winter, with tortoises, rattlesnakes, Gila monsters, and packrats sharing the same dens (Repp 1998). Although this October she returns to the same burrow under the boulder she used last winter, she has used several other hibernacula during her lifetime. This burrow is not as deep as some of the ones used by males, who usually hibernate for longer periods in shelters extending deeper underground (Averill-Murray et al., ch. 7 of this volume). Here she will pass her thirty-seventh winter, dormant and largely sedentary. If unusually wet conditions occur, she and others may occasionally emerge on sunny days in midwinter to bask and escape uncomfortably soggy hibernacula. So completes another year in the life of our female tortoise.

Our ideal Arizona Upland female has spent her entire life within about 1 km² of the nest where she hatched. Everything that she needs for shelter, food, and mates can be found in a relatively small area in the complex Arizona Upland habitat. Her population lives on an isolated granite mountain, although other tortoise populations are found in mountains formed of volcanic (andesite, rhyolite, basalt), metamorphic (gneiss), and sedimentary (lake deposits, limestone; Averill-Murray et al., ch. 7 of this volume) rocks. The population occupies the entire mountain above the bajadas. For most of its 40-km² area, there are 30 to 100 tortoises per square kilometer, for a total of some 2600 tortoises in the population. However, in her home range of 10 hectares (100,000 m², 0.1 km²), she only encounters two females and three males of reproductive or near-reproductive age. A couple of males with larger home ranges occasionally wander by. There are also likely another four or five subadults living in the area. As the eighteen or so hatchlings emerge each year, the number of tortoises in her area briefly triples. The five males who could potentially father her offspring at any time represent less than 1 percent of the seven hundred or so breeding males in the entire population. Considering the natural turnover in a dynamic population, her genes were likely mixed with other males during her twenty-eight reproductive years.

This discussion of the natural history of a model tortoise serves well to illustrate the interactions of individuals on a home-range scale and their contribution to the population on the scale of an isolated mountain. It was based on the average home ranges, densities, sex ratios, and demographic structures for Sonoran tortoise plots in Arizona (Averill-Murray et al., chs. 6 and 7 of this volume). Summarizing the natural history of the Sonoran desert tortoise, a long-lived animal, presents difficulties because important aspects are poorly known or not known at all, especially in the following areas. The distances that hatchling Sonoran tortoises disperse from the nests are not known, but are likely much less in the complex Arizona Upland habitats than in the Mohave Desert flatlands. Although adult tortoises have on occasion been recorded making exceptionally long-distance movements (Scott J. Bailey, pers. comm. 1999), the importance of long-distance dispersal to isolated populations is unknown. Because few tortoise plots have been studied more than a season or two, the lifetime movements of individual tortoises are not known. If home ranges expand or contract — tracking decadal climatic fluctuations, health condition, or even age — the number of other tortoises encountered, including potential mates, would vary. Tortoises have been studied in square-mile or square-kilometer plots but never on the scale of a mountain population. If the number of tortoises were variable in ideal-looking habitat, extrapolation of the numbers of tortoises from plot densities to larger areas would be exaggerated.

Most Sonoran desert tortoise populations live on rocky slopes and adjacent upper bajadas on isolated hills, where the most shelters and food resources are. On the lower bajadas and along major washes below the rock outcrops, there are fewer tortoises. Here, tortoises use packrat houses, soil burrows dug by other animals or themselves, or small caves in caliche exposed in arroyo embankments as shelters. *Caliche* is the term for the hard calcium carbonate layers that are common in desert soils in Arizona Upland. Caliche was formed in the wetter climates of the Pleistocene and is mostly eroding away today. Cryptically concealed tortoises in such low-density populations are not easily located by desert tortoise biologists.

Habitats and Dispersal

For the most part, Sonoran tortoises are absent from, or present in low densities, in habitats in the fine-grained silt and clay soils of the valleys adjacent to their rocky citadels. Lowe (1990) interpreted this as a genetically fixed behavior that evolved from the high mortality costs of eggs and young during the periodic (annual or more often) sheet flooding during intense rains on the floodplains of the great tropical rivers (Ríos Yaqui, Mayo, and Fuerte) in southern Sonora, while the upland, ridge-based populations thrived. In Arizona, the summer rains are normally not as intense as in more tropical areas, although periodic late summer–fall *chubascos* (the Sonoran term for tropical storms) can flood the valleys. In the Arizona Upland region, winter storms are occasionally intense enough to flood the valley bottoms. The Mohave tortoise trait of living in the bottoms of broad desert valleys is apparently a newly derived character only possible because of the absence of summer rainfall (Lowe 1990; Van Devender, ch. 2 of this volume). Only the enhanced protection of the extensive burrows of the Mohave tortoises allows them to survive the physiological stresses of extended summer drought and the inability to rehydrate and feed during the warm season (Peterson 1996). Questions arise about the valley flood hypothesis in the Mohave Desert in southern Nevada, which periodically receives heavy summer rains (Mitchell 1976), and in the more arid portions of southwestern Arizona, southeastern California, and adjacent Baja California and Sonora, where summer floods are less frequent. In these hyperarid areas, the distribution and abundance of tortoises may be controlled by regular droughts lasting eighteen to thirty-six months and very high ambient temperatures.

Most of the desert tortoise populations on isolated desert mountains in Arizona do not appear to be in decline or show unusual genetic variability (Averill-Murray et al., ch. 6 of this volume). This raises questions about the importance of cross-valley dispersal for the genetics and conservation of isolated montane animal populations (Averill-Murray et al., ch. 6 of this volume; Howland and Rorabaugh, ch. 14 of this volume). Are enough individuals moving between isolated populations to prevent inbreeding in the populations? Are they maintaining genetic diversity in these relatively small populations? In many species, including many desert snakes, the individuals most likely to leave the population are young males (Phillip Rosen, pers. comm. 1999). Male tortoises appear to have larger home ranges than females (Averill-Murray et al., ch. 7 of this volume). Dispersing males can carry new genetic material to a neighboring population, but cannot alone establish new populations on outlying hills.

An interesting area to consider is north of Tucson between the north-northeast-oriented Santa Catalina, Black, and Tortilla Mountains and the Picacho Mountains to the west (essentially a northern extension of Avra Valley). Tortoise populations have been studied in the Desert Peak, Granite Hills, Picacho Mountains, and Tortolita Mountains (Barrett 1990; Martin 1995; Averill-Murray et al., ch. 6 of this volume). In the Picacho Mountains on the western side of the valley, Barrett (1990) found tortoises in the paloverde-saguaro community but not in the lower creosotebush *(Larrea divaricata)*-triangleleaf bursage desertscrub. On the eastern side of the valley, Arizona Upland merges into desert grassland at about 1,000-m elevation. Much of the upper bajada at about 790- to 915-m elevation are relatively rockless areas dominated by foothills paloverde–saguaro communities.

Throughout this area tortoises may be present at low densities and have been overlooked by surveyors, being cryptically concealed in soil burrows, vegetation, or packrat houses. Urbanization and road paving are tangible threats to tortoises on the higher Arizona Upland bajadas (Howland and Rorabaugh, ch. 14 of this volume). Tortoise populations in Arizona Upland are especially vulnerable to the ecological devastation resulting from the introduction of exotic grasses. The spring annual red brome *(Bromus rubens)* and the African perennial buffelgrass *(Pennisetum ciliare)* produce abundant fine fuel and subsequent fires (Esque et al., ch. 13 of this volume).

Dispersal of tortoises between populations is increasingly difficult and less likely through sparse desertscrub in very hot, dry valleys in the lower Colorado River Valley. Populations at lower elevations such as the Eagletail, Maricopa, and Sand Tank Mountains have likely been completely isolated for a long time. Surprisingly, a late Pleistocene fossil of a very large desert tortoise (carapace 500 mm long, 420 mm maximum width) was found in McClellan Wash on the eastern base of the Picacho Mountains (McCord 1994). The inside of the shell is full of the sandy clay typical of ponded areas in valley bottoms. This fossil indicates that desert tortoises formerly attained larger adult sizes than today and that they were more widespread in the valley habitats, although the paleovegetation was likely more mesic than the modern creosotebush desertscrub (Van Devender, ch. 2 of this volume). Paleoclimatic reconstructions of the late Wisconsin glacial age in the northern Sonoran Desert reveal much cooler summers, greatly increased winter precipitation, and summer rainfall greatly reduced or absent.

The Southern Limits: Thornscrub Tortoises

In the Sonoran Desert in northern Sonora, the Arizona Upland sub-division gives way to the Plains of Sonora. The transition from Sonoran desertscrub to foothills thornscrub begins on desert inselbergs north of Hermosillo and higher areas east of the Sonoran Desert (Turner and Brown 1982; Búrquez M. et al. 1999). Arizona Upland is more similar to foothills thornscrub than to the arid creosotebush communities of the lower Colorado River Valley. Indeed, Turner and Brown (1982) suggested that the Arizona Upland could be considered a drier and more temperate thornscrub rather than a subdivision of the Sonoran Desert. The distribution, natural history, and conservation of tortoises in Mexico are discussed in Bury et al. (ch. 5 of this volume). The distributions of Sonoran and Sinaloan tortoises, best recognized by mitochondrial DNA (Lamb and McLuckie, ch. 4 of this volume), are not well known but likely break at the desertscrub-thornscrub transition just south of Guaymas. In the summer, tortoises from the Sonoran Desert in Arizona south to thornscrub in southern Sonora share a similar natural history, one that is essentially unchanged from that of their ancestors dwelling in tropical communities before the Miocene formation of the Sonoran Desert (Van Devender, ch. 2 of this volume). The desert tortoise reaches its southern terminus in northern Sinaloa as thornscrub grades into tropical deciduous forest. It is unclear what factors set the southern limits, but they may be related to the growth of pathogens and fungi in burrows unable to dry out during the summer tropical rainy season (Bury et al., ch. 5 of this volume).

Life at the Edge: Tortoises in the Lower Colorado River Valley

The transition from Arizona Upland to the lower Colorado River Valley in southwestern Arizona is at about 550-m elevation near Gila Bend, extending across the Colorado River to below sea level in the Salton Basin of southeastern California (Turner and Brown 1982). The vegetation in the lower Colorado River Valley is less diverse, is adapted to more arid conditions, and grows in low densities compared to the Arizona Upland. Saguaros, blue paloverdes, and ironwood become riparian and restricted to usually dry desert washes; unlike in Arizona Upland, they do not grow on the slopes above. The valley vegetation is a simple desertscrub community dominated by creosotebush, white bursage *(Ambrosia dumosa)*, and big galleta grass *(Pleuraphis rigida)*. This is the hottest and most arid region in North America (Turner and Brown 1982). Plant remains in ancient packrat middens document that creosotebush desertscrub communities were present 12,000 years ago in the late Wisconsin glacial age in the Picacho Peaks, California (just north of Yuma, Arizona; Van Devender 1990). The lower Colorado River Valley has been the core desert of North America for the entire 2.4 million years of the Pleistocene or longer.

Tortoises in lower elevation ranges, such as the Maricopa and Sand Tank Mountains, are living in habitats with relatively sparse vegetation. Due to extreme and prolonged heat and reduced rainfall with longer, more frequent droughts, the climate is more stressful than in the Arizona Upland. Only the extraordinary low metabolic rates of inactive tortoises allow them to survive (Peterson 1996). Rains may be sufficient to rehydrate tortoises, allowing them to extract quality food from dried plants but not enough to produce fresh summer herbs and annuals. Grasses are less common and frequently absent in the lower Colorado River Valley, and tortoises are more dependent on shrubs for food (Wirt and Holm 1997; Van Devender et al., ch. 8 of this volume). The tortoise study plot in the Eagletail Mountains, a low desert range in Maricopa County about 115 km west of Phoenix, is an interesting example (Averill-Murray et al., ch. 6 of this volume). The tortoises are common in a rough boulder field at 475-m elevation on the northwestern base of Double Eagle Peak in an area shaded by the steep vertical cliff, but are absent on sunny slopes and the adjacent creosotebush–white bursage desertscrub flats (fig. 1.4).

The distribution and abundance of tortoises in the lower areas of southwestern Arizona are poorly known. Tortoises may be absent from the hyperarid sandy valleys of the Gila and lower Colorado Rivers but exist in very low densities in such desert mountain ranges as the Gila, Mohawk, and Tinajas Altas in Arizona, where only unconfirmed sign, not actual tortoises, have been found (Johnson et al. 1990). For example, widespread granite boulder habitats in the Tinajas Altas Mountains (fig. 1.5) are similar to those in Arizona Upland populations, where tortoises are abundant (e.g., Four Peaks, Granite Hills, Little Shipp Wash, and the Rincon Mountains); yet, here tortoises are rare. Presumably, tortoise distribution, population density, and dispersal in the lower Colorado River Valley are limited by heat and aridity and not by the availability of physical habitat.

Indeterminate fossil shell fragments on sandy windswept surfaces suggest that desert tortoises may have been more widespread in the lowland valleys of southwestern Arizona in the winter-rainfall climates of the Pleistocene (McCord, ch. 3 of this volume; Van Devender, ch. 2 of this volume).

Life at the Edge: Tortoises and Cold

Emergent sky-island mountains aside, Arizona can be generally be divided into low-elevation Sonoran and Mohave desertscrub in the west and southwest, the middle-elevation desert grasslands and Chihuahuan Desert in the southeast, and the highlands of the Colorado Plateau in the Great Basin Desert to the north and northeast. The massive Mogollon Rim rises above the deserts in a broad band from southeast to northwest across Arizona. The presence of Sonoran Desert plants and animals, including desert tortoises, in higher-elevation communities are limited by colder winter temperatures (Lowe 1964; Van Devender et al. 1994) and wildfires (Esque et al., ch. 13 of this volume). Sonoran desertscrub, thornscrub, and tropical deciduous forest are frost sensitive and severely impacted by fires. In higher-elevation desert grassland, chaparral, woodland, and forest, winters are colder and fire plays an important role in shaping vegetation structure and composition. Tortoises enter the lower edges of some of these communities, but are not typical residents.

Desert Grassland

Desert grassland is a grass- or shrub-dominated landscape positioned between warmer, drier desertscrub below and colder, wetter communities at higher elevations in the southwestern United States and northern Mexico (Brown 1982; McClaran and Van Devender 1995). In Arizona, the transition between the Sonoran Desert and desert grassland is at approximately 1,100- to 1,400-m elevation. Such frost-sensitive subtropical desert plants as saguaro and foothills paloverde reach their upper elevational limits in desert grassland, whereas ecologically widespread and freeze-resistant plants such as velvet mesquite thrive.

Figure 1.4. Eagletail Mountains Wilderness Area, west of Phoenix, Maricopa County, Arizona. Left: A Sonoran desert tortoise population lives in a pocket of Arizona Upland desertscrub at 475-m elevation in this boulder field on the northwest base of Double Eagle Peak. Right: Sparse lower Colorado River Valley

creosotebush desertscrub and hot, dry climates on exposed ridges and in adjacent valley bottoms effectively isolate the tortoise population. (Photographs by Thomas Van Devender)

Figure 1.5. Tinajas Altas Mountains, Yuma County, southwestern Arizona, in the lower Colorado River Valley subdivision of the Sonoran Desert. Although the granite boulder areas are similar to prime Sonoran desert tortoise habitats in central Arizona, sparse vegetation and hot, dry climates eliminate tortoises or reduce populations to extremely low levels. (Photograph by Thomas Van Devender)

The desert tortoise enters the western edge of desert grassland from the adjacent Arizona Upland (Martin 1995). Although food abundance and generous summer rainfall in desert grassland would seem to be conducive for this species, it is likely that the severity and extended season of freezes and regular wildfires inhibit it from ranging farther into higher-elevation, colder, and more fire-prone communities (Esque et al., ch. 13 of this volume).

Curiously, a few isolated individual tortoises have been reported in southeastern Arizona in desert grassland or Chihuahuan desertscrub (Hulse and Middendorf 1979; Johnson et al. 1990). Considering that tortoises have been transported from place to place for food or pets for a long time, there is always suspicion that these particular individuals were captured in the Sonoran Desert and were released or escaped. However, other Sonoran amphibians and reptiles, including the Arizona coral snake *(Micruroides eury-xanthus)*, Gila monster, Sonoran Desert toad, and tiger rattlesnake, are locally present in southeastern Arizona and southwestern New Mexico (Stebbins 1985; Phillip Rosen, pers. comm. 1999), and likewise may be the desert tortoise. These species are widespread in foothills thornscrub and may have entered southeastern Arizona via the Río Bavispe Valley in Sonora rather than eastward from the Sonoran Desert. Additionally, tortoise shells and epidermal scutes of desert tortoises found in ancient packrat middens and cave deposits in late Pleistocene pinyon-juniper woodlands in the northern Chihuahuan Desert in New Mexico and Texas (Van Devender et al. 1976; Thompson et al. 1980; McCord, ch. 3 of this volume) indicate that at some time tortoises lived in the area between the Sonoran and Chihuahuan Deserts.

Below the Mogollon Rim

In the Rincon Mountains near Tucson, oak woodland grows just above desert grassland. One of the few desert tortoise populations found in oak woodland is in a south-facing basin at 1,520-m elevation in Chiminea Canyon. To the northwest of Tucson in the mountains below the Mogollon Rim, desert grassland is replaced by interior chaparral (Lowe 1964). This is a winter rainfall-adapted shrub community with strong affinities to similar communities in southern California. Tortoise plots at Bonanza Wash, Four Peaks, and Little Shipp Wash are in Arizona Upland desertscrub not far below chaparral. Similar to desert grassland, tortoises likely live in the lower edges of chaparral but have not been studied there. The charred tortoise in figure 13.4 was from a similar habitat in the foothills of the Bradshaw Mountains, Yavapai County, Arizona (Esque et al., ch. 13 of this volume).

The Arizona Upland, the prime habitat of the Sonoran tortoise, parallels the Mogollon Rim from the Tucson area northwestward until it merges with the Mohave Desert in western Arizona south of Kingman (Turner and Brown 1982). The northwestern limits of the Sonoran Desert coincide with the edges of the geographic ranges of foothills paloverde and saguaro, while the range of the desert tortoise extends into the Mohave Desert in California, Nevada, and Utah (Germano et al. 1994).

The Mohave Tortoise: An Evolutionary Youngster

If our ideal female tortoise were transplanted from her hibernaculum in Arizona Upland to a similar burrow under a boulder on a rocky slope in the Mohave Desert, when she emerged in the spring she might be alone or in a population with fewer tortoises. However, a thriving population of active, feeding tortoises, both males and females, could be found in the valley below. Mohave tortoises are not just marginal populations on the periphery of the Sonoran tortoise's distribution, but are well adapted to that environment. Although this chapter and book are about the Sonoran tortoise, the marked differences between the extensively studied Mohave tortoise (Bury and Germano 1994) and the two southern races (Sonoran and Sinaloan tortoises, which are recognized but undescribed subspecies; see Lowe 1990) need to be clearly stated. The information on the Sonoran tortoise summarized in this book allows us to do that.

Lamb and McLuckie (ch. 4 of this volume) discerned substantial differences in mitochondrial DNA among Sinaloan, Sonoran, and Mohave tortoises, reflecting evolution that progressed due to mutations at the molecular level in the absence of natural selection. Mohave tortoise populations are primarily valley based but also live on surrounding mountain slopes, resulting in widespread and continuous habitat use with potential widespread gene flow between populations (Bury and Germano 1994). Within its usually rockless valley habitat, the Mohave desert tortoise relies heavily on self-constructed burrows and natural fissures in eroded dry stream beds (arroyos) for shelter (Luckenbach 1982; Bury and Germano 1994). The ability to excavate deep burrows along desert washes and on surrounding desert flats essentially frees the populations from sheltersite limitations and allows them to reach high densities in large areas.

The Mohave tortoise is the only North American tortoise that lives in a uniseasonal winter-rainfall climate with little summer rainfall. In the biseasonal rainfall regime of the Sonoran Desert, droughts would rarely be longer than nine months, because rain would fall in the next season. In the Mohave Desert, tortoises will experience periods of drought extending eighteen months or longer during their lives. Mohave desert tortoises are on the edge physiologically and are vulnerable to catastrophic population declines during severe droughts (Peterson 1996). In northwestern Baja California, beyond the range of desert tortoises, extended droughts of thirty-six months have been recorded (Tony L. Burgess, pers. comm. 1985). Such extensive droughts apparently exclude desert tortoises from the hotter Sonoran Desert lowland areas.

The general activity period of the Mohave tortoise (active in spring, inactive in summer), especially in the western Mohave Desert, is essentially opposite of those of Sonoran and Sinaloan tortoises (less active in spring, active in summer). In the Mohave tortoise, summer estivation merges with winter hibernation. Mohave tortoises primarily feed in the spring on fresh annuals even though they cannot achieve a positive energy balance (Peterson 1996). They need to eat dried grasses later to produce enough excess energy to be stored as fat and used the next spring. The inability to rehydrate and eliminate concentrated salts from the bladder during the summer has important physiological consequences, especially in the inhibition of digestion of dried plants (Oftedal, ch. 9 of this volume). Female Mohave tortoises have multiple clutches of eggs, laying in late May, late June, and August. The multiple clutches may contain about the same number of eggs per year as the single clutch of the Sonoran tortoise (Jeff Lovich, pers. comm. 1998; Averill-Murray et al., ch. 6 of this volume).

The listing of the Mohave population as threatened by the U.S. Fish and Wildlife Service in 1990 was hastened by extensive mortality in the western Mohave Desert due to upper respiratory tract disease caused by *Mycoplasma agassizii* (Brown et al. 1994; Berry 1997). Mycoplasmosis has not been shown to be a major threat to Sonoran tortoise populations, which appear to maintain low levels of infection without developing the disease (Dickinson et al., ch. 10 of this volume). Apparently, the Mohave tortoise has lost this natural resistance to *M. agassizii*.

I propose that the ecology, behavior, and physiology of Sonoran and Sinaloan tortoises are more like those of their tropical ancestors (Van Devender, ch. 2 of this volume). Most of the differences between Sonoran and Mohave tortoises are newly derived characters in the Mohave tortoise that evolved in response to the development of winter-rainfall climates at the beginning of the Pleistocene.

ACKNOWLEDGMENTS

This manuscript is a literary device to summarize the wealth of knowledge about desert tortoises reflected in the chapters in this book and to salute the author-biologists who love them. Brent Martin helped in many ways with his insight and experience. Roy Averill-Murray's thoughtful and thorough editing of the manuscript was very helpful.

LITERATURE CITED

Barrett, S. L. 1990. Home range and habitat use of the desert tortoise *(Xerobates agassizii)* in the Picacho Mountains, Pinal County, Arizona. *Herpetologica* 46:202–6.

Berry, K. H. 1997. Demographic consequences of disease in two desert tortoise populations in California, USA. Pp. 91–99 *in* J. Van Abbema, ed. *Proceedings of the international conference on conservation, restoration, and management of tortoises and turtles.* New York Turtle and Tortoise Society, New York.

Brown, D. E. 1982. Semidesert grassland. *Desert Plants* 4:123–31.

Brown, M. B., I. M. Schumacher, P. A. Klein, K. Harris, T. Correll, and E. R. Jacobson. 1994. *Mycoplasma agassizii* causes upper respiratory tract disease in the desert tortoise. *Infection and Immunity* 62:4580–86.

Búrquez M., A., A. Martínez Y., R. S. Felger, and D. Yetman. 1999. Vegetation and habitat diversity at the southern edge of the Sonoran Desert. Pp. 36–67 *in* R. H. Robichaux, ed. *Ecology of Sonoran Desert plants and plant communities.* University of Arizona Press, Tucson.

Bury, R. B., and D. J. Germano, eds. 1994. *Biology of North American tortoises.* Fish and Wildlife Research no. 13, U.S. Dept. of the Interior National Biological Service, Washington, D.C.

Dimmitt, M. A. 2000. Biomes and communities of the Sonoran Desert region. Pp. 3–18 *in* S. J. Phillips and P. W. Comus, eds. *A natural history of the Sonoran Desert.* Arizona-Sonora Desert Museum Press, Tucson.

Germano, D. J., R. B. Bury, T. C. Esque, T. H. Fritts, and P. A. Medica. 1994. Range and habitats of the desert tortoise. Pp. 73–84 *in* R. B. Bury and D. J. Germano, eds. *Biology of North American tortoises.* Fish and Wildlife Research no. 13, U.S. Dept. of the Interior National Biological Service, Washington, D.C.

Hulse, A. C., and G. A. Middendorf. 1979. Notes on the occurrence of *Gopherus agassizii* (Testudinidae) in extreme eastern Arizona. *Southwestern Naturalist* 24:545–46.

Johnson, T. B., N. M. Ladehoff, C. R. Schwalbe, and B. K. Palmer. 1990. Summary of literature on the Sonoran Desert population of the desert tortoise. Report to U.S. Fish and Wildlife Service, Phoenix.

Lowe, C. H. 1964. *The vertebrates of Arizona.* University of Arizona Press, Tucson.

———. 1990. Are we killing the desert tortoise with love, science, and management? Pp. 84–102 *in* K. R. Beaman, F. Chaporaso, S. McKeown, and M. D. Graff, eds. *Proceed-*

ings of the first international symposium on turtles and tortoises: Conservation and captive husbandry, Chapman University, Chapman, Calif.

Luckenbach, R. A. 1982. Ecology and management of the desert tortoise (Gopherus agassizii) in California. Pp. 1–37 in R. B. Bury, ed. North American tortoise conservation and ecology. Wildlife Research Report no. 12, U.S. Fish and Wildlife Service, Washington, D.C.

Martin, B. E. 1995. Ecology of the desert tortoise (Gopherus agassizii) in a desert-grassland community in southern Arizona. M.S. thesis, University of Arizona, Tucson.

McClaran, M. P., and T. R. Van Devender. 1995. The desert grassland. University of Arizona Press, Tucson.

McCord, R. D. 1994. Fossil tortoises of Arizona. Pp. 83–89 in D. Boaz, M. Dornan, and S. Bolander, eds. Proceedings of the fossils of Arizona symposium. Vol. 2. Mesa Southwest Museum and Southwest Paleontological Society, Mesa, Arizona.

Mitchell, V. L. 1976. The regionalization of climate in the western United States. Journal of Applied Meteorology 15:920–27.

Nagy, K. A., B. T. Henen, and D. B. Vyas. 1998. Nutritional quality of native and introduced food plants of wild desert tortoises. Journal of Herpetology 32:260–67.

Peterson, C. C. 1996. Ecological energetics of the desert tortoise (Gopherus agassizii): effects of rainfall and drought. Ecology 77:1831–44.

Repp, R. 1998. Wintertime observations on five species of reptiles in the Tucson area: shelter-site selection/fidelity to sheltersites/notes on behavior. Bulletin of the Chicago Herpetological Society 33:49–56.

Rondeau, R., T. R. Van Devender, C. D. Bertelsen, P. Jenkins, R. K. Wilson, and M. A. Dimmitt. 1996. Annotated flora of the Tucson Mountains, Pima County, Arizona. Desert Plants 12:3–46.

Secretaría de Desarrollo Social. 1994. Poder ejecutivo. Diario Oficial de la Federación. Tomo 488 no. 10, Mexico, D.F., lunes 16 de mayo de 1994.

Stebbins, R. C. 1985. A field guide to western reptiles and amphibians. Houghton Mifflin, Boston.

Thompson, R. S., T. R. Van Devender, P. S. Martin, T. Foppe, and A. Long. 1980. Shasta ground sloth (Nothrotheriops shastense Hoffstetter) at Shelter Cave, New Mexico: environment, diet and extinction. Quaternary Research 14:360–76.

Turner, R. M., and D. E. Brown. 1982. Sonoran desertscrub. Desert Plants 4:121–81.

Van Devender, T. R. 1990. Late Quaternary vegetation and climate of the Sonoran Desert, United States and Mexico. Pp. 134–65 in J. L. Betancourt, T. R. Van Devender, and P. S. Martin, eds. Packrat middens: The last 40,000 years of biotic change. University of Arizona Press, Tucson.

Van Devender, T. R., K. B. Moodie, and A. H. Harris. 1976. The desert tortoise (Gopherus agassizii) in the Pleistocene of the northern Chihuahuan Desert. Herpetologica 32:298–304.

Van Devender, T. R., C. H. Lowe, and H. E. Lawler. 1994. Factors influencing the distribution of the Neotropical vine snake (Oxybelis aeneus) in Arizona and Sonora, Mexico. Herpetological Natural History 2:25–42.

Van Devender, T. R., A. C. Sanders, R. K. Wilson, and S. A. Meyer. 2000. Flora and vegetation of the Río Cuchujaqui, a tropical deciduous forest near Alamos, Sonora. Pp. 36–

101 *in* R. H. Robichaux and D. Yetman, eds. *The tropical deciduous forest of the Alamos: Biodiversity of a threatened ecosystem.* University of Arizona Press, Tucson.

Wirt, E. B., and P. A. Holm. 1997. Climatic effects on survival and reproduction of the desert tortoise *(Gopherus agassizii)* in the Maricopa Mountains, Arizona. Report to Arizona Game and Fish Department, Phoenix.

CHAPTER 2

Cenozoic Environments and the Evolution of the Gopher Tortoises (Genus *Gopherus*)

THOMAS R. VAN DEVENDER

The Sonoran Desert is considered to be the most tropical of the four North American deserts, both because of its climate, with mild virtually frost-free winters and summer monsoonal rainfall from the tropical oceans, and its physical connections with more tropical communities to the south in Baja California and Sonora, Mexico (Turner and Brown 1982). The transition from temperate biomes to the New World tropics in western Mexico is between 31° and 28° N latitude in Sonora. Winter freezes decrease in frequency and intensity, and the absolute amounts and percentages of summer precipitation increase toward the tropics (Van Devender et al. 1994). The distribution of the desert tortoise *(Gopherus agassizii)* spans this transition and more, extending into the winter-rainfall and cold-winter climates of the Mohave Desert in northwestern Arizona, southeastern California, southwestern Utah, and southern Nevada.

In the northeastern Sonoran Desert in Arizona, desert tortoises are closely associated with the foothills paloverde *(Parkinsonia [Cercidium] microphylla)*–saguaro *(Carnegiea gigantea)* desertscrub typical of rocky slopes in the Arizona Upland subdivision (Van Devender, ch. 1 of this volume). Most of the visual dominants in the saguaro forests of southern Arizona, including saguaro, foothills paloverde, ironwood *(Olneya tesota)*, and creosotebush *(Larrea divaricata)*, do not grow in tropical deciduous forests near Alamos in southern Sonora (Van Devender et al. 2000). Indeed, the ranges of only a few dominant Sonoran Desert plant communities, such as *brea (Parkinsonia praecox)*, guayacán *(Guaiacum coulteri)*, organpipe cactus *(pitahaya; Stenocereus thurberi)*, and tree ocotillo *(Fouquieria macdougalii)*, extend into dry tropical forests near Alamos in southern Sonora (Van Devender 1999). Moreover, many tropical species such as coral bean *(Erythrina flabelliformis)*, manihot *(Manihot davisae)*, and brown vine snake *(Oxy-*

belis aeneus) reach their northern limits in southern Arizona in the desert grassland–oak woodland ecotone at 1,220- to 1,525-m elevation above the Sonoran Desert (Van Devender et al. 1994).

To understand the tropicalness of the Sonoran Desert and the desert tortoise, we must look into their distant pasts. The fossil record of North American tortoises is summarized by McCord (ch. 3 of this volume). Here, I present a summary of the vegetation and climates of the Cenozoic for the southwestern United States and explore the evolutionary implications for gopher tortoises *(Gopherus)*.

Early Tertiary

The Cenozoic era is the geological time period following the extinction of the dinosaurs at the end of the Cretaceous period of the Mesozoic era (66.4 million years ago; table 2.1, modified from Phillips and Comus 2000). It is divided into the Tertiary and Quaternary periods. Major subdivisions of the Tertiary are the Paleocene, Eocene, Oligocene, Miocene, and Pliocene epochs, originally defined on the evolutionary stages of marine invertebrates in European sediments. The Quaternary, beginning 1.8 million years ago, includes the Pleistocene and Holocene epochs.

The Paleocene (66.4 to 57.8 million years ago) climates of North America were warm with humid forests with strong Asian affinities and primitive ferns *(Anemia)*, cycads *(Dioon, Zamia)*, and palms as far north as Alaska (Wolfe 1977). Palms grew at latitude 70° N in Greenland, and recognizable tropical rain forests appeared in the southern United States for the first time (Graham 1999). Broad-leaved temperate or tropical rain forests spanned the continent with little east–west differentiation.

Tropical climates and forests continued into the Eocene (57.5 to 36.6 million years ago). In the western United States, studies of fossil leaf sizes and shapes indicate that the climates of the Eocene became warmer and drier (Wolfe and Hopkins 1967; Axelrod and Bailey 1969). Deciduous trees became increasingly more important through the Eocene as the tropical dry season, and thus tropical deciduous forest, developed (Graham 1993; Graham and Dilcher 1995).

The similarities of Eocene small mammal faunas between North America and western Europe are strong indicators of a corridor between

TABLE 2.1. Geologic time scale. Time spans are in millions of years except for thousands of years (ka) in the Quaternary.

Era	Period	Epoch	Time Span
CENOZOIC	Quaternary	Holocene	11.0 ka–present
		Pleistocene	1.8–11.0 ka
	Tertiary	Pliocene	5.3–1.8
		Miocene	23.7–5.3
		Oligocene	36.6–23.7
		Eocene	57.8–36.6
		Paleocene	66.4–57.8
MESOZOIC	Cretaceous		144–66.4
	Jurassic		208–144
	Triassic		245–208

the continents. A search for a land route in the Arctic, the now-sundered landmass of Euramerica (Dawson et al. 1976; Graham 1993), yielded a remarkable fauna from the early Eocene Eureka Sound Formation on Ellesmere Island in the Canadian Arctic Archipelago (78° N). Not only were the typical mammals present but also an extinct alligator *(Allognathosuchus)*, tortoise *(Hadrianus)*, softshell turtle *(Trionyx)*, varanid lizard (monitor lizards and relatives, Varanidae), and ground boa (Estes and Hutchison 1980). Plant fossils from the Eocene of Alaska have been interpreted as a paratropical rain forest with a remarkable mixture of species with lowland Malaysian affinities, temperate evergreen trees, palms, deciduous plants including ginkgo *(Ginkgo)*, legumes, and riparian trees in extant genera (*Alnus, Juglans, Platanus, Prunus, Salix*, and *Sorbus*; Wolfe 1972, 1977; see discussion in Graham 1993).

The concept of a major environmental change at the end of the Eocene has been controversial. Wolfe (1992) envisioned a "big chill," a cold period of 1 to 2 million years in length in the earliest Oligocene that had a profound impact on the flora. Other climatic reconstructions for the Eocene-Oligocene boundary, primarily based on the climatic relationships of surviving taxa, were very different. For example, analyses of fossil reptiles and amphibians by Hutchison (1992) and pollen and leaf floras by Leopold et al. (1992) indicated increasing aridity and seasonality with little cooling.

Miocene Revolution

Geologic factors have been extremely important in the evolution of plants, animals, and biotic communities throughout the Earth's history (Tiffany 1985). A series of enormous volcanic eruptions from the late Oligocene to the middle Miocene (about 30 to 15 million years ago) changed the climates and established the modern biogeographic provinces of North America (Axelrod 1979). The Rocky Mountains were uplifted at least 1525 m near Florissant, Colorado, and more than 2,135 m of volcanic rocks were deposited in the Jackson Hole area of west-central Wyoming (Leopold and MacGinitie 1972). In the Sierra Madre Occidental in northwestern Mexico, extensive Oligocene (36.6 to 23.7 million years ago) rhyolitic ash-flow tuffs were deposited on top of Laramide (90 to 40 million years ago) andesites in combined layers as much as 2 km thick, forming the modern high plateaus of the Sierra Madre Occidental (Swanson and Wark 1988; Cochemé and Demant 1991; Roldán and Clark 1992). Basaltic volcanism predominated in the northern Sierra Madre (24 to 12 million years ago; Cochemé and Demant 1991) accompanied by basin and range faulting and extension (Henry and Aranda 1992), especially from 15 to 5 million years ago (Menges and Pearthree 1989). As the mountains continued uplifting, they interrupted the upper flow of the atmosphere for the first time in the Tertiary. Tropical moisture from both the Pacific Ocean and the Gulf of Mexico was blocked from the midcontinent, thus drying out the modern Great Plains and Mexican Plateau. Harsher climates segregated drought- and cold-tolerant species into new environmentally limited biomes, including tundra, modern conifer forests, and grasslands, which were distributed along elevational and latitudinal environmental gradients (Van Devender 1995). The Miocene Revolution also initiated major evolutionary radiations in many of today's successful groups. Pollen of the Compositae appeared in abundance for the first time at the Oligocene-Miocene boundary (Graham 1993).

In the Miocene, tropical forests were restricted by the rising Sierra Madres to lowland ribbons along the coasts from about Sonora and Tamaulipas southward. Evidence for tropical deciduous forest on the eastern side of the continent (southern Mexico and Central America) was not found in the pollen record until the middle Pliocene (4 million years ago; Graham and Dilcher 1995), 8 to 10 million years later than postulated for the Pacific

Coast. Apparently, thornscrub, at least the modern northwestern Mexican type, was also a new vegetation type that formed on the lower, drier edges of tropical deciduous forest in the early Miocene. Thornscrub may well have been the regional vegetation covering the drier areas that are now Sonoran Desert.

Paleoelevations of the Rocky Mountains

A new method of estimating paleoelevations has challenged the above scenario of middle Tertiary modernization of the continent. Wolfe (1971) used correlations between leaf morphology and climate in modern floras to estimate paleotemperatures for Tertiary fossil floras. Recently, he expanded his analyses to include regressions between foliar morphology and various aspects of climate and developed a method of estimating the elevations of fossil floras at the time of deposition (Wolfe 1993). In this approach, the physiological tolerances and limits of the living populations or closest relatives of the fossil taxa are not considered, because the latter are extinct species that might have been different than the extant species. The leaf morphology–climate relationship based on worldwide floras is thought to be a better indicator of climate than the physiological tolerances of the living relatives of the fossil taxa. The first studies using this methodology reached dramatically different paleoelevation estimates than previous studies based on floristic affinities. For example, MacGinitie (1953) inferred a paleoelevation of 915 m (now at 2,500 m) for the latest Eocene (35 million years ago, formerly called the Oligocene) Florissant Beds in Colorado based on a paleoflora closely allied with the highlands of northeastern Mexico. The flora was a mixture of plants now found in tropical and montane areas and a sequoia *(Sequoia affinis)*. In contrast, Gregory (1994), using Wolfe's (1993) multivariate climate analysis techniques, estimated that the Florissant Beds were at 2,300- to 3,300-m elevation. The climatic implications of the additional 1,385- to 2,385-m elevation of the Rocky Mountains in the late Eocene are profound, especially the inferences about cold temperatures. All of the ecological, evolutionary, and biogeographic changes in the biota discussed for the Oligocene–middle Miocene in the Axelrod (1979) model should have occurred earlier if Gregory were correct. It is difficult to accept this in light of the persistence of tropical plants such as cedar *(Cedrela)*, palms, and *piocha*

(Trichilia) into the early Oligocene, or the gradual modernization of the Rocky Mountain flora from the Oligocene to the early Miocene (Leopold and MacGinitie 1972).

A reexamination of Gregory's (1994) study of the Florissant flora is enlightening. The living relatives of at least 34 percent of the twenty-nine taxa used in her paleoclimatic analysis are today restricted to elevations and latitudes lower than Florissant. Cedar, hopbush *(Dodonaea)*, mesquite *(Prosopis)*, and soapberry *(Sapindus)* are genera with tropical affinities whose extant species live in areas with higher mean annual temperatures than the upper elevations of the Rockies. *Trichilia*, in particular, is a tropical genus in the Meliaceae that reaches its northern limit in southern Sonora at 27° N latitude, where two piochas *(T. americana, T. hirta)* grow in tropical deciduous forest below about 1,000-m elevation. The inescapable conclusions of a paleoelevation of 2,300–3,300 m for the Florissant Beds are that a third of the flora had greater cold tolerances than their living relatives and that cold-adapted species were more vulnerable to extinction than their tropical descendants.

If Wolfe's (1993) multivariate climate analyses of leaf floras systematically underestimate mean annual temperatures, then paleoelevations are overestimated, resulting in questionable landscape and paleoclimatic reconstructions. I feel that important paleoecological signals from plant and animal taxa in these fossil deposits must be considered. For the present, the Axelrod (1979) model of landscape evolution is the most useful, although it is susceptible to revisions in the timing of the uplift of the Sierra Madre Occidental.

Evolution of the Sonoran Desert

The Sonoran Desert did not form as part of the late Oligocene–early Miocene orogeny, but as result of a drying trend beginning in the middle Miocene (15 to 8 million years ago; Axelrod 1979). The dominant species in the new desertscrub communities were segregated out of thornscrub. However, many of the nearest relatives of important Sonoran Desert plants, especially the columnar cacti, are not in southern Sonora or northern Sinaloa, but in seasonally dry vegetation on the Pacific slopes of central and southern Mexico (Cornejo 1994), presumably reflecting pre–Sonoran

Desert floristic connections from a time when the Sierra Madre Occidental and the Western Mexican Volcanic Belt were not such formidable geographic barriers. A notable example is the saguaro and its closest relatives, *Neobuxbaumia*, a group of columnar cacti found in central Mexico from the Balsas Basin southeastward. Grismer (1994) reported that the closest relatives of several reptiles endemic to Baja California (e.g., *Bipes biporus, Cnemidophorus hyperythrus, Phyllodactylus unctus, Pituophis vertebralis, Urosaurus nigricaudus*) are found on mainland Mexico from Colima southeast to Guerrero and Oaxaca, reflecting continuous distributions prior to the formation of the Gulf of California in the Miocene.

Baja California had a different history than the mainland Sonoran Desert. About 12 million years ago, in the Miocene, the proto–Gulf of California opened and several large islands stocked with tropical plants and animals drifted in splendid isolation northwestward to meet California. The actual timing of the formation of the present Gulf of California along the San Andreas Fault and the biological and evolutionary importance of the proto-gulf have been controversial, depending on interpretations of the geological record (Axelrod 1979; Morafka et al. 1992; Grismer 1994). Natural selection shaped the plants isolated on Baja California into many unique endemics including boojum tree *(cirio; Fouquieria columnaris)*, elephant tree *(Pachycormus discolor)*, and many other plants and reptiles.

Thus, the habitats of the desert tortoise in the Sonoran Desert were in existence by the late Miocene (8 to 5 million years ago), the result of geologic events that shaped the landscape and altered regional climates mostly in the early to middle Miocene. The speciation of many important species of plants and animals occurred earlier in thornscrub or tropical deciduous forest. In contrast, the Mohave Desert continued to develop well into the Pleistocene.

High Sea Levels, Capybaras, Iguanas, and Tropical Climates

After the Miocene, changing climates and immigration—rather than evolution—were the major impacts on the biota, dramatically shifting species ranges and community compositions. During the late Miocene–Pliocene (5 to 1.8 million years ago) and again briefly in the Pleistocene, there were reversals to more tropical climates. Evidence of tropical species in the Sonoran Desert and much higher sea levels indicate warmer global temperatures. Warmer ocean water results in enhanced monsoonal summer rainfall at higher latitudes and northward expansions of tropical deciduous forest and thornscrub.

During the latest Miocene–early Pliocene (centered about 5 million years ago), the sea level rose enough that the newly formed Gulf of California expanded into the Salton Trough of southeastern California to deposit the marine sediments of the Imperial Formation. Later, when the Colorado River connected with the Gulf of California, deltaic sediments separated the Salton Basin (see discussions in Cassiliano 1994). The extensive sediments of the roughly contemporaneous Bouse Formation in the lower Colorado River Valley Basin in Arizona and California have been interpreted as either estuarine, based on invertebrate fossils including barnacles, or as freshwater lakes, based on strontium isotopic analyses (see discussion and references in Spencer and Patchett 1997). Fossils of in situ mollusc shells and fish bones record later incursions of marine water into the Fish Creek–Vallecitos area of the Anza Borrego Desert State Park of southern California in the Pliocene and early Pleistocene. These incursions were mostly tectonically controlled because there was, in general, an equilibrium between sedimentation and subsidence keeping the Salton Basin near sea level. When subsidence was greater than deposition, the area was reconnected with the Gulf of California (Johnson et al. 1983). The vertebrate fossil record provides some indication that Pliocene environments and some interglacials in the generally cool Pleistocene were more tropical than the present interglacial, the Holocene — that is, a frost-free climate with much greater rainfall in the warm season and higher humidity.

Anza-Borrego, California

In Anza-Borrego Desert State Park in southern California, the Pliocene–early Pleistocene terrestrial sediments of the Palm Springs Formation overlie the late Miocene Imperial Formation. However, the area is west of the San Andreas Fault and has moved approximately 300 km to the northwest in the last 5.5 million years (Winker 1987). Today the area 300 km to the southwest is on the Gulf of California at about the latitude of Desemboque, Sonora (30°40' N). Both Anza-Borrego and Desemboque are in the arid lower Colorado River Valley subdivision of the Sonoran Desert (Turner and Brown 1982). At a rate of 54.5 mm/year, younger sediment deposits moved progressively lesser distances.

Vertebrate fossils, including a great variety of extinct mammals, are common in the Palm Springs Formation (Cassiliano 1994). A fossil lizard skull from the lower portion of the Palm Springs Formation (about 4.3 to 2.5 million years ago, Pliocene), described as *Pumilia novaceki*, is closely related to the extant *Iguana* (Norell 1989). The modern green iguana *(I. iguana)* is a tropical lizard that today ranges no farther north than southern Sinaloa, about 1,500 km to the southeast. Placing the new species in *Pumilia* rather than *Iguana* obscured its tropical origins. Indeed, the lizard was used along with the desert iguana *(Dipsosaurus dorsalis)* as an indicator of relatively dry environments, even though most of the associated animals in the fauna reflected more mesic environments than Anza-Borrego today (Cassiliano 1994; Jolly 2000). Today, Pacific pond turtles *(Clemmys marmorata)*, mud turtles *(Kinosternon* cf. *sonoriense)*, and sliders *(Trachymys scripta)* are aquatic animals; the southern alligator lizard *(Gerrhonotus multicarinatus)* is common in winter-rainfall chaparral on the Pacific Coast; and desert iguana and kangaroo rats *(Dipodomys* spp.) are desert dwellers.

Fossil wood (the Carrizo local flora) from the lower Palm Springs Formation was mostly from widespread riparian trees including ash *(Fraxinus caudata)*, cottonwood *(Populus* cf. *alexanderi)*, walnut *(Juglans pseudomorpha)*, and willows *(Salix gooddingii, Salix* sp.; Remeika 1994). Reconstructions of a wet winter and dry summer paleoclimate based mainly on a buckeye *(Aesculus* sp.) and a bay *(Umbellularia salicifolia)*, relatives of coastal southern California species (Cassiliano 1994; Remeika 1994), are untenable. All genera of riparian trees in the flora as well as juniper (*Juniperus* sp.) are widespread in biseasonal rainfall regimes all across the continent. The palm *(Sabal* sp.) in the flora is a tropical species reaching its northwestern limits in summer-rainfall-dominated climates in central Sonora (Felger and Joyal 1999). Moreover, the distributions of the modern species of many of the animals in the vertebrate fauna are restricted to summer-rainfall areas. This includes the slider and cotton rats *(Sigmodon)*, grasshopper mice

(Onychomys), mud turtle, pocket gophers *(Geomys)*, and pygmy mouse *(Baiomys)*. Clearly, summer rainfall was important in the early Pliocene climate of the Anza-Borrego area, whereas the presence of cotton rats, the palm, and the slider suggests that temperatures were warm throughout the year.

Late Pliocene and early Pleistocene fossils from Anza-Borrego continued to indicate environments more mesic than today with the addition of some eastern or tropical summer-rainfall species including antelope jackrabbit *(Lepus* cf. *alleni)*, a capybara *(Hydrchoerus* sp.), eastern cottontail *(Sylvilagus floridanus)*, Gambel quail *(Lophortyx gambeli)*, jaguar *(Felis* cf. *onca)*, tortoises *(Geochelone* sp.), peccary *(Platygonus* sp.), and tapir *(Tapirus merriami)*. Gopher tortoise *(Gopherus* sp.) was found in all Pliocene and Pleistocene levels of the Palm Springs Formation (Cassiliano 1994).

El Golfo, Sonora

Early Pleistocene (Irvingtonian Land Mammal Age [LMA], beginning 1.8 million years ago) sediments from El Golfo de Santa Clara in northwestern Sonora yielded the first fossil record of the giant anteater *(Myrmecophaga tridactyla)* in North America (Lindsay 1984; Shaw and McDonald 1987). The nearest populations of this large tropical mammal are 3,000 km to the southeast in the humid, tropical lowlands of Central America. As for many large mammals, its historical distribution may not accurately reflect the potential range because of human predation in the last 11,000 years (Martin 1984). The El Golfo fauna included many extinct mammals. These included antelope *(Tetrameryx* sp.), a bear *(Tremarctos* cf. *floridanus)*, beaver *(Castor* cf. *californicus)*, camels *(Camelops* sp., *Hemiauchenia* spp., *Titanotylopus* sp.), a jaguar *(Felis* cf. *onca)*, horses *(Equus* spp.), proboscidians *(Cuvieronius* sp., *Mammuthus imperator)*, and a tapir *(Tapirus* sp.). Extant species in the fauna included boa constrictor *(Boa constrictor)*, slider, Sonoran Desert toad *(Bufo alvarius)*, and jaguar *(Felis* cf. *onca)*. Boa constrictors and sliders are found today in Sonora in wetter, more tropical areas to the southeast. El Golfo is at the head of the Gulf of California in the lower Colorado River Valley subdivision of the Sonoran Desert (Turner and Brown 1982). Today, this hyperarid desert is too dry to support any of these animals, although historically the delta of the Colorado River was a very wet area with abundant beaver *(Cas-*

tor canadensis) and extensive cottonwood *(Populus fremontii)* gallery forests (Davis 1982).

RANCHO LA BRISCA, SONORA. The late Pleistocene (Rancholabrean LMA) Rancho la Brisca fauna from a riparian canyon north of Cucurpe, north-central Sonora, was dominated by the Sonoran mud turtle *(Kinosternon sonoriense)* and fish, reflecting a wet *ciénega* paleoenvironment (Van Devender et al. 1985). The tropical sabinal frog *(Leptodactylus melanonotus)* was 240 km north of its northernmost population on the Río Yaqui. The presence in the fauna of a large extinct bison *(Bison* sp.), which only immigrated to North America from Siberia in the late Pleistocene between about 170,000 and 150,000 years ago (C. A. Repenning, pers comm. 1984), with tropical species indicated the Rancho la Brisca sediments were deposited about 80,000 years ago in the last interglacial (the Sangamon).

Mohave Desert

The Mohave Desert apparently formed much later than the Sonoran Desert. The middle Miocene (17 million years ago) Tehachapi flora on the westernmost edge of the Mohave Desert in California recorded a vegetation with woodland-forest trees and shrubs including madrone *(Arbutus)*, oaks *(Quercus* spp.), California bay *(Umbellularia)*, and toyón *(Heteromeles)*. Madroño and oaks are widespread in the western United States and the Sierra Madres in Mexico, whereas California bay and toyón are Californian species restricted to the West Coast. Tropical species in the flora reflecting summer rainfall included avocado *(Persea)*, bebelama *(Sideroxylon [Bumelia])*, cacachila *(Karwinskia)*, corcho *(Clethra)*, fig *(Ficus)*, hopbush, kidneywood *(Eysenhardtia)*, mesquite *(Prosopis)*, palms *(Brahea* and *Sabal)*, palo chino *(Havardia [Pithecellobium])*, papache *(Randia)*, and torote *(Bursera;* Axelrod 1979, 1983). Axelrod interpreted these species as indicators of summer rainfall and the development of semidesert (Sonoran Desert) vegetation. However, all of them, with the exception of the avocado, grow in tropical deciduous forest and thornscrub in southeastern Sonora (Martin et al. 1998). All but avocado and corcho have species that grow in the Sonoran Desert. The flora did not have any Mohave Desert indicators. The fossil record of *Clethra* in the Clethraceae in California is especially interesting. There are six or seven species in the genus that are found in montane forests of Central America and Mexico (Standley 1924). *Clethra mexicana* grows from Central America northwest to Durango and Sonora, Mexico. The nearest populations are near Yécora in the Sierra Madre Occidental of eastern Sonora about 1,140 km southeast of the Tehachapi Mountains (Ana L. Reina G., unpubl. data). The trees live at 1,100- to 1,450-m elevations in the pine-oak forests with an annual rainfall of 900–1,000 mm, with 60–70 percent falling between June and October.

Pollen floras from the Coso Formation on the northern border of the Mohave Desert and at localities eastward to Panamint Valley indicate the presence of a mixed-conifer forest in the Pliocene (3.5 to 2.0 million years ago). The flora included alder *(Alnus)*, beech *(Betula)*, dogwood *(Cornus)*, Douglas fir *(Pseudotsuga)*, fir *(Abies)*, giant sequoia *(Sequouiadendron)*, incense cedar *(Calocedrus)*, and pine *(Pinus)*. Except for giant sequoia and incense cedar, these plants are typical of mixed-conifer forests at high elevations in the Sierra Nevada and Rocky Mountains.

More than these floras, the exact timing of the development of the Mohave Desert is not well understood. Aridity in the Mohave Desert is primarily the result of the coastal mountains blocking moisture from the Pacific Ocean. Fully two-thirds of the elevation of the Sierra Nevada arose in the last 2 million years. The range was already high by about 5 million years ago, reached 2,100 m by 3 million years ago, and has risen another 950 m since (Huber 1981). Each increment of the uplift intensified the rain shadow, progressively drying the Mohave Desert. Uplift of the Mohave block during the Quaternary separated the area climatically from the Sonoran Desert. Continued cooling of the Pacific Ocean and enhancement of the California current completed the shift to the winter-rainfall dominance of the West Coast of North America. The distributions of the summer-rainfall tropical species contracted into central and southern Sonora. Thus, the Mohave Desert is the youngest biotic province in North America. The Pleistocene climatic fluctuations discussed below and the immigration of creosotebush from South America, presumably in the Pleistocene, had further impacts on the Mohave Desert.

Ice Age Environments

The warmth of the Pliocene ended at the beginning of the Pleistocene as the Earth entered a new climatic era that far surpassed the middle Miocene in cool, continental conditions. Traditionally, four ice ages, or glacial periods, were recognized based on terrestrial sedimentary deposits in North America and widely correlated between Europe and South America. However, recent studies of isotopic climatic indicators in continuous sediment cores from the ocean floors record fifteen to twenty glacial periods in the last 2.4 million years, with ice ages about five to ten times as long as the 10,000- to 20,000-year interglacials (Imbrie and Imbrie 1979).

In the last glacial period (the Wisconsin), the massive Laurentide ice sheet covered most of Canada and extended as far south as New York and Ohio. Boreal forest with spruce (*Picea* spp.) and jack pine *(Pinus banksiana)* moved southward and displaced mixed-deciduous forest in much of the eastern United States (Delcourt and Delcourt 1993). Glaciers covered the tops of the Rocky Mountains and the Sierra Nevada in the western United States and the Sierra Madre del Sur in south-central Mexico. Now-dry playa lakes in the Great Basin were full. Enough water was tied up in ice on land to lower sea level about 100 m.

Plant remains in ancient packrat (*Neotoma* spp.) middens document the expansion of woodland trees and shrubs into desert elevations from 45,000 to 11,000 years ago (radiocarbon years before 1950; Van Devender 1990). Woodlands with singleleaf pinyon *(Pinus monophylla)*, junipers (*Juniperus* spp.), shrub live oak *(Quercus turbinella)*, and Joshua tree *(Yucca brevifolia)* were widespread in the present Arizona Upland subdivision of the Sonoran Desert in Arizona (Van Devender 1990) and the southern Mohave Desert in California (Spaulding 1990). Ice age climates with greater winter rainfall from the Pacific and reduced summer monsoonal rainfall from the tropical oceans likely favored woody cool-season shrubs with northern affinities and spring annuals (Neilson 1986), rather than the summer-rainfall trees, shrubs, and cacti of tropical forests and subtropical deserts. Warm desertscrub communities dominated by creosotebush were restricted to below 300-m elevation in the lower Colorado River Valley in the Sonoran Desert and to the southern Chihuahuan Desert (Van Devender 1990).

In the Puerto Blanco Mountains in Organ Pipe Cactus National

Monument, saguaro and brittlebush *(Encelia farinosa)* returned from Sonora soon after the beginning of the Holocene, about 11,000 years ago (Van Devender 1990). Sonoran desertscrub formed about 9,000 years ago when the last woodland-chaparral plants retreated upslope. However, relatively modern desertscrub communities did not develop until about 4,500 years ago with the arrival of foothills paloverde, ironwood, and organpipe cactus.

Similar successional stages likely occurred during each of fifteen to twenty interglacials. Although the late Holocene desertscrub communities likely resembled the original late Miocene Sonoran Desert, relatively modern communities were only developed for about 5–10 percent of the 2.4 million years of the Pleistocene (Porter 1989; Winograd et al. 1997); ice age woodlands were in desert lowlands for about 90 percent of this period.

Tertiary Environments and the Evolution of the Gopher Tortoises

Tortoises apparently evolved from the batagurines, either a family (Bataguridae) or a subfamily of the Emydidae (pond turtles), now restricted to southeastern Asia and the New World tropics (McCord, ch. 3 of this volume). Today, one species, the Central American wood turtle *(Rhinoclemmys pulcherrima)*, ranges from South and Central America to southern Sonora. *Hadrianus*, the first North American tortoise, apparently immigrated from Eurasia through the high-latitude paratropical rain forest in the early Eocene (McCord, ch. 3 of this volume). In the Eocene, many plant and animal groups, including tortoises, began adaptive radiations in response to more open, diverse, and variable habitats available with the development of tropical deciduous forest. By the early Oligocene (37 million years ago), the descendants of *Hadrianus* had diverged into the genera *Gopherus, Hesperotestudo* (formerly a subgenus of *Geochelone*), and *Stylemys* (McCord, ch. 3 of this volume).

The divergence into the modern desert tortoise *(G. agassizii)* and Bolson and gopher tortoises *(Gopherus flavomarginatus* and *G. polyphemus)* lineages occurred as the North American herpetofauna was modernizing in the late Oligocene–early Miocene, in part by immigration (Holman 1995). The modern colubrids (Colubridae) appeared in North America in the late Oligocene, followed by the coral snakes (Elapidae) and pit vipers (Viperidae) in the early Miocene. All of these groups had earlier fossil records in Europe. Presumably the Bering Strait region, as in later dispersals, was the primary intercontinental route.

James L. Jarchow (pers. comm. 1999) suggested that the present southern limits of the desert tortoise in Sinaloa are probably related to the growth of disease-causing bacteria, other pathogens, or fungi in soil in the shelters in response to the extended hot, wet, humid conditions during the tropical summer rainy season. Desert tortoises are very susceptible to infections caused by the bacterium *Aeromonas hydrophila*, which thrives in moist soil. It first attacks the tortoise as a skin infection, but can spread quickly throughout the body via the blood stream, causing serious damage to various body systems and even death. Based on limited bacteriological surveys, *A. hydrophila* does not appear to be a major pathogen in the gopher tortoise (*G. polyphemus*; Elliott R. Jacobson, pers. comm. 2000), which lives in humid tropical areas in the southeastern United States. The evolutionary divergence of the Bolson-gopher tortoise lineage in the early Miocene was interpreted as adaption for more efficient burrowing, especially in the shoulder girdle (Bramble 1982). Fossils in this lineage are typically relatively complete, often with limb or skull elements preserved, as if they died in their burrows. It is interesting to speculate that the development of resistance to soil pathogens might have helped in their success as burrowers and that the susceptibility of the desert-Texas tortoise lineage *(Xerobates)* might have led to active selection to keep burrow use to the minimum needed for protection from environmental extremes and predators.

The east–west distributions of the modern desert and Texas *(Gopherus berlandieri)* tortoises are similar to those of various other plants and animals. Some of them reflect the uplift of the Sierra Madres in the early to middle Miocene, whereas others reflect the expansions of woodlands in the Pleistocene (Van Devender 2001). However, the fossil record of the desert-Texas tortoise lineage, which is characterized by primitive characters relative to the Bolson-gopher tortoise lineage, is poor and provides a confusing evolutionary picture. Lamb and Lydeard's (1994) molecular phylogeny of the gopher tortoises simply implies that desert and Texas tortoises evolved from a common ancestor.

The earliest fossil record in the desert-Texas tortoise lineage that has been well studied was a middle Pleistocene (likely 500,000 years or less in age) fossil Texas tortoise from Aguascalientes, an interior Mexican state, west of its modern range along the Gulf of Mexico (Mooser 1972). However, the oldest report for the desert tortoise is from the upper portion of the Palm Springs Formation in Anza Borrego Desert State Park of California (Cassiliano 1994). The fossils are in the Arroyo Seco and Vallecito Creek local faunas in the late Blancan-Irvingtonian LMA (3.1 to 0.75 million years ago). The picture is further confused by fossils of a small late Pleistocene tortoise from the coast of the Gulf of California in Sonora within the desert tortoises' modern range that are very similar to the Texas tortoise (Dennis M. Bramble cited in Lamb et al. 1989) and others from several middle to late Pleistocene sites in southern Arizona (Kevin B. Moodie, pers. comm. 1999). Study of these fossils or discoveries of new fossils, especially from Miocene or Pliocene sites in central or northern Mexico, would very likely precipitate revision and new interpretations of the evolution and biogeography of the desert tortoises.

Activity Periods

The activity periods of tortoises have changed dramatically through time. In the early Eocene tropical rain forests at low latitudes, seasons were not especially differentiated. Tortoises likely only used shelters to avoid predators or to dry out. The fossil records of tropical plants and animals indicate that equable climates with essentially frost-free winters extended to Arctic latitudes (Estes and Hutchison 1980). Their presence at latitudes with six-month-long nights suggests that polar inactivity periods initiated by light and dark cycles developed earlier than the hot-dry (estivation), cold (hibernation), or photoperiodic (deciduousness) triggers of lower latitudes. The dispersal of *Hadrianus* from Asia to North America through the paratropical rain forest may have prepared it and its descendants to the seasonal climates of tropical deciduous forest, thornscrub, and desertscrub that were to develop at lower latitudes much later.

In the Eocene, lower-latitude tropical forests developed wet and dry seasons. Leaves fell at the beginning of the dry season, as a cold winter did not yet exist. Tortoises may have begun to estivate during the warm, dry season, initiating the use of shelters to buffer environmental conditions as well as avoid predators. Bramble (1971) suggested that early *Gopherus* were more accomplished burrowers than contemporaneous *Stylemys*, although the abundance of complete specimens of the latter would argue that both were burrowers.

In the Miocene, regional uplift established the modern climatic and biotic regimes. Cold was likely a new environmental factor that affected tortoise activity, at least at higher elevations. As the deserts developed in the middle and late Miocene, estivation intensified as the summer rainy season was shortened, the arid foresummer developed, and climate became more variable (i.e., had periodic extended droughts). Tortoises were still primarily active in the warm, summer rainy season and inactive in the spring dry season. Shelters became increasingly important in buffering low humidity and high temperatures.

In the Pleistocene, there were shifts to winter rainfall and reductions in summer rainfall in the modern Mohave and Sonoran Deserts. In the Sonoran Desert, tortoises were probably still inactive for first half of year even if spring annuals were available. Hibernation developed as winters become colder. Shelters still primarily buffered extremes in humidity but also cold and hot temperatures. Biseasonal rainfall shifted southward in Sonora, possibly extending the range of the desert tortoise southward in Sinaloa. Pleistocene fossils should be looked for in Sinaloa and Nayarit. Considering fossils from the McKittrick Asphalt in California (Miller 1942) and caves and packrat midden sites in New Mexico and Texas (Van Devender et al. 1976), the desert tortoise's distribution was considerably larger in the latest Pleistocene than today. Unlike the modern desert tortoises, the now-extinct desert tortoises in the northern Chihuahuan Desert were living in pinyon-juniper-oak woodland.

Evolution of the Mohave Tortoise

Lamb and McLuckie (ch. 4 of this volume) put the cladogenesis (lineage separation) of Sonoran and Mohave tortoises at 5 million years ago based on mitochondrial DNA data, supposedly linked to the east–west separation caused by the lakes or marine waters of the Bouse Formation in the lower Colorado River Valley between Arizona and California. However, this event did not reflect the timing of the evolution of the Mohave tortoise because warmer-than-modern climates in the late Miocene and Pliocene would have reinforced the characteristics of Sonoran or Sinaloan tortoises rather than selecting new ones related to the uniseasonal winter-rainfall climate of the Mohave Desert. Several factors including the movement of land west of the San Andreas Fault hundreds of kilometers to the northwest, the Pleistocene formation of the Mohave Desert, the expansion of woodlands in the Mohave Desert lowlands for 80–90 percent of the last 2.4 million years, and the Pleistocene immigration of creosotebush from South America suggest that the Bouse Formation scenario needs to be reconsidered.

I suggest that the evolution of the Mohave tortoise occurred in the Pleistocene (2.4 million years ago). A population can respond to a major climatic change in one of three ways: extinction, speciation (adapting to the new climate), or range adjustments. The packrat midden record for the last major glacial-interglacial climate change at about 11,000 years ago demonstrated that most species simply adjust their geographical and elevational ranges during a climate change. In the Mohave Desert bioregion, the Pleistocene was a time of evolution of new adaptations to the uniseasonal winter-rainfall Mohave Desert bioregion. The evolution of the diverse spring annual flora from temperate herbaceous perennial ancestors was an important development in the biota.

With loss of summer rainfall in the northern Sonoran Desert during glacial periods, most summer-rainfall species simply retracted their ranges to the southeast. However, the desert tortoise population apparently evolved into the Mohave tortoise. It is easy to visualize geographic isolation at the local population scale best suited for the selection of new adaptive characters. With the reduction of summer rainfall, most tortoise populations in the modern Mohave Desert would have declined dramatically or disappeared. Only individuals with spring characteristics survived. It is intriguing to think that the evolution of the Mohave tortoise might have been, in part, due to the establishment of juvenile characters of their tropical ancestors in the breeding population. Certainly, hatchling and juvenile tortoises are active at relatively cool temperatures, spend a great deal of time in burrows or shelters, and have a diet dominated by protein-rich herbs including spring and summer annuals (Morafka 1994).

In Mohave desert tortoises, the activity period shifted to spring, and estivation began in early summer and extended through the warm season to merge with fall-winter hibernation. The excavation of extensive burrows allowed Mohave tortoises to survive the extended inactivity periods including colder winter temperatures than tortoises experience in the Sonoran Desert. The diet shifted to primarily fresh spring annuals with less dry grasses and mallows. Survivorship of juveniles was enhanced by laying eggs in several clutches in May, late June, and August, although the number of eggs per year was essentially the same as in the Sonoran tortoise (Averill-Murray et al., ch. 7 of this volume). The packrat midden record (Spaulding 1990) suggests that the general vegetation of the Mohave Desert was a pinyon-juniper woodland or chaparral rather than a desertscrub for 80–90 percent of last 2.4 million years of the Pleistocene. For much of this time, the Mohave tortoise has been a woodland tortoise rather than a desert tortoise.

If the Mohave tortoise evolved as winter-rainfall climates and the Mohave Desert formed in the Pleistocene, it raises the question of the 3.2 million–year lag between the time of lineage separation estimated from the DNA and the development of winter-rainfall climates. The discrepancy could well be that the DNA data reflects evolution at a molecular level that is not subject to natural selection, whereas the behavioral and physiological characteristics that make the Mohave tortoise distinctive are environmentally linked, and thus the result of natural selection. Apparently, the connections between DNA evolution and the phenotypic expressions of morphology, behavior, and physiology that are so important at population and species levels remain unclear.

ACKNOWLEDGMENTS

I thank Roy Averill-Murray and David Morafka for their careful and stimulating reviews of the manuscript. Kevin Moodie provided access to the fossil tortoise from near Picacho Peak, Arizona, in the University of Arizona Paleontological Collection. Jim Jarchow and Elliott Jacobson shared their insight into the possible role of infectious bacteria in limiting desert tortoises in tropical habitats.

LITERATURE CITED

Axelrod, D. I. 1979. Age and origin of the Sonoran Desert. *California Academy of Sciences Occasional Paper* 132:1–74.

————. 1983. Paleobotanical history of the western desert. Pp. 113–29 *in* S. G. Wells and D. R. Haragan, eds. *Origin and evolution of deserts.* University of New Mexico Press, Albuquerque.

Axelrod, D. I., and H. P. Bailey. 1969. Paleotemperature analysis of Tertiary floras. *Palaeography, Palaeoclimatology, Palaeoecology* 6:163–95.

Bramble, D. M. 1971. Functional morphology, evolution, and paleoecology of gopher tortoises. Ph.D. diss., University of California, Berkeley.

————. 1982. *Scaptochelys:* Generic revision and evolution of gopher tortoises. *Copeia* 1982: 852–67.

Cassiliano, M. L. 1994. Paleoecology and taphonomy of vertebrate faunas from the Anza-Borrego Desert of California. Ph.D. diss., University of Arizona, Tucson.

Cochemé, J. J., and A. Demant. 1991. Geology of the Yécora area, north Sierra Madre, northern Sierra Madre Occidental, Mexico. *Geological Society of America Special Paper* 254:81–94.

Cornejo, D. O. 1994. Morphological evolution and biogeography of Mexican columnar cacti, tribe Pachycereeae, Cactaceae. Ph.D. diss., University of Texas, Austin.

Davis, G. P., Jr. 1982. *Man and wildlife in Arizona: The American exploration period, 1824–1865.* Arizona Game and Fish Department, Phoenix.

Dawson, M. R., R. M. West, W. Langston Jr., and J. H. Hutchinson. 1976. Paleogene terrestrial vertebrates: northernmost occurrence, Ellesmere Island, Canada. *Science* 192:781–82.

Delcourt, P. A., and H. R. Delcourt. 1993. Paleoclimates, paleovegetation, and paleofloras during the late Quaternary. Pp. 71–94 *in* Flora of North America Editorial Committee, eds. *Flora of North America north of Mexico.* Oxford University Press, New York.

Estes, R., and J. H. Hutchison. 1980. Eocene lower vertebrates from Ellesmere Island, Canadian arctic archipelago. *Palaeogeography, Palaeoclimatology, Palaeoecology* 30:325–47.

Felger, R. S., and E. Joyal. 1999. The palms (Arecaceae) of Sonora, Mexico. *Aliso* 18:1–18.

Graham, A. 1993. History of the vegetation: Cretaceous (Maastrichian)-Tertiary. Pp. 57–70 *in* Flora of North America Editorial Committee, eds. *Flora of North America north of Mexico.* Oxford University Press, New York.

————. 1999. *Late Cretaceous and Cenozoic history of North American vegetation: North of Mexico.* Oxford University Press, New York.

Graham, A., and D. Dilcher. 1995. The Cenozoic record of tropical dry forests in northern Latin America and the southern United States. Pp. 124–45 *in* S. H. Bullock, H. A. Mooney, and E. Medina, eds. *Seasonally dry tropical forests.* Cambridge University Press, New York.

Gregory, K. M. 1994. Palaeoclimate and palaeoelevation of the 35 ma Florissant flora, Front Range, Colorado. *Palaeoclimates* 1:23–57.

Grismer, L. L. 1994. The origin and evolution of the peninsular herpetofauna of Baja California. *Herpetological Review* 2:51–106.

Henry, C. D. and J. J. Aranda. 1992. The real southern Basin and Range: mid- to late Cenozoic extension in Mexico. *Geology* 20:701–4.

Holman, J. A. 1995. *Pleistocene amphibians and reptiles of North America.* Oxford University Press, New York.

Huber, N. K. 1981. Amount and timing of the Cenozoic uplift and tilt of the Sierra Nevada,

California: Evidence from the upper San Joaquin Basin. *U.S. Geological Survey Professional Paper* 1197:1–28.

Hutchison, J. H. 1992. Western North American reptile and amphibian record across the Eocene/Oligocene boundary and its climatic implications. Pp. 451–63 *in* D. R. Prothero and W. A. Berggren, eds. *Eocene-Oligocene climatic and biotic evolution*. Princeton University Press, Princeton, N.J.

Imbrie, J., and K. P. Imbrie. 1979. *Ice ages: Solving the mystery*. Enslow, Hillside, N.J.

Johnson, N. M., C. B. Officer, N. D. Opdyke, G. D. Woodward, P. D. Zeither, and E. H. Lindsay. 1983. Rates of late Cenozoic tectonism in the Vallecitos–Fish Creek Basin, western Imperial Valley, California. *Geology* 11:664–67.

Jolly, D. W. 2000. Fossil turtles and tortoises of Anza-Borrego Desert State Park, California. M.S. thesis, Northern Arizona University, Flagstaff.

Lamb, T., and C. Lydeard. 1994. A molecular phylogeny of the gopher tortoise, with comments on familial relationships within the Testudinoidea. *Molecular Phylogenetics and Evolution* 3:283–91.

Lamb, T., J. C. Avise, and J. W. Gibbons. 1989. Phylogenetic patterns in mitochondrial DNA of the desert tortoise *(Xerobates agassizii)*, and evolutionary relationships among the North American gopher tortoises. *Evolution* 43:76–87.

Leopold, E. B., and H. D. MacGinitie. 1972. Development and affinities of Tertiary floras in the Rocky Mountains. Pp. 147–200 *in* A. Graham, ed. *Floristics and paleofloristics of Asia and eastern North America*. Elsevier, Amsterdam.

Leopold, E. B., G. Liu, and S. Clay-Poole. 1992. Low-biomass vegetation in the Oligocene. Pp. 399–420 *in* D. R. Prothero and W. A. Berggren, eds. *Eocene-Oligocene climatic and biotic evolution*. Princeton University Press, Princeton, N.J.

Lindsay, E. H. 1984. Late Cenozoic mammals from northwestern Mexico. *Journal of Vertebrate Paleontology* 4:208–15.

MacGinitie, H. D. 1953. *Fossil plants of the Florissant Beds, Colorado*. Publication no. 599, Carnegie Institution of Washington, Washington, D.C.

Martin, P. S. 1984. Pleistocene overkill: The global model. Pp. 354–403 *in* P. S. Martin and R. Klein, eds. *Pleistocene extinctions*. University of Arizona Press, Tucson.

Martin, P. S., D. Yetman, M. Fishbein, P. Jenkins, T. R. Van Devender, and R. K. Wilson. 1998. *Gentry's Río Mayo plants: The tropical deciduous forest and environs of northwest Mexico*. University of Arizona Press, Tucson.

Menges, C. M., and P. A. Pearthree. 1989. Late Cenozoic tectonism in central Sonora and its impact on regional landscape evolution. *Arizona Geological Society Digest* 17:649–80.

Miller, L. H. 1942. A Pleistocene tortoise from the McKittrick Asphalt. *Transactions of the San Diego Society of Natural History* 9:439–42.

Mooser, O. 1972. A new species of Pleistocene fossil tortoise, genus *Gopherus*, from Aguascalientes, Aguascalientes, Mexico. *Southwestern Naturalist* 17:61–65.

Morafka, D. J. 1994. Neonates: Missing links in the life histories of North American tortoises. Pp. 161–73 *in* R. B. Bury and D. J. Germano, eds. *Biology of North American tortoises*. Fish and Wildlife Research no. 13, U.S. Dept. of the Interior National Biological Survey, Washington, D.C.

Morafka, D. J., G. A. Adest, L. M. Reyes, G. Aguirre L., and S. S. Lieberman. 1992. Differentiation of North American deserts: A phylogenetic evaluation of a variance model. *Tulane Studies in Zoology and Botany* 1(Suppl.):195–226.

Neilson, R. P. 1986. High-resolution climatic analysis and Southwest biogeography. *Science* 232:27–34.

Norell, M. A. 1989. Late Cenozoic lizards of the Anza-Borrego Desert, California. *Natural History Museum of Los Angeles County Contributions in Science* 414:1–31.

Phillips, S. J., and P. W. Comus, eds. 2000. *A natural history of the Sonoran Desert*. Arizona-Sonora Desert Museum Press, Tucson.

Porter, S. C. 1989. Some geological implications of average Quaternary glacial conditions. *Quaternary Research* 32:245–61.

Remeika, P. 1994. Lower Pliocene angiosperm hardwoods from the Vallecitos–Fish Creek Basin, Anza Borrego Desert State Park: Deltaic stratigraphy, paleoclimate, paleoenvironment, and phytographic significance. *San Bernardino County Museum Association Quarterly* 41:26–27.

Roldán, J., and K. F. Clark. 1992. An overview of the geology and mineral deposits of the northern Sierra Madre Occidental and adjacent areas. Pp. 39–65 *in* K. F. Clark, J. Roldán, and R. H. Schmidt, eds. *Geology and mineral resources of the northern Sierra Madre Occidental, Mexico*. El Paso Geological Society, El Paso, Tex.

Shaw, C. A., and H. G. McDonald. 1987. First record of giant anteater (*Xenartha*, Myrmecophagidae) in North America. *Science* 26:186–88.

Spaulding, W. G. 1990. Vegetational and climatic development of the Mojave Desert: The last glacial maximum to the present. Pp. 166–69 *in* J. L. Betancourt, T. R. Van Devender, and P. S. Martin, eds. *Packrat middens: The last 40,000 years of biotic change*. University of Arizona Press, Tucson.

Spencer, J. E., and P. J. Patchett. 1997. Sr isotope evidence for a lacustrine origin for the upper Miocene to Pliocene Bouse Formation, lower Colorado River trough, and implications for the uplift of the Colorado Plateau. *Geological Society of America Bulletin* 109:767–78.

Standley, P. C. 1924. Trees and shrubs of Mexico (Passifloraceae-Scrophulariaceae). *Contributions from the United States National Herbarium* 23:849–1312.

Swanson, E., and D. Wark. 1988. Mid-Tertiary silic volcanism in Chihuahua, Mexico. Pp. 229–39 *in* K. F. Clark, P. C. Goodell, and J. M. Hoffer, eds. *Stratigraphy, tectonics and resources of parts of the Sierra Madre Occidental Province, Mexico*. El Paso Geological Society, El Paso, Tex.

Tiffany, B. H. 1985. Geological factors and the evolution of plants. Pp. 1–10 *in* B. H. Tiffany, ed. *Geological factors and the evolution of plants*. Yale University Press, New Haven, Conn.

Turner, R. M., and D. E. Brown. 1982. Sonoran desertscrub. *Desert Plants* 4:121–81.

Van Devender, T. R. 1990. Late Quaternary vegetation and climate of the Sonoran Desert, United States and Mexico. Pp. 134–65 *in* J. L. Betancourt, T. R. Van Devender, and P. S. Martin, eds. *Packrat middens: The last 40,000 years of biotic change*. University of Arizona Press, Tucson.

———. 1995. Desert grassland history: changing climates, evolution, biogeography, and community dynamics. Pp. 68–99 *in* M. P. McClaran and T. R. Van Devender, eds. *The Desert Grassland*. University of Arizona Press, Tucson.

———. 1999. Evolution of the Sonoran Desert. Pp. 221–28 *in* D. Vasquez del Castillo,

M. Ortega N., and C. A. Yocupicio C., eds. *Symposium Internacional sobre la Utilización y Aprovechamiento de la Flora Silvestre de Zonas Áridas*. Universidad de Sonora, Hermosillo.

———. 2001. Deep history and biogeography of *La Frontera*. Pp. 56–83 *in* G. L. Webster and C. J. Bahre, eds. *Vegetation and flora of* la frontera: *Vegetation change along the United States–Mexican boundary*. University of New Mexico Press, Albuquerque.

Van Devender, T. R., K. B. Moodie, and A. H. Harris. 1976. The desert tortoise *(Gopherus agassizii)* in the Pleistocene of the northern Chihuahuan Desert. *Herpetologica* 32:298–304.

Van Devender, T. R., A. M. Rea, and M. L. Smith. 1985. The Sangamon interglacial vertebrate fauna from Rancho la Brisca, Sonora. *Transactions of the San Diego Society of Natural History* 21:23–55.

Van Devender, T. R., C. H. Lowe, and H. E. Lawler. 1994. Factors influencing the distribution of the Neotropical vine snake *(Oxybelis aeneus)* in Arizona and Sonora. *Herpetological Natural History* 2:25–42.

Van Devender, T. R., A. C. Sanders, R. K. Wilson, and S. A. Meyer. 2000. Flora and vegetation of the Río Cuchujaqui, a tropical deciduous forest near Alamos, Sonora, México. Pp. 36–101 *in* R. H. Robichaux and D. Yetman, eds. *The tropical deciduous forest of the Alamos: Biodiversity of a threatened ecosystem*. University of Arizona Press, Tucson.

Winker, C. D. 1987. Neogene stratigraphy of the Fish Creek–Vallecito section, southern California: Implications for early history of the northern Gulf of California. Ph.D. diss., University of Arizona, Tucson.

Winograd, I. J., J. M. Landwehr, K. R. Ludwig, T. B. Coplen, and A. C. Riggs. 1997. Duration and structure of the past four interglaciations. *Quaternary Research* 48:141–54.

Wolfe, J. A. 1971. Tertiary climatic fluctuations and methods of analysis of Tertiary floras. *Palaeogeography, Palaeontology, and Palaeoecology* 9:27–57.

———. 1972. An interpretation of Alaskan Tertiary floras. Pp. 201–33 *in* A. Graham, ed. *Floristics and paleofloristics of Asia and eastern North America*. Elsevier, Amsterdam.

———. 1977. *Paleogene floras from the Gulf of Alaska region*. Professional Paper no. 997, U.S. Geological Survey, Washington, D.C.

———. 1992. Climatic, floristic, and vegetational changes near the Eocene/Oligocene boundary in North America. Pp. 421–36 *in* D. R. Prothero and W. A. Berggren, eds. *Eocene-Oligocene climatic and biotic evolution*. Princeton University Press, Princeton, N.J.

———. 1993. *A method of obtaining climatic parameters from leaf assemblages*. Professional Paper no. 1964, U.S. Geological Survey, Washington, D.C.

Wolfe, J. A., and D. Hopkins. 1967. Climatic changes recorded by Tertiary land floras in northwestern North America. Pp. 67–76 *in* K. Hatai, ed. *Proceedings of the Tertiary correlations and climatic changes in the Pacific symposium*, 11th Pacific Scientific Congress, Tokyo, Japan.

CHAPTER 3

Fossil History and Evolution of the Gopher Tortoises (Genus *Gopherus*)

ROBERT D. McCORD

The history of the desert tortoise *(Gopherus agassizii)* and its ancestors spans nearly the entire Cenozoic. This tortoise evolved in a world far different from that of today, one with different vegetation, climate, and other environmental factors, and on a continent with a large diversity of tortoise species. An appreciation of the desert tortoise's history is essential to understanding its place in the world today.

Evolution and Immigration of Tortoises

The story of the desert tortoise begins with the origin of the truly terrestrial turtles, the tortoises (Testudinidae). Morphological (Hirayama 1984; Gaffney and Meylan 1988) and chromosomal (Bickman and Baker 1976; Carr and Bickman 1986) evidence suggests an origin of the testudinids from the batagurines, a family of mostly Asian aquatic turtles formerly included within the pond turtles (Emyidae). *Rhinoclemmys pulcherrima* (painted wood turtle) is a New World batagurine whose range extends from South and Central America northward into southern Sonora, where it is sympatric with the desert tortoise. Tortoises likely had their origin in Asia, today's center of diversity for the batagurines. In fact, possible Paleocene tortoises have been reported from Asia (Yeh 1974, 1979).

In the early Eocene (50 million years ago) tortoises of the genus *Hadrianus*, considered by some to be the same genus as the extant *Manouria*, suddenly appeared on the European and North American continents (de Broin 1977; Hutchison 1980; McCord 1996). Today *Manouria* is represented by the Asian brown tortoise *(M. emys)* and the impressed tortoise *(M. impressa)* of Indochina. This sudden appearance in the fossil record without suitable local antecedents strongly suggests that *Hadrianus* immigrated from

Asia. Moreover, the identification of a particular genus as being the ancestor to the entire subsequent radiation is a rare paleontological event. Today, *Manouria* inhabits tropical regions in the Indochina Peninsula, and it is likely that these Eocene *Hadrianus* also entered a largely tropical American continent. In fact, Eocene *Hadrianus* have been discovered about 10° south of the northern paleopole, along with such typically tropical creatures as crocodiles, monitor lizards (Varanidae), and the flying lemurs (dermopterans; Dawson et al. 1976; Estes and Hutchison 1980; McKenna 1980). These fascinating faunas, from the early and middle Eocene of Ellesmere Island, permit us a glimpse of the connections between North America, Europe, and Asia at that time. Many of the mammal taxa had affinities to Europe, but some seemed to have originated in Asia. Another interesting turtle known from Ellesmere Island is an anosteirine carettochelyid, an extinct subfamily of the pig-nose turtles (Anosterinae: Carettochelyidae). Today, this family is restricted to Asia, but has fossil records from Asia and North America.

By the late Eocene and early Oligocene, tortoises were diverse in North America, including several species of *Stylemys* (whose abundant, well-preserved remains are often seen for sale in rock shops) and tortoises in the genus *Hesperotestudo*. The latter were formerly classified as the North American representatives of *Geochelone*, but today are recognized as a discrete separate lineage of tortoises.

Stylemys was represented by several species of generally primitive tortoises that lived from the late Eocene to middle Miocene (40 to 10 million years ago). Size and presumably habitat varied greatly between the species. The lack of burrowing adaptations or burrow burial and varying abundance suggests a differentiation of a more mesic-adapted *Stylemys* and xeric-adapted *Gopherus* in the late Eocene and early Oligocene (34 million years ago; Bramble 1971).

Hesperotestudo were the giant tortoises of North America and bore broad similarities to the extant *Geochelone*. As with *Geochelone*, the many species varied greatly in size and morphology, with some species being by no means giant. *Hesperotestudo* lived from the Oligocene (25 million years ago) to late Pleistocene (12,000 years ago, radiocarbon years before 1950). In fact, one specimen discovered in Florida was killed with a sharpened wooden stake dated at 12,000 years ago (Clausen et al. 1979; Holman and Clausen 1984). No evidence of burrowing adaptation is evident in the genus, but some forms had extensive dermal armoring. Although this armor has generally been interpreted as an antipredator defense, it has also been suggested as an adaptation to resist desiccation and temperature fluctuation. *Hesperotestudo* may have been a more xeric-adapted, nonburrowing form than *Stylemys* (Bramble 1971).

Gopher Tortoise Evolution

The gopher tortoises *(Gopherus)* evolved in North America, with these many other tortoises from its origin to the end of the Pleistocene. From whence did this early *Gopherus* spring? Ultimately, the answer must be *Hadrianus*. Bramble (1971, 1982) suggested *Hadrianus utahensis* and *"Testudo" uintensis* (Gilmore 1915), likely synonyms, as candidates. The fossils are both known from the middle Eocene (45 million years ago) of Utah. These forms have some characters reminiscent of those in *Gopherus*, but on a whole were very primitive tortoises having much in common with the earlier *H. majusculus*. Moreover, *"T." uintensis* has been suggested as the ancestor of *Stylemys* (Auffenberg 1962). These inferred relationships have not yet been rigorously examined but seem to be the only candidates to date. Although many fossil tortoises are known, the appropriate portions that permit precise phylogenetic placement of those tortoises (e.g., skulls, complete shells, limbs) are frequently missing.

Studies have revealed a long history of association between *Gopherus* and *Stylemys* (Hay 1908; Williams 1950, 1952; Auffenberg 1964). Recent phylogenies of North American testudines (Bramble 1971; Crumly 1994; McCord 1997) do not preclude a close relationship (fig. 3.1), but do preclude a close, unique common ancestor. The shared character of a premaxillary ridge, often used to unite *Gopherus* and *Stylemys*, deserves special comment. Although certainly a derived (newly evolved) character for tortoises (Crumly 1994), the premaxillary ridge may well be a primitive character in North American tortoises. The condition in North American *Hadrianus* has not been reported, although the recent description (Parkham and Hutchison 1998) of a *Hadrianus* dentary does suggest that a ridge was present. One of the living *Manouria (M. emys)* possesses an incipient ridge, whereas the other does not (Crumly 1994). Furthermore, a primitive *Hesperotestudo* from the

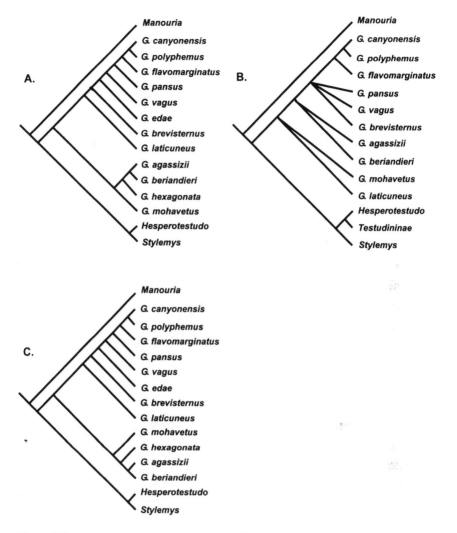

Figure 3.1. (A) The *Gopherus* phylogeny of Bramble (1971) presented as a dendrogram to facilitate comparison. (B) The Adams consensus tree of Crumly (1994). (C) An Adams consensus tree from the stratocladistic analysis of McCord (1997).

early Oligocene (Whitneyan, 31 million years ago) also has a partial premaxillary ridge (Hutchison 1996), thus strengthening the concept that the premaxillary ridge was a primitive character for the North American tortoises.

The Land Mammal Ages (LMA; e.g., Whitneyan), following the Oligocene here and elsewhere deserve some comment. LMAs are discrete times of varying duration characterized by discrete mammalian faunas. Mammals are employed in this scheme due to their high diversity and relatively rapid evolution. These LMAs are commonly used as a fine-scale, relative time scale by vertebrate paleontologists.

By the Oligocene (34 million years ago), the gopher tortoises had also appeared, either directly from *Hadrianus* or from an as-yet-unknown ancestor. Auffenberg (1966, 1976) recognized that the living gopher tortoises and many of the fossil species can be put into two groups based on similarity: the *flavomarginatus-polyphemus* group and the *agassizii-berlandieri* group (fig. 3.1). These groups have variously been recognized as the genera *Gopherus* and *Xerobates*, subgenera of *Gopherus* (Lowe 1990), and/or as a nonphylogenetic grade classification (i.e., primitive versus advanced gopher tortoises) not suitable for taxonomic recognition. To avoid confusion, I will hereafter refer to both lineages and their unique predecessors as "the gopher tortoises." Tortoises uniquely leading to the *flavomarginatus-polyphemus* group will be referred to as the "subgenus *Gopherus*."

There is general consensus (Bramble 1971; Crumly 1987, 1994; Lamb et al. 1989; Lamb and Lydeard 1994; McCord 1997) that the Bolson tortoise *(G. flavomarginatus)* and the gopher tortoise *(G. polyphemus)* form a holophyletic group or clade (evolutionary branches containing the ancestor and all descendants; fig. 3.1), although opinions vary as to which fossil species belong to the clade.[1] The Bolson tortoise has a good fossil record in the Pleistocene of the southwestern United States, including Arizona (McCord 1982; Archer 1989) and Texas (Van Devender and Bradley 1994), but not, apparently, in California. Unfortunately, dates are lacking on most of these fossil Bolson tortoises, but it is likely that they coexisted with the desert tortoise. The gopher tortoise is present today in Florida and adjacent states. Its fossil record suggests a range greater than that found today, but still absent from the Southwest. The lineage, which ultimately and uniquely led to the modern *G. polyphemus* and *G. flavomarginatus*, is our subgenus *Gopherus* and should not be confused with the more restrictive clade of the common ancestor of the modern species and all of its descendants, of which only one extinct species, *G. canyonensis*, is known.

I will now return to the history of *Gopherus* and its first certain representative. *Gopherus laticuneus* is known from late Eocene (Chadronian LMA, 35 million years ago) and early Oligocene (Orellan LMA, 33 million years ago) beds of Colorado, Wyoming, and South Dakota, in some cases existing in at least the same county and formation as *Stylemys* (Hutchison 1996). However, recent cladistic (Crumly 1994) and stratocladistic (McCord 1997) phylogenetic analyses place *G. laticuneus* as a primitive offshoot of the lineage ultimately leading to the common ancestor of today's *G. polyphemus* and *G. flavomarginatus*. Hutchison (1996) noted that *G. laticuneus* possesses several derived characters shared with the restricted subgenus *Gopherus* that were not employed in the cladistic and stratocladistic analyses; thus, he erected a new subgenus, *Oligopherus*, to contain *G. laticuneus*. This is an interesting hypothesis that deserves further study, but should not be accepted uncritically. The characters stated by Hutchison have not been subject to rigorous phylogenetic analysis; that is, *all* characters have not been examined to determine *Oligopherus*'s phylogenetic relationships. At least one of the characters (medial maxillary triturating ridge occasionally joining the median premaxillary ridge) is not present in the extant *G. flavomarginatus*. If *G. laticuneus* is a member of the restricted subgenus *Gopherus*, then the *Xerobates* lineage or lineages have an even longer unaccounted history than is currently thought. Incidentally, Bramble (1982) classified *G. laticuneus* as an early *Xerobates*, an interpretation perhaps based more on phenetic than phylogenetic grounds, as he earlier placed it within the lineage of subgenus *Gopherus* (Bramble 1971).

The restricted subgenus *Gopherus* is generally accepted to include the Miocene *G. brevisterna* (Hemingfordian LMA, 18 million years ago), *G. vaga* (Barstovian, 14 million years ago), and *G. pansa* (Barstovian LMA; Bramble 1971, 1982; Crumly 1994; McCord 1997). The *G. polyphemus* lineage apparently diverged from the *G. flavomarginatus* lineage by the Pliocene, with the Blancan *G. huecoensis* viewed as a synonym of *G. flavomarginatus* (Bramble 1971, 1982; Auffenberg 1974; Crumly 1994) and the Blancan *G. canyonensis* viewed as more closely related to *G. polyphemus* than *G. flavomarginatus* (Bramble 1971; Crumly 1994; McCord 1997).

The fossil histories of *G. agassizii* and *G. berlandieri* (Texas tortoise) are much less clear. These have been regarded as a single lineage, separate from the restricted subgenus *Gopherus* (Bramble 1971; McCord 1997) and classified as the genus *Scaptochelys* (Bramble 1982), later noted by Bour and Dubois (1984) to be a junior synonym of *Xerobates*. Alternately, the desert tortoise and the Texas tortoise as primitive grades of gopher tortoises, with no shared-derived characters uniting them (Crumly 1987, 1994), are best considered, along with the other extant North American tortoises, in the genus *Gopherus* (Crumly 1987). Although not defended in current literature, I am sure that there are some who would be comfortable recognizing a *Xerobates* grade based on primitive characters and a *Gopherus* grade based on derived characters, given their considerable phenetic difference (Bramble 1982).

The difficulties resolving the phylogeny of *"Xerobates"* are evident. First, the genus *Xerobates* is not defined by what is but rather by what is not (Bramble 1982). *Xerobates* is a gopher tortoise lacking the skull, cervical, and limb specializations, chiefly for burrowing, of the restricted subgenus *Gopherus*. Second, *Xerobates* has a poor fossil record. Regardless of what phylogeny is accepted for these tortoises, we would expect far older fossils.

Bramble (1971) hypothesized that *G. mohavetus* (as *mohavense*) of the middle Miocene of California represents the first recognized taxon belonging to the *Xerobates* clade. Subsequent cladistic (Crumly 1994) and stratocladistic (McCord 1997) analyses have not supported this interpretation and reveal *G. mohavetus* as an outgroup to the extant gopher tortoises.

Crumly (1994) analyzed the gopher tortoises *G. berlandieri* and *G. agassizii* cladistically and concluded that there was no evidence for (or against) uniting them as a clade. McCord (1997) reanalyzed Crumly's (1994) dataset stratocladistically and concluded that, on grounds of stratigraphic parsimony (the simplest explanation considering the ages of the fossils), the extant *Xerobates* may well form a clade, but that this clade excludes the known fossil species. Stratigraphic parsimony favors phylogenetic hypotheses that reconcile the observed and expected order of temporal occurrence (Fisher 1991, 1992). In other words, ancestors should precede descendants.

If *Xerobates* is a clade, then it is missing a substantial portion of its fossil history. If *G. laticuneus* does indeed belong to the restricted subgenus *Gopherus*, this would imply that its sister-group *Xerobates* is at least as old, namely the late Eocene or early Oligocene. More conservatively, if *G. brevisterna* is the oldest known member of the restricted subgenus

Gopherus, then *Xerobates* would be present in the early Miocene, about 19 million years ago.

Conversely, the oldest known member of the *Xerobates* group would be at about 14 million years ago, in the middle Miocene, if *G. mohavetus* is included. However, *G. berlandieri* (published as *"G. auffenbergi"* from Aguascalientes, an interior state in Mexico; Mooser 1972), which is probably not older than 0.5 million years, is the oldest secure record of *Xerobates*. Curiously, the range of *G. berlandieri* extends as far south as Aguascalientes today, but is restricted to the Atlantic coastal region along the Gulf of Mexico, far east of this fossil locality. It is also interesting to note that undated fossils resembling a "robust *G. berlandieri*" have been discovered near Desemboque, Sonora, near the coast of the Gulf of California (Lamb et al. 1989; Dennis Bramble, pers. comm. 1999).

Several observations are appropriate at this time. First, the likely cladogenesis (i.e., separation of lineages) of the two gopher tortoise clades, based on fossil evidence at about 19 million years ago, agrees reasonably well with those of 18 to 17 million years ago predicted on molecular grounds (Lamb and Lydeard 1994). It should be noted that in the discussion of these and subsequent cladogenetic events, all the traits that will ultimately characterize the lineage are not yet present; in fact, very little differentiation would be expected. Second, the existence of *G. berlandieri* in the late Pleistocene of Aguascalientes and Sonora, Mexico, suggests that an important and as yet largely unknown area of gopher tortoise evolution is in Mexico. Finally, an 18 million–year or more gap in *Xerobates* evolution may be present in the record. If *Xerobates* is considered a primitive grade of gopher tortoise rather than a clade, then this gap becomes even more significant because we then have to account for at least two lineages with no fossil records. It is tempting to speculate that the absence of specimens for *Xerobates* suggests a more upland and less burrowing habitus for this group, but this would be drawing conclusions from purely negative evidence.

Fossil Desert Tortoises

Our long journey has finally arrived at the desert tortoise itself. Numerous fossils of *G. agassizii* have been discovered in California. Sites there include Antelope Cave, Invanpah Mountains (11,080 years ago; Jefferson 1991); Cool Water Coal Gasification Site, Daggett (Rancholabrean Age, 500,000 to 11,000 years ago; Reynolds and Reynolds 1985); Kokoweef Cave (San Bernardino County; 9,850 years ago; Jefferson 1991); Ludlow Cave (San Bernardino County; Rancholabrean Age; Jefferson 1989); McKittrick Asphalt (Kern County; 38,000 years ago; Miller 1942); Mescal Range (Rancholabrean Age; Brattstrom 1958); Mitchell Caverns, Providence Mountains (Rancholabrean Age and Holocene; Jefferson 1991); Newberry Cave (San Bernardino County; Rancholabrean Age and Holocene; Davis and Smith 1981); Panamint Crater (Inyo County; Rancholabrean Age; Jefferson 1989); Schuiling Cave, Newberry Mountains (12,500 years ago; Downs et al. 1959); and Solar One Generation Station, Daggett (Rancholabrean Age and early Holocene; Reynolds and Reynolds 1985). In Arizona, *G. agassizii* fossils have been discovered in the Grand Canyon at Vulture Cave in radiocarbon-dated packrat *(Neotoma)* middens (30,000 to 10,250 years ago; Van Devender et al. 1977; Mead and Phillips 1981); in undated sediments at Picacho Peak (McCord 1994); at Rampart Cave (Grand Canyon) in association with packrat middens (19,000 to 12,230 years ago; Van Devender et al. 1977; Mead and Phillips 1981); and in packrat middens at Wellton Hills (Yuma County; 8,750 years ago; Van Devender and Mead 1978). From New Mexico, desert tortoise fossils have been reported from Conkling's Cavern and Shelter Cave on Bishop's Cap (12,520 to 11,130 years ago; Brattstrom 1961, 1964; Thompson et al. 1980); Dry Cave (Eddy County; 33,590 to 25,160 years ago; Van Devender et al. 1976); and Robledo Cave (Doña Ana County; no date; Van Devender et al. 1976). Even Texas has provided desert tortoise fossils in a Hueco Mountains midden (34,000 years ago; Van Devender and Moodie 1977).

Several things may be noted about this fossil record. First, desert tortoises formerly extended their range as far east as Dry Cave in eastern New Mexico and as far north and west as the McKittrick Asphalt in California, although not necessarily at the same time. Second, the fossil record of desert tortoises does not extend with certainty beyond the Rancholabrean LMA and, indeed, beyond the last 34,000 years.

Recently, tortoises reported as *G. agassizii* have been found from the Blancan LMA (Pliocene) and Irvington LMA (Pleistocene) sediments of the Anza-Borrego Desert State Park in California (Jolly 2000). The remains were isolated elements of variable reliability, with some elements best re-

ferred to a gopher tortoise and others having remarkable phenetic similarity to *G. agassizii*. Jolly did not indicate which elements were recovered at
which horizons. Although entoplastra were recovered, the confirmation of
the interclavicular keel, important in differentiating *Xerobates* from the restricted subgenus *Gopherus*, was not reported. None of these elements, especially singularly, can be taken uneqivocably to represent *G. agassizii*. Nevertheless, the findings provide a stimulating possibility for recovering some of
G. agassizii's missing record.

Other evidence of an earlier origin for the desert tortoise is provided
by the Aguascalientes *G. berlandieri* that, if it is the sister-species, pushes
the *G. agassizii* lineage to perhaps over 1 million years. But where are the
fossils? If *"Xerobates"* is indeed a paraphyletic grouping, the origin of the
desert tortoise lineage would be more than 19 million years ago, the age of
certain establishment of the restricted subgenus *Gopherus*. If this is the case,
we have an even greater time span with no fossils, and two separate lineages
(G. agassizii and *G. berlandieri)* missing.

Molecular phylogeny does predict a far older age for the *G. agassizii* lineage and some objective evidence for a large fossil gap (Lamb and
Lydeard 1994). If we assume Lamb and Lydeard's divergence rate of 0.4 percent per million years, then, based on genetic distance, *G. agassizii* should
have diverged from *G. berlandieri* about 10.5 million years ago in the middle
Miocene.

Sonoran versus Mohave Tortoises

Finally, we reach the focus of this book: the Sonoran Desert populations of the desert tortoises. Attempts to distinguish the morphology of
Mohave Desert from the Sonoran Desert populations with recent specimens
have met with limited success (Germano 1993; McLuckie et al. 1999). My
attempts to extend morphometric methodology to fossils have so far failed.
Estimates based on divergence of mitochondrial DNA (Lamb and Lydeard
1994) put this cladogenetic event at 6 million years ago (Lamb and Lydeard
1994) or 5 million years ago (Lamb and McLuckie, ch. 4 of this volume),
dates far older than the fossil record of the species. Such remarkable ancient divergence times for forms thought to be within the same species is
not without precedence. The type specimen to the extant subspecies *Kino-*

sternon flavescens arizonense is a Pliocene fossil found near Benson, Arizona (Iverson 1979).

Regardless of the timing of the divergence of the Mohave and Sonoran Desert populations, there can be no denying that a great deal of genetic distance is now present between them. What drove this differentiation? At least three scenarios have been proposed. The first and most obvious, but the least probable, is that the differentiation occurred as a result of the varying rainfall regimes within the ranges of the two populations, with the Mohave Desert populations existing within a winter-dominated rainfall system and the Sonoran Desert populations existing within a summer-dominated system (Van Devender, ch. 2 of this volume).

The second related hypothesis (these ideas can, do, and should intermix) as to the cause of genetic distance between the Mohave and Sonoran desert tortoise populations is that the Sonoran tortoise is a relic of the tropical deciduous forest–evolved population, such as the living so-called Sinaloan tortoises in southern Sonora and northern Sinaloa (Lowe 1990). First, this hypothesis postulates that the desert tortoise should be viewed as adapted to thornscrub, not desert—an observation that has been reinforced physiologically. Second, thornscrub tortoises do not inhabit valley bottoms due to heavy flooding during summer rains. Third, thornscrub tortoises live in a largely frost-free environment and have no need to construct extensive burrows. With this hypothesis, the Sonoran Desert populations continue a thornscrub-tortoise lifestyle because they can. The thermally stressed, relatively flood-free Mohave populations have adopted a more derived ecology.

A third hypothesis has the advantage of being the least refuted, but perhaps also currently has the least evidence to support it. This hypothesis is that the Sonoran desert tortoises' nonextensive burrowing and upland-dwelling habit is a result of competitive displacement or ecological exclusion by the Bolson tortoise. To put it another way, the Mohave population's ecology is determined by a lack of Bolson tortoises. This idea is suggested by the similar habits of the Bolson tortoise and the Mohave populations and the fact that the Bolson tortoise was widespread in Arizona during the Pleistocene (McCord 1982, 1994; Archer 1989) and absent in California (Jefferson 1991).

NOTE

1. *Holophyletic* is a term used to describe that portion of a phylogenetic tree that includes a common ancestor and all its descendants. *Paraphyletic* refers to a taxonomic grouping that includes some descendants of a common ancestor, but not all (Ashlock 1971).

LITERATURE CITED

Archer, B. 1989. Quarternary fossil tortoises of the Phoenix Basin. M.S. thesis, Arizona State University, Tempe.

Auffenberg, W. 1962. *Testudo amphithorax* Cope referred to *Stylemys. American Museum Novitiates* 2120:1–10.

———. 1964. A redefinition of the fossil tortoise genus *Stylemys* Leidy. *Journal of Paleontology* 38:316–24.

———. 1966. The carpus of land tortoises (Testudinidae). *Bulletin of the Florida State Museum* 10:159–91.

———. 1974. Checklist of fossil land tortoises (Testudinidae). *Bulletin of the Florida State Museum* 18:121–251.

———. 1976. The genus *Gopherus* (Testudinidae) osteology and relationships of extinct species. *Bulletin of the Florida State Museum* 20:47–110.

Bickman, J. W., and R. J. Baker. 1976. Chromosome homology and evolution of emydid turtles. *Chromosoma* 54:201–19.

Bour, R., and A. Dubois. 1984. *Xerobates* Agassiz, 1857, synonyme plus ancien de *Scaptochelys* Bramble, 1982 (Reptilia, Chelonii, Testudinidae). *Bulletin Mensuel Société Linnéenne de Lyon* 53:30–32.

Bramble, D. M. 1971. Functional morphology, evolution, and paleoecology of gopher tortoises. Ph.D. diss., University of California, Berkeley.

———. 1982. *Scaptochelys*: generic revision and evolution of gopher tortoises. *Copeia* 1982:852–67.

Brattstrom, B. H. 1958. New records of Cenozoic amphibians and reptiles from California. *Bulletin of the Southern California Academy of Sciences* 57:5–13.

———. 1961. Some new fossil tortoises from western North America with remarks on zoogeography and paleoecology of tortoises. *Journal of Paleontology* 35:543–60.

———. 1964. Amphibians and reptiles from cave deposits in south-central New Mexico. *Bulletin of the Southern California Academy of Science* 63:93–103.

Carr, J. L., and J. W. Bickman. 1986. Phylogenetic implications of karyotypic variation in the Batagurinae (Testudines: Emydidae). *Genetica* 70:89–106.

Clausen, C. J., A. D. Cohen, C. Emiliani, J. A. Holman, and J. J. Stipp. 1979. Little Salt Spring, Florida: A unique underwater site. *Science* 203:609–14.

Crumly, C. R. 1987. The genus name for North American gopher tortoises. *Proceedings of the Desert Tortoise Council Symposium* 1984:147–48.

————. 1994. The phylogenetic systematics of North American tortoises (genus *Gopherus*): Evidence for their classification. Pp. 7–32 *in* R. B. Bury and D. J. Germano, eds. *Biology of North American tortoises*. Fish and Wildlife Research no. 13, U.S. Dept. of the Interior National Biological Survey, Washington, D.C.

Davis, C. A., and G. A. Smith. 1981. *Newberry Cave*. San Bernardino County Museum Association, Redlands, Calif.

Dawson, M. R., R. M. West, W. Langston, and J. H. Hutchison. 1976. Paleocene terrestrial vertebrates: Northernmost occurrence, Ellesmere Island, Canada. *Science* 192:781–82.

de Broin, F. 1977. Contribution à l'étude des Chèloniens: Chèloniens continentaux du Crétacé et du Tertiare de France. *Mémoires des Muséum National d'Histoire Naturalle (new series) C* 39:1–366.

Downs, T., H. Howard, T. Clements, and G. A. Smith. 1959. Quarternary animals from Schuiling Cave in the Mojave Desert, California. *Los Angeles County Museum Contributions in Science* 29:1–21.

Estes, R., and J. H. Hutchison. 1980. Eocene lower vertebrates from Ellesmere Island, Canadian arctic archipelago. *Palaeogeography, Palaeoclimatology, Palaeoecology* 30:325–47.

Fisher, D. C. 1991. Phylogenetic analysis and its application in evolutionary paleobiology. Pp. 103–22 *in* N. L. Gilinsky and P. W. Signor, eds. *Analytical paleobiology: Short courses in paleontology*. Paleontological Society, Knoxville, Tenn.

————. 1992. Stratigraphic parsimony. Pp. 124–29 *in* W. P. Maddison and D. R. Maddison, eds. *MacClade*. Ver. 3.0. Manual. Sinauer Associates, Sunderland, Mass.

Gaffney, E. S., and P. A. Meylan. 1988. A phylogeny of turtles. Pp. 157–219 *in* M. J. Benton, ed. *The phylogeny and classification of the tetrapods*. Vol. I, *Amphibians, reptiles, birds*. Clarendon Press, Oxford, U.K.

Germano, D. J. 1993. Shell morphology of North American tortoises. *American Midland Naturalist* 129:319–35.

Gilmore, C. W. 1915. Fossil turtles from the Unita Formation. *Memoirs of the Carnegie Museum* 7:101–61.

Hay, O. P. 1908. *The fossil turtles of North America*. Publication no. 75, Carnegie Institute of Washington, Washington, D.C.

Hirayama, R. 1984. Cladistic analysis of batagurine turtles (Batagurinae: Emydidae: Testudinoidea): A preliminary result. *Studia Geologica Salmanticiensia Special Vol.* 1:141–57.

Holman, J. A., and C. J. Clausen. 1984. Fossil vertebrates associated with Paleo-Indian artifact at Little Salt Spring, Florida. *Journal of Vertebrate Paleontology* 4:146–54.

Hutchison, J. H. 1980. Turtle stratigraphy of the Willwood Formation, Wyoming: Preliminary results. *University of Michigan Papers on Paleontology* 24:115–18.

————. 1996. Testudines. Pp. 337–53 *in* D. R. Prothero and R. J. Emry, eds. *The terrestrial Eocene-Oligocene transition in North America*. Cambridge University Press. Cambridge, U.K.

Iverson, J. B. 1979. A taxonomic reappraisal of the yellow mud turtle, *Kinosternon flavescens* (Testudines: Kinosternidae). *Copeia* 1979:212–25.

Jefferson, G. T. 1989. Late Pleistocene and earliest Holocene fossil localities and vertebrate taxa from the western Mojave Desert. Pp. 27–40 *in* R. E. Reynolds, ed. *The west-central Mojave Desert: Quarternary studies between Kramer and Afton Canyon*. Special Publication, San Bernardino County Museum Association, Redlands, Calif.

————. 1991. *A catalogue of late Quarternary vertebrates from California*. Part 1. *Nonmarine lower vertebrates and avian taxa*. Technical Report no. 5, Natural History Museum of Los Angeles County, Los Angeles.

Jolly, D. W. 2000. Fossil turtles and tortoises of Anza-Borrego Desert State Park, California. M.S. thesis, Northern Arizona University, Flagstaff.

Lamb, T., and C. Lydeard. 1994. A molecular phylogeny of the gopher tortoises, with comments on familial relationships within the Testudinoidea. *Molecular Phylogenetics and Evolution* 3:283–91.

Lamb, T., J. C. Avise, and J. E. Gibbons. 1989. Phylogeographic patterns in mitochondrial DNA of the desert tortoise *(Xerobates agassizii)*, and evolutionary relationships among the North American gopher tortoises. *Evolution* 43:76–87.

Lowe, C. H. 1990. Are we killing the desert tortoise with love, science, and management? Pp. 84–106 *in* K. R. Beaman, F. Caporaso, S. McKeown, and M. D. Graff, eds. *Proceedings of the first international symposium on turtles and tortoises: Conservation and captive husbandry*, Chapman University, Chapman, Calif.

McCord, R. D. 1982. Fossil *Gopherus* from northern Yuma County, Arizona. *Arizona State University Anthropological Research Papers* 28:401–5.

————. 1994. Fossil tortoises of Arizona. Pp. 83–89 *in* D. Boaz, M. Dornan, and S. Bolander, eds. *Proceedings of the fossils of Arizona symposium*. Vol. 2. Mesa Southwest Museum and Southwest Paleontological Society, Mesa, Ariz.

————. 1996. Turtle biostratigraphy of late Cretaceous and early Tertiary continental deposits, San Juan Basin, New Mexico. Pp. 135–53 *in* D. Boaz, P. Dierking, M. Dornan, R. McGeorge, and B. J. Tegowski, eds. *Proceedings of the fossils of Arizona symposium*. Vol. 4. Mesa Southwest Museum and Southwest Paleontological Society, Mesa, Ariz.

————. 1997. Preliminary stratocladistic analysis of North American tortoises (Genus *Gopherus*). Pp. 87–89 *in* B. Anderson, D. Boaz, and R. D. McCord, eds. *Proceedings of the Southwest paleontological symposium*. Vol. 1. Mesa Southwest Museum and Southwest Paleontological Society, Mesa, Ariz.

McKenna, M. C. 1980. Eocene paleoaltitude, climate, and mammals of Ellesmere Island. *Paleogeography, Paleoclimatology, Paleoecology* 30:349–62.

McLuckie, A. M., T. Lamb, C. R. Schwalbe, and R. D. McCord. 1999. Genetic and morphometric assessment of an unusual tortoise *(Gopherus agassizii)* population in the Black Mountains of Arizona. *Journal of Herpetology* 33:36–44.

Mead, J. I., and A. M. Phillips. 1981. The late Pleistocene and Holocene fauna and flora of Vulture Cave, Grand Canyon, Arizona. *Southwestern Naturalist* 26:257–88.

Miller, L. H. 1942. A Pleistocene tortoise from the McKittrick Asphalt. *Transactions of the San Diego Society of Natural History* 9:439–42.

Mooser, O. 1972. A new species of Pleistocene fossil tortoise, genus *Gopherus*, from Aguascalientes, Mexico. *Southwestern Naturalist* 17:61–65.

Parkham, J. F., and J. H. Hutchison. 1998. New evidence on the origin and relationships of New World testudinids. *Journal of Vertebrate Paleontology* 18(Suppl.):69A.

Reynolds, R. E., and R. L. Reynolds. 1985. Mid-Pleistocene faunas of the Mojave Desert. Pp. 175–91 *in* R. Reynolds, ed. *The west-central Mojave Desert: Quarternary studies between Kramer and Afton County*. Special Publication, San Bernardino County Museum Association, Redlands, Calif.

Thompson, R. S., T. R. Van Devender, P. S. Martin, T. Foppe, and A. Long. 1980. Shasta ground sloth (*Nothrotheriops shastense* Hoffstetter) at Shelter Cave, New Mexico: Environment, diet, and extinction. *Quaternary Research* 14:360–76.

Van Devender, T. R., and G. L. Bradley. 1994. Late Quaternary amphibians and reptiles from Maravillas Canyon Cave, Texas, with discussion of the biogeography and evolution of the Chihuahuan Desert herpetofauna. Pp. 23–53 *in* P. R. Brown and J. W. Wright, eds. *Herpetology of the North American deserts.* Special Publication no. 5., Southwestern Herpetologists Society, Van Nuys, Calif.

Van Devender, T. R., and J. I. Mead. 1978. Early Holocene and late Pleistocene amphibians and reptiles in Sonoran Desert packrat middens. *Copeia* 1978:464–75.

Van Devender, T. R., and K. B. Moodie. 1977. The desert tortoise in the late Pleistocene with comments about its earlier history. *Proceedings of the Desert Tortoise Council Symposium* 1977:41–45.

Van Devender, T. R., K. B. Moodie, and A. H. Harris. 1976. The desert tortoise *(Gopherus agassizii)* in the Pleistocene of the northern Chihuahuan Desert. *Herpetologica* 32:298–304.

Van Devender, T. R., A. M. Phillips, and J. I. Mead. 1977. Late Pleistocene reptiles and small mammals from the lower Grand Canyon of Arizona. *Southwestern Naturalist* 22:49–66.

Williams, E. E. 1950. *Testudo cubensis* and the evolution of the Western Hemisphere tortoises. *American Museum of Natural History Bulletin* 95:7–36.

———. 1952. A new fossil tortoise from Mona Island, W.I., and a tentative arrangement of the tortoises of the world. *American Museum of Natural History Bulletin* 99:545–60.

Yeh, H. K. 1974. Cenozoic chelonian fossils from Nanhsiung, Kwantung. *Vertebrata Palasiatica* 12:26–37.

———. 1979. Paleocene turtles from Anhui. *Vertebrata Palasiatica* 7:49–56.

CHAPTER 4

Genetic Differences among Geographic Races of the Desert Tortoise

TRIP LAMB AND ANN M. McLUCKIE

Over the past decade, a disparate array of researchers have generated an equally diverse body of data geared toward a common goal: understanding the biology of the desert tortoise *(Gopherus agassizii)*. Through these efforts, entailing research on ecology, physiology, behavior, and conservation, the desert tortoise ranks as one of the more intensively studied turtles in the world. This chapter focuses on population genetics, emphasizing geographic variation and evolutionary history. We elaborate on molecular genetic approaches used to reconstruct the tortoise's past and stress how these same techniques may help safeguard its future. Given our central theme—relating geographic pattern to genetic process—we offer a brief overview of the term *geographic race* to provide some perspective on geographic variation and its evolutionary significance.

Geographic Races: A Brief History and Contemporary Perspective

Efforts to describe subgroups within species trace to Linnaeus (1751), who used the term *variety* to define distinguishable forms ranging from individual variants within a population to aggregates of similar populations occupying distinct geographic subdivisions of the species' range (i.e., geographic races). The breadth and ambiguity associated with varietal classification prompted the inception of another taxonomic category, *subspecies*, a term used to describe "geographical forms which cannot rank as full species" (Rothschild et al. 1894). The concept was quickly embraced, and geographic race became largely synonymized with subspecies. Subspecies are assigned formal taxonomic names, extending the Linnaean binomial system of nomenclature (e.g., *Crotalus viridis*, western rattlesnake) to include trinomials (e.g., *C. viridis cerberus*, Arizona black rattlesnake).

During the early 1900s, recognition of geographic variation pervaded taxonomy, and considerable effort was directed toward the identification and formal description of infraspecific variation. This preoccupation, in its extreme form, resulted in as many as 150 trinomial assignments within a species (Goldman 1935). The trend was fueled in part by the mindset of late-nineteenth-century taxonomists, who often based full species descriptions on minor morphological distinctions observed in microgeographic settings. For example, in 1886 C. Hart Merriam described eighty-six separate species of grizzly and brown bears on the basis of subtle skeletal differences among specimens in the U.S. National Museum; three species are currently recognized.

Today biologists must contend with the nomenclatural legacies left by the zealous taxonomic activities at the turn of the twentieth century. Current philosophical views now stress that classifications must reflect what can we can recover from evolutionary history (Wiley 1981). Moreover, analytical methodologies provide standard, objective approaches with which to infer evolutionary relationships among organisms and avoid nonhistorical (i.e., nonevolutionary) classifications (Wiley 1981). Accompanying the general acceptance of this classification paradigm, termed *phylogenetic systematics*, are refined concepts of species and, by extension, different views on infraspecific variation traditionally encompassed by the taxonomic categories subspecies and geographic race (Frost and Hillis 1990; Frost et al. 1992). Species are defined as distinct evolutionary lineages, entities with clear, nonreticulate bases; infraspecific variation, although recognized, seldom receives formal taxonomic recognition.

Many traditionally described subspecies whose diagnoses involved characters of marginal significance have since had their trinomials subsumed. In other cases, subspecific designations appear legitimate upon reappraisal, as distinctive character complexes with long-term adaptive or biogeographic bases are uncovered. (The latter examples are often elevated to specific status.) Studies of infraspecific geographic variation are still widely conducted, but tend to emphasize the historical processes that have shaped observed geographic patterns. Most of these studies are concerned with genetic variation, and one particularly fruitful approach for animal species involves analysis of a special form of DNA called mitochondrial DNA (mtDNA), which lies within the cell's energy-producing organelle, the mitochondrion.

Mitochondrial DNA

A Molecular Profile

The merit of animal mtDNA as a high-resolution genetic marker for assessing geographic variation is well founded (Avise 1994). Studies commonly document extensive variation among populations, often with some degree of geographical structure. Although several characteristics of mtDNA make it amenable to surveys of this nature, we stress the two more important factors. First, it is a rapidly evolving DNA molecule. MtDNA's principal method of evolutionary change involves substitution, a process whereby one of the four nucleotides (adenine [A], cytosine [C], guanine [G], and thymine [T]) — the basic building blocks of DNA — simply replaces another. This rapid rate of nucleotide substitution accounts for the wealth of mtDNA variation that typifies many animal species.

A second and far more significant feature of animal mtDNA is maternal inheritance. With very few exceptions (Zouros et al. 1994; Liu et al. 1996), mitochondria and hence their DNA are transmitted from the female parent to her offspring via the egg. The male parent contributes only nuclear DNA, not mtDNA. As a result, mtDNA is the maternal analogue of surname inheritance, where sons and daughters assume their father's surname, but only sons pass the surname on to the next generation. Biologically, both sons and daughters inherit their mother's mtDNA, but only daughters can transmit it to their offspring. Surname inheritance allows one to trace family lines and construct genealogies with relative ease. Similarly, the maternal transmission of mtDNA across generations greatly simplifies genetic bookkeeping. MtDNA genotypes function like names that can be traced from generation to generation as well as from location to location, providing an organismal family tree.

The Concept of Phylogeography

Because it is easy to track mtDNA genotypes across generations, mtDNA lineages can be viewed as extended pedigrees. Collectively, these extended pedigrees constitute a species' genealogical history, or entire matrilineal phylogeny, and provide a look at the evolutionary pathways taken by different groups of populations. It is not unusual for a species' mtDNA phylogeny to display some degree of geographic structure. For example, populations may become separated by a major historical event such as sea-level rise or mountain formation. As these isolated populations begin to diverge over time, genetic differences soon become apparent in their rapidly evolving mtDNA. The profound effect these events may impose on a species' evolutionary history, either through population isolation or integration, underscores the ongoing interaction between organismal history and Earth history at the infraspecific level. John Avise and coworkers at the University of Georgia coined a term that describes this relationship, *intraspecific phylogeography* (Avise et al. 1987; Avise 2000). We define intraspecific phylogeography as the historical (e.g., geologic, paleoclimatic) processes governing the geographical distributions of mtDNA genealogical lineages within a species.

Detecting Mitochondrial DNA Variation

The aforementioned characteristics of rapid evolution and maternal inheritance provide the rationale for examining mtDNA variation to infer associations between organismal history and Earth history. How does one go about identifying distinct mtDNA genotypes within a species? Two routine molecular approaches are commonly employed. Both capitalize on detecting nucleotide substitutions, the main source of variation distinguishing one mtDNA genotype from another.

The first approach is called restriction site analysis. In this procedure the small, circular mtDNA molecule (about 16,500 nucleotides in *G. agassizii*) is cut into linear fragments using restriction enzymes. Each restriction enzyme recognizes a short but specific nucleotide sequence (e.g., AGATCT) called a restriction site; the enzyme will sever the DNA molecule wherever these sequences occur. Normally, two to ten DNA fragments (one for each restriction site) are produced and can be sorted by size. Because mtDNA from isolated populations undergoes change by nucleotide substitutions, the number of restriction sites present on these mtDNA molecules is apt to change as well. Figure 4.1 illustrates the restriction fragment profiles of *G. agassizii* from a series of geographic localities. Note that the mtDNA

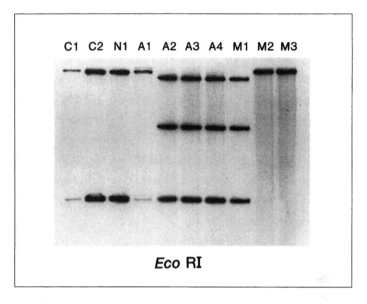

Figure 4.1. Restriction fragment profiles of desert tortoise mitochondrial DNA generated by the restriction enzyme *Eco* RI. Each vertical lane represents the mitochondrial profile of an individual tortoise. The three distinct profile patterns, from left to right, characterize tortoises from the Mohave (two fragments; samples C1, C2, N1, A1), Sonoran (three fragments; A2–4, M1), and Sinaloan (one fragment; M2, M3) assemblages (see fig. 4.2). C, California; N, Nevada; A, Arizona; M, Sonora, Mexico.

profiles vary from one to three fragments, reflecting one to three restriction sites present on the mtDNA molecule.

Fifteen to twenty different enzymes are used in a typical restriction site analysis, generating sixty to one hundred restriction sites with which to examine population variation. By quantifying the number of identical restriction sites shared between populations, one can assess degrees of relatedness among populations. Closely related populations share substantially more restriction sites than do distantly related populations. Often, populations in close geographic proximity will be characterized by similar, if not identical, mtDNA genotypes.

The second technique for examining mtDNA variation involves DNA sequencing, a more sophisticated molecular approach in which a gene (or gene segment) is isolated and its particular sequence of nucleotides is determined. Thus, DNA sequencing directly reveals nucleotide substitutions. Typically, DNA fragments ranging from three hundred to two thousand nucleotides in length are analyzed. The technical aspects of sequencing, both in terms of equipment costs and expertise, initially hindered its application in population-level surveys. However, advances in automated sequencing have made population-level research quite feasible. Recent studies of turtle population genetics have demonstrated the great potential mtDNA sequencing holds for enhanced resolution of genetic variation in future phylogeographic surveys (Norman et al. 1994; Lahanas et al. 1998; Lenk et al. 1999; Roman et al. 1999; Weisrock and Janzen 2000).

Evolutionary Genetics of the Desert Tortoise

Phylogeographic Patterns among Tortoise Populations

Detecting phylogeographic patterns for a given species requires a detailed geographic assessment of the genetic variation harbored within that species. Thus, the best survey strategy involves sampling individuals from a series of different populations throughout the species' range. In their phylogeographic survey of *G. agassizii*, Lamb et al. (1989) examined fifty-six individuals representing twenty-two populations (fig. 4.2). Populations were carefully selected to include animals from the northern, southern, eastern, and western termini of the tortoise's range, as well as an even distribution of sample locales across the center of the range. Under such a sampling regime, any genetic variation of geographical significance should be disclosed.

Using restriction site analysis, Lamb et al. (1989) resolved five different mtDNA genotypes among the fifty-six tortoises surveyed. Although the total number of genotypes is low compared to typical vertebrate surveys, the phylogeographic structure observed among tortoise populations is striking, yielding three well-defined genetic assemblages (fig. 4.2). The first assemblage consists of three closely related genotypes found in populations north and west of the Colorado River. Included within this assemblage are populations from Utah (the northernmost representatives of the tortoise's range) and the Virgin Mountains of northwestern Arizona. The distribution

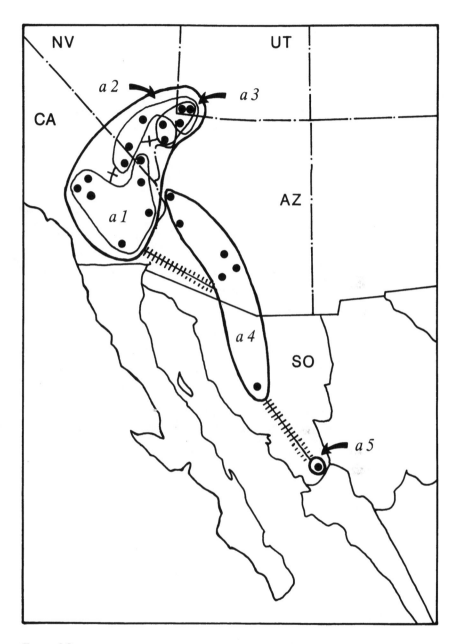

Figure 4.2. Phylogeographic patterns in the desert tortoise superimposed over the collection locales (solid circles) from Lamb et al. (1989). The mitochondrial DNA genotypes *a1–3* form the Mohave assemblage, whereas genotypes *a4* and *a5* represent the Sonoran and Sinaloan assemblages, respectively. Tick marks on the solid lines connecting mitochondrial genotypes indicate the numbers of restriction site changes responsible for genotype differences.

of this western assemblage largely coincides with the Mohave Desert and is henceforth designated the "Mohave assemblage."

The second major genetic assemblage lies south and east of the Colorado River and is represented by a single genotype observed in tortoises from central Arizona south to central Sonora (fig. 4.2). Largely coincident with the Arizona Upland and upper portions of the lower Colorado River Valley subdivisions of the Sonoran Desert (sensu Turner and Brown 1982), this eastern mtDNA genotype is referred to as the "Sonoran assemblage." In southern Sonora, near Alamos, tortoises are characterized by another highly distinct genotype, constituting a third genetic assemblage that presumably extends south 100–150 km to the terminus of the species' range in northern Sinaloa. Tortoises here occupy very different habitats: foothills thornscrub and tropical deciduous forest (Fritts and Jennings 1994; this vol., ch. 5). The Alamos genotype is designated the "Sinaloan assemblage."

Comparisons among genotypes across the three assemblages show appreciable genetic divergence. For example, the Sonoran genotype is distinguished from the three Mohave genotypes by a minimum of seventeen restriction site changes. The Sinaloan genotype differs from all others by a comparable number of restriction site changes. Not unexpectedly, genetic distances (quantitative values derived from the number of site changes observed between tortoise genotypes) are large, especially for a within-species comparison. The pronounced phylogeographic structure (and attendant genetic distances) exhibited by *G. agassizii* suggests remarkably deep separations in its mtDNA genealogy. If the observed phylogeographic structure is indeed a function of historical events, which historical environmental features of the southwestern landscape might be responsible? In the following sections, we detail how these contemporary genetic patterns can be used to infer some of the tortoise's evolutionary history.

Converting Genetic Distance to Time: The Tortoise and the Hare

The genetic distances (5.1–5.6 percent) observed between Mohave and Sonoran genotypes of desert tortoise are substantial; they are, in fact, significantly higher than distance values reported for any other turtle species (Walker and Avise 1998). How and to what degree of certainty can these genetic distances be converted to time, allowing a date to be assigned to the isolation event responsible for such high genetic divergence? The evolutionary record of population separation inscribed in DNA is marked by a clocklike rate of change (Avise 1994). Under the simplest mode of molecular evolution, nucleotide substitution, DNA accrues mutational changes in a time-dependent fashion. Molecular clocks seldom tick in a regular, metronomic manner, however; instead they display "stochastically constant behavior, not unlike radioactive decay" (Ayala 1986). Neither are molecular clocks evolving at some universal rate. Rates of change vary markedly from one gene to the next, differing in extreme cases by orders of magnitude.

As noted earlier, animal mtDNA evolves rather rapidly. Even though certain mtDNA genes evolve at a pace considerably slower than others, the overall rate of mtDNA evolution is about five to ten times faster than nuclear DNA. Initially, the average rate for animal mtDNA evolution was calibrated at 2 percent sequence divergence per million years for pairs of lineages isolated less than 10 million years (Brown et al. 1979). This calibration, based on primates (Brown et al. 1982), was extended to other groups and led to the assumption of a uniform molecular clock in vertebrates (Wilson et al. 1985). However, more rigorous calibration surveys have revealed substantive rate heterogeneity in mtDNA evolution among vertebrates. For example, shark mtDNA evolves seven to eight times slower than that of primates or ungulates (Martin et al. 1992). Other lines of evidence point to significant rate reductions for mtDNA evolution in a second major vertebrate group, the turtles (Avise et al. 1992).

Rate inference for a particular taxonomic group requires a well-represented fossil record and/or established biogeographic barriers, which serve as temporal reference points for calibrating the molecular clock. Using these criteria, estimates of mtDNA evolution for six species of turtles (representing six genera and three families) demonstrate two- to fourteen-fold reductions in rates relative to the conventional calibration of 2 percent divergence per million years derived for mammals (Avise et al. 1992). One explanation offered to account for this rate discrepancy involves metabolism: Higher rates of mtDNA evolution in mammals are perhaps causally related to higher metabolic activities associated with endothermy, the "warm-blooded" condition in mammals (Martin and Palumbi 1993). Other factors such as body size and generation time probably also come into play. Whatever the reasons, mtDNA evolution in turtles is slow compared to mammals, providing an interesting molecular perspective on the race between the tortoise and the hare.

Avise et al.'s (1992) proposed slow rate of mtDNA evolution in turtles has garnered additional support from refined rate estimates based on a particular mitochondrial gene, cytochrome *b*, DNA sequence comparisons for sea turtles (Chelonioidea; Bowen et al. 1993), map turtles (*Graptemys*, Lamb et al. 1994), and the genus *Gopherus* (Lamb and Lydeard 1994) all indicate a sequence divergence rate of 0.4–0.5 percent per million years for this mtDNA gene. Cytochrome *b* sequence comparisons between Mohave and Sonoran lineages of *G. agassizii* revealed a divergence value of 2.4 percent (Lamb and Lydeard 1994). Assuming the putative 0.4–0.5 percent rate for turtle cytochrome *b*, the Mohave and Sonoran mtDNA lineages appear to have diverged some 5 million years ago, tracing to a common maternal ancestor in the late Miocene epoch.

The Bouse Formation: A Possible Isolation Event

The genetically distinct Mohave and Sonoran assemblages abut along the Colorado River, a geographic setting that implicates this general region as an historical impediment to gene flow. While physiographic constraints are evident in the steep canyon areas of the upper Colorado River, it is difficult to envision the river's lower region as a serious barrier to tortoise dispersal. To appreciate how the lower Colorado may have influenced desert tortoise evolution, it is necessary to consider the middle and late Cenozoic history of this drainage within the surrounding Basin and Range Province of southern California and western Arizona.

Within the southern Basin and Range Province, two major incursions of waters centered about the Miocene-Pliocene boundary (5 million years ago) were important in the formation of the lower Colorado River. The first was a marine embayment, known as the Imperial Formation, that covered much of California's lower Imperial Valley along a geologic feature called the Salton Trough as a result tectonic activity in the late Miocene (Luchitta 1979; Winker and Kidwell 1986). The second geologic event, known as the Bouse Formation, involved extensive inundation from the Yuma area north to southern Nevada. Deposits from the Bouse Formation indicate a shallow, broad (40–50 km) body of water, which was originally interpreted as estuarine influx from the Gulf of California (Smith 1970). However, a recent study involving strontium isotope analysis provides an altogether different interpretation of the Bouse Formation (Spencer and Patchett 1997). The strontium isotope measurements offer little support for marine inundation, but are consistent with freshwater deposition in a chain of lakes fed by an early Colorado River. Centered at this debate are various explanations for the presence of marine barnacles and clams in Bouse Formation sediments near Parker, Arizona (Smith 1970; Spencer and Patchett 1997). Nonetheless, the inundation responsible for the Bouse Formation, whether freshwater or marine, occupied a geographic realm sufficient to serve as a potential geographic barrier.

Although the Bouse Formation is geographically congruent with the genetic break observed between Mohave and Sonoran tortoise assemblages, an equally important issue concerns timing. Is the level of genetic divergence between tortoise mtDNA lineages temporally consistent with the Bouse Formation deposition? This appears to be the case. Age assignments for the Bouse Formation range from late Miocene to early Pliocene (9 to 4 million years ago; Spencer and Patchett 1997), but a date of 5.7 to 5.3 million years ago based on potassium-argon estimates is considered the most reliable (Shafiqullah et al. 1980; Buising 1990). This date closely approximates the 5 million–year estimate for lineage isolation in tortoises. Thus, the Mohave and Sonoran assemblages may have descended from a common ancestral lineage sundered by the abrupt inundation responsible for the Bouse Formation.

As noted earlier, the Sinaloan mtDNA genotype exhibits genetic distances of significant magnitude from both the Sonoran (4.2 percent) and Mohave (5.1 percent) mtDNA assemblages, suggesting estimated separation times of about 3 to 5 million years between the Sinaloan assemblage and the other two assemblages. The absence of any current or notable historical barrier in Sonora makes resolution of the event(s) responsible for the genetic divergence of the Sinaloan genotype difficult. The paucity of sample localities in Sonora (fig. 4.2) also complicates matters, because the exact geographic position delimiting the genetic break between Sonoran and Sinaloan mtDNA assemblages remains unknown. Although speculative, the potential role Baja California's biogeographic history (Grismer 1994) might have played in desert tortoise evolution as an isolating mechanism for *any* of the mtDNA assemblages cannot be dismissed.

Conservation Genetics

East versus West: Genetic Markers for Mohave and Sonoran Tortoises

The presence of Sonoran and Mohave tortoise assemblages partitioned roughly east and west of the Colorado River begs the question, "To what degree has dispersal occurred between lineages?" A major phylogenetic break in mtDNA genealogy does not necessarily negate the possibility of subsequent and/or contemporary gene flow at points along a historical barrier. Opportunities for tortoise dispersal across the river basin seem possible, even plausible, during regional conditions associated with the Pleistocene and Holocene (Van Devender 1990). Like the tortoise, two major geographic assemblages of the pocket gopher *(Thomomys bottae)* meet along the Colorado River. Yet allozyme analysis revealed that certain gopher populations adjacent to the river are very similar genetically and are likely products of dispersal events (Smith and Patton 1980). A recent study of an unusual tortoise population in the Black Mountains of Arizona has brought to the fore the possibility of similar genetic patterns (McLuckie 1995).

Tortoises inhabiting the Black Mountains of western Arizona exhibit behaviors more like Mohave animals than those that typify Arizona populations. Specifically, they occupy *bajadas* dominated by creosotebush *(Larrea divaricata)* and white bursage *(Ambrosia dumosa)*, as opposed to rocky hillsides (Barrett 1990; McLuckie 1995). The tortoise's habitat preferences, coupled with the fact that the Black Mountains lie within the Mohave Desert, suggest that the Black Mountain tortoises may be Mohavean. We conducted a detailed study of these tortoises, in which a primary research issue concerned the genetic status of the Black Mountain population (McLuckie 1995; McLuckie et al. 1999). Is the population represented by Mohave mtDNA genotypes, Sonoran genotypes, or a combination thereof? To address this population genetics question, we developed a set of mtDNA markers that distinguish the Mohave versus Sonoran assemblages (McLuckie et al. 1999).

Our mtDNA marker system involves an analytical protocol that is quick, economical, and minimally invasive. As a result, large numbers of individuals from natural populations can be sampled relatively easily and without harm. A key feature to the approach is the use of a powerful molecular technique called the polymerase chain reaction, which synthesizes millions of copies of a gene or gene fragment from a just a few DNA molecules, generating ample material for further analysis. Thus, tissue samples from which the DNA is isolated can be accordingly small, for example, a few drops of blood or a fragment of shed skin.

To establish our mtDNA markers, we compared DNA sequences of Mohave and Sonoran tortoises previously reported for two mtDNA segments: a portion of the cytochrome *b* gene (Lamb and Lydeard 1994) and the adjacent control region (Osentoski and Lamb 1995). Through these comparisons we identified diagnostic restriction enzyme sites (again, restriction sites present on one assemblage mtDNA type but not the other) to serve as unique Mohave and Sonoran markers. Once diagnostic sites were identified, we genetically typed eleven Black Mountain tortoises using the assay procedure depicted in figure 4.3. Ten of the eleven Black Mountain animals possessed Mohave mtDNA genotypes, thus confirming the presence of a Mohavean tortoise population on the east side of the Colorado River.

The mechanism responsible for the establishment of the Black Mountain population is less clear; active dispersal (recent or historical), river meander (where a land parcel becomes part of the opposite bank upon a shift in the river channel), or human transport (early or modern peoples) are possible explanations. Evidence for different regional vegetation and climatic patterns in the Sonoran Desert during the Wisconsin glacial of the late Pleistocene provides an interesting environmental setting in which to consider tortoise dispersal. Plant remains from ancient packrat (*Neotoma* spp.) middens indicate that central Arizona was characterized by a woodland-chaparral community (including many Mohave Desert species) that received predominantly winter rainfall (Van Devender 1990). A primary environmental difference between Mohave and Sonoran desert tortoises involves the latter's dependency on summer rainfall; populations in the western Mohave Desert seldom experience summer rainfall (Peterson 1996; Henen et al. 1998). Van Devender (ch. 2 of this volume) suggests that the Mohavean tortoise is perhaps the only North American tortoise ever to be tied ecologi-

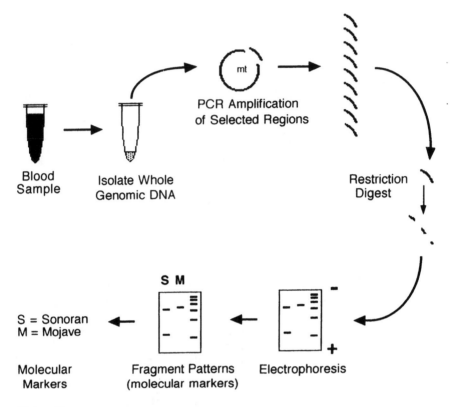

Figure 4.3. The mitochondrial DNA marker protocol involves the following steps (presented clockwise, starting at the upper left): (1) DNA is isolated from blood; (2) a specific segment of the mitochondrial DNA molecule (cytochrome *b* and control region) is amplified via the polymerase chain reaction (PCR); (3) the amplified segments are cut using a restriction enzyme; (4) the resulting fragments are sorted by electrophoresis; and (5) the tortoise is identified as having a Mohave or Sonoran mitochondrial DNA genotype based on the diagnostic fragment profile.

cally to a winter-rainfall climate. Thus, the Black Mountain population may represent a relictual component of Mohavean tortoise dispersal eastward into Arizona during glacial winter-rainfall climates of the Pleistocene.

Our mtDNA markers consistently typed respective Mohave and Sonoran populations in the vicinity of the Black Mountains, but they may also be broadly applicable for the genetic typing of Mohave versus Sonoran tortoises throughout their respective ranges. We are currently examining a series of populations in Arizona and California to confirm the markers' general utility, and thus far, they appear to hold. Such a marker system should prove useful for future tortoise conservation and population management, especially in resolving the superposition of recent dispersal upon patterns of historical isolation.

Tortoise Mitochondrial DNA Lineages as Units of Conservation Management

The identification of phylogeographic pattern within a species provides an important baseline in formulating conservation strategy. However, geographic populations can be genetically structured across a range of ecological (shallow) and evolutionary (deep) levels. Sound management practice must take this spectrum into account. Even though the tortoise lineages exhibit pronounced genealogical depth, mtDNA variation is but a small component of the genetic legacy of this species. Correspondence between geographic patterns in mtDNA variation and those observed for other aspects of organismal evolution would bolster arguments for the conservation of the tortoise assemblages. Fortunately, Mohave and Sonoran tortoise populations also demonstrate concordant geographic variation for certain nuclear genes (Jennings 1985; Glenn et al. 1990; Britten et al. 1997), morphology (Weinstein and Berry 1987), and behavior (Barrett 1990). The high degree of phylogenetic concordance across these independent markers (mitochondrial, nuclear, as well as genetically based complex traits like behavior and morphology) serves as "a useful if not necessary criterion for distinguishing deep population genetic structures from those that are evolutionarily shallow" (Avise 1996).

Recently, information on the degree and depth of phylogeographic structure has been integrated in conservation practice as a method to identify populations of high conservation value (Dizon et al. 1992; Moritz 1994; Vogler and DeSalle 1994; Mayden and Wood 1995; Avise and Hamrick 1996). Populations or groups of populations historically isolated from one another, and thus representing deep phylogenetic subdivisions within species, constitute evolutionarily significant units, or ESUs (sensu Vogler and DeSalle 1994). Although ESUs stem largely from modern systematic concepts, they are perceived as diagnostic units of conservation more so than taxonomic entities (e.g., subspecies or varieties). Each ESU comprises a substantial fraction of a species' overall genetic diversity and, from this perspective, warrants serious conservation effort.

Given their geographic distribution, genealogical depth, and concordant suite of characters, the Mohave, Sonoran, and Sinaloan tortoise assemblages clearly qualify as ESUs. Although none of the assemblages have received formal taxonomic recognition, the same criteria would appear to identify them as equally valid geographic races or, perhaps, *evolutionary species* (sensu Frost and Hillis 1990). Ironically, by using ESUs conservation biologists attempt to define infraspecific geographic variation in a manner reminiscent of the subspecies concept of the early twentieth century. However, there are satisfying distinctions: Whereas the subspecies concept is fraught with subjective ambiguity, the ESU is an evolutionarily sound, scientifically objective entity, serving more to conserve than to codify biodiversity.

LITERATURE CITED

Avise, J. C. 1994. *Molecular markers, natural history, and evolution.* Chapman and Hall, New York.

———. 1996. Toward a regional conservation genetics perspective: Phylogeography of faunas in the southeastern United States. Pp. 431–70 *in* J. C. Avise and J. L. Hamrick, eds. *Conservation genetics: Case histories from nature.* Chapman and Hall, New York.

———. 2000. *Phylogeography: The history and formation of species.* Harvard University Press, Cambridge, Mass.

Avise, J. C., and J. L. Hamrick, eds. 1996. *Conservation genetics: Case histories from nature.* Chapman and Hall, New York.

Avise, J. C., J. Arnold, R. M. Ball, E. Bermingham, T. Lamb, J. E. Neigel, C. A. Reeb, and N. C. Saunders. 1987. Intraspecific phylogeography: The mitochondrial DNA bridge between population genetics and systematics. *Annual Review of Ecology and Systematics* 18:489–522.

Avise, J. C., B. W. Bowen, T. Lamb, A. B. Meylan, and E. Bermingham. 1992. Mitochondrial DNA evolution at a turtle's pace: Evidence for low genetic variability and reduced microevolutionary rate in the Testudines. *Molecular Biology and Evolution* 9:457–73.

Ayala, F. J. 1986. On the virtues and pitfalls of the molecular evolutionary clock. *Journal of Heredity* 77:226–35.

Barrett, S. L. 1990. Home range and habitat of the desert tortoise *(Xerobates agassizii)* in the Picacho Mountains of Arizona. *Herpetologica* 46:202–6.

Bowen, B. W., W. S. Nelson, and J. C. Avise. 1993. A molecular phylogeny for marine turtles: Trait mapping, rate assessment, and conservation relevance. *Proceedings of the National Academy of Sciences USA* 90:5574–77.

Britten, H. B., B. R. Riddle, P. F. Brussard, R. Marlow, and T. E. Lee. 1997. Genetic delineation of management units for the desert tortoise, *Gopherus agassizii*, in the northeastern Mojave Desert. *Copeia* 1997:523–30.

Brown, W. M., M. George Jr., and A. C. Wilson. 1979. Rapid evolution of animal mitochondrial DNA. *Proceedings of the National Academy of Sciences USA* 76:1967–71.

Brown, W. M., E. M. Prager, A. Wang, and A. C. Wilson. 1982. Mitochondrial DNA sequences of primates: Tempo and mode of evolution. *Journal of Molecular Evolution* 18:225–39.

Buising, A. V. 1990. The Bouse Formation and bracketing units, southeastern California and western Arizona: Implications for the evolution of the proto–Gulf of California and the lower Colorado River. *Journal of Geophysical Research* 95:20111–32.

Dizon, A. E, C. Lockyer, W. F. Perrin, D. P. Demaster, and J. Sisson. 1992. Rethinking the stock concept: A phylogeographic approach. *Conservation Biology* 6:24–36.

Fritts, T. H., and R. D. Jennings. 1994. Distribution, habitat use, and status of the desert tortoise in Mexico. Pp. 49–56 *in* R. B. Bury and D. J. Germano, eds. *Biology of North American tortoises*. Fish and Wildlife Research no. 13, U.S. Dept. of the Interior National Biological Survey, Washington, D.C.

Frost, D. R., and D. M. Hillis. 1990. Species in concept and practice: Herpetological applications. *Herpetologica* 46:87–104.

Frost, D. R., A. G. Kluge, and D. M. Hillis. 1992. Species in contemporary herpetology: Comments on phylogenetic inference and taxonomy. *Herpetological Review* 23:46–54.

Glenn, J. L., R. C. Straight, and J. W. Sites Jr. 1990. A plasma protein marker for population genetic studies of the desert tortoise *(Xerobates agassizii)*. *Great Basin Naturalist* 50:1–8.

Goldman, E. A. 1935. Pocket gophers of the *Thomomys bottae* group in the United States. *Proceedings of the Biological Society of Washington* 48:153–58.

Grismer, L. L. 1994. The origin and evolution of the peninsular herpetofauna of Baja California, Mexico. *Herpetological Natural History* 2:51–106.

Henen, B. T., C. C. Peterson, I. R. Wallis, K. H. Berry, and K. A. Nagy. 1998. Desert tortoise field metabolic rates and water fluxes track local and global climatic conditions. *Oecologia* 117:365–73.

Jennings, R. D. 1985. Biochemical variation of the desert tortoise, *Gopherus agassizii*. M.S. thesis, University of New Mexico, Albuquerque.

Lahanas, P. N., K. A. Bjorndal, A. B. Bolten, S. Encalata, M. M. Miyamoto, and B. W. Bowen. 1998. Genetic composition of a green turtle feeding ground population: Evidence for multiple origins. *Marine Biology* 130:345–52.

Lamb, T., and C. Lydeard. 1994. A molecular phylogeny of the gopher tortoises, with comments on familial relationships within the Testudinoidea. *Molecular Phylogenetics and Evolution* 3:283–91.

Lamb, T., J. C. Avise, and J. W. Gibbons. 1989. Phylogeographic patterns in mitochondrial DNA of the desert tortoise *(Xerobates agassizii)*, and evolutionary relationships among the North American gopher tortoises. *Evolution* 43:76–87.

Lamb, T., C. Lydeard, R. B. Walker, and J. W. Gibbons. 1994. Molecular phylogeny of map turtles *(Graptemys)*: A comparison of mitochondrial restriction site versus sequence data. *Systematic Biology* 43:543–59.

Lenk, P., U. Fritts, U. Joger, and M. Winks. 1999. Mitochondrial phylogeny of the European pond turtle, *Emys orbicularis* (Linnaeus 1758). *Molecular Ecology* 8:1911–22.

Linnaeus, C. 1751. *Philosophia Botanica*. Stockholm, Sweden.

Liu, H.-P., J. P. Mitton, and S.-K. Wu. 1996. Paternal mitochondrial DNA differentiation far exceeds maternal mitochondrial DNA and allozyme differentiation in the freshwater mussel *Anodonta grandis grandis*. *Evolution* 50:952–57.

Luchitta, I. 1979. Late Cenozoic uplift of the southwestern Colorado Plateau and adjacent lower Colorado River region. *Tectonophysics* 61:63–95.

Martin, A. P., and S. R. Palumbi. 1993. Body size, metabolic rate, generation time, and the molecular clock. *Proceedings of the National Academy of Sciences USA* 90:4087–91.

Martin, A. P., G.J.P. Naylor, and S. R. Palumbi. 1992. Rates of mitochondrial DNA evolution in sharks are slow compared with mammals. *Nature* 357:153–55.

Mayden, R. L., and R. M. Wood. 1995. Systematics, species concepts, and the evolutionary significant unit in biodiversity and conservation biology. *American Fisheries Society Symposium* 17:58–113.

McLuckie, A. M. 1995. Genetics, morphology, and ecology of the desert tortoise *(Gopherus agassizii)* in the Black Mountains, Mohave County, Arizona. M.S. thesis, University of Arizona, Tucson.

McLuckie, A. M., T. Lamb, C. R. Schwalbe, and R. D. McCord. 1999. Genetic and morphometric assessment of an unusual tortoise *(Gopherus agassizii)* population in the Black Mountains of Arizona. *Journal of Herpetology* 33:36–44.

Moritz, C. 1994. Applications of mitochondrial DNA analysis in conservation: A critical review. *Molecular Ecology* 3:401–11.

Norman, J. A., C. Mortiz, and C. J. Limpus. 1994. Mitochondrial DNA control region polymorphisms: Genetic markers for ecological studies of marine turtles. *Molecular Ecology* 3:363–73.

Osentoski, M. F., and T. Lamb. 1995. Intraspecific phylogeography of the gopher tortoise, *Gopherus polyphemus*: RFLP analysis of amplified mtDNA segments. *Molecular Ecology* 4:709–18.

Peterson, C. C. 1996. Ecological energetics of the desert tortoise *(Gopherus agassizii)*: Effects of rainfall and drought. *Ecology* 77:1831–44.

Roman, J., S. D. Santhuff, P. E. Moler, and B. W. Bowen. 1999. Population structure and cryptic evolutionary units in the alligator snapping turtle. *Conservation Biology* 13:135–42.

Rothschild, W., E. Hartert, and K. Jordan. 1894. Note of the editors. *Novitates Zoologicae* 1:1.

Shafiqullah, M., P. E. Damon, D. J. Lynch, S. J. Reynolds, W. A. Rehrig, and R. H. Raymond. 1980. K-Ar geochronology and geologic history of southwestern Arizona and adjacent areas. *Arizona Geological Society Digest* 12:201–60.

Smith, B. P. 1970. New marine evidence for a Pliocene marine embayment along the lower Colorado River area, California and Arizona. *Geological Society of America Bulletin* 81:1411–20.

Smith, M. F., and J. L. Patton. 1980. Relationships of pocket gopher *(Thomomys bottae)* populations of the lower Colorado River. *Journal of Mammalogy* 61:681–96.

Spencer J. E., and P. J. Patchett. 1997. Sr isotope evidence for a lacustrine origin for the upper Miocene to Pliocene Bouse Formation, lower Colorado River Trough, and implications for timing of Colorado Plateau uplift. *Geological Society of America Bulletin* 109:767–78.

Turner, R. M., and D. E. Brown. 1982. Sonoran desertscrub. *Desert Plants* 4:181–221.

Van Devender, T. R. 1990. Late Quaternary vegetation and climate of the Sonoran Desert, United States and Mexico. Pp. 134–65 *in* J. L. Betancourt, T. R. Van Devender, and P. S. Martin, eds. *Packrat middens: The last 40,000 years of biotic change.* University of Arizona Press, Tucson.

Vogler, A. P., and R. DeSalle. 1994. Diagnosing units of conservation management. *Conservation Biology* 8:354–63.

Walker, D., and J. C. Avise. 1998. Principles of phylogeography as illustrated by freshwater and terrestrial turtles in the southeastern United States. *Annual Review of Ecology and Systematics* 29:23–58.

Weinstein, M. N., and K. H. Berry. 1987. *Morphometric analysis of desert tortoise populations.* Report no. CA950-CT7-003, Bureau of Land Management, Riverside, Calif.

Weisrock, D. W., and F. J. Janzen. 2000. Comparative molecular phylogeography of North American softshell turtles *(Apalone)*: Implications for regional and wide-scale historical evolutionary forces. *Molecular Phylogenetics and Evolution* 14:152–64.

Wiley, E. O. 1981. *Phylogenetics: The theory and practice of phylogenetic systematics.* John Wiley and Sons, New York.

Wilson, A. C., R. L. Cann, S. M. Carr, M. George Jr., U. B. Gyllenstein, K. M. Helm-Bychowski, R. H. Huguchi, S. R. Palumbi, E. M. Prager, R. D. Sage, and M. Stoneking. 1985. Mitochondrial DNA and two perspectives on evolutionary genetics. *Biological Journal of the Linnaean Society* 26:375–400.

Winker, C. D., and S. M. Kidwell. 1986. Paleocurrent evidence for lateral displacement of the Pliocene Colorado River Delta by the San Andreas Fault system, southeastern California. *Geology* 14:788–91.

Zouros, E., A. O. Ball, C. Saaverda, and K. R. Freeman. 1994. A new type of mitochondrial DNA inheritance in the blue mussel *Mytilus. Nature* 359:412–14.

CHAPTER 5

The Desert Tortoise in Mexico

Distribution, Ecology, and Conservation

R. BRUCE BURY, DAVID J. GERMANO, THOMAS R.
VAN DEVENDER, AND BRENT E. MARTIN

Although more than one-third of the geographic range of the desert tortoise *(Gopherus agassizii)* is in northwestern Mexico, most of our knowledge on its biology stems from studies in the Mohave Desert (see Berry 1989; Bury et al. 1994; Germano and Bury 1994) and the Sonoran Desert (Burge 1979; Barrett 1990; Bailey et al. 1995; Murray et al. 1996; Averill-Murray et al., chs. 6 and 7 of this volume) in the United States. The desert tortoise is listed as *amenazada* ("threatened"; Secretaría de Desarrollo Social 1994) in Mexico, yet its distribution and ecology in Mexico are the least known of any major portion of the tortoise's range (Germano and Bury 1994). The desert tortoise is also listed in Appendix 1 of the Convention on International Trade in Endangered Species of Wild Fauna and Flora, and thus people need permits to transport the animal between member nations (e.g., Mexico and the United States).

In Sonora the desert tortoise is called *tortuga del monte* ("scrub tortoise") or *tortuga de los cerros* ("hill tortoise") in rural areas and *tortuga del desierto* (likely a translation of the English "desert tortoise") in cities. It lives in desertscrub, thornscrub, and tropical deciduous forest, where the tortoise is one of the largest herbivorous reptiles, with adults reaching a mean carapace length of 26–28 cm and a biomass of 3–4 kg (Bury and Germano 1994).

Auffenberg (1969) provided the first substantial field observations on the species in Sonora. Surveys and basic ecological research on the tortoise were later conducted in the fall and early winter on Isla Tiburón in the Sonoran portion of the Gulf of California (Bury et al. 1978; Reyes O. and Bury 1982). Germano (1994) reviewed the life-history traits of the desert tortoise across its range, but there was little new information from Mexico. Fritts and Jennings (1994) and Treviño et al. (1994) added appreciably to our knowledge of the desert tortoise in Mexico. Here, we provide a current review of its distribution, general ecology, and conservation in Mexico.

Distribution

The desert tortoise ranges widely across northwestern Mexico from northern Sinaloa and probably western Chihuahua through northwestern Sonora and then into the southwestern United States as far north as southern Nevada and southwestern Utah (Van Devender, ch. 1 of this volume). The number of locality records of tortoises in Mexico remained fairly low for decades: twenty-two (Auffenberg and Franz 1978), twenty-five (Patterson 1982), and forty-three (Smith and Smith 1979). Many of these were repetitive records of the same sites. In Fritts and Jennings's (1994) review of the tortoise's distribution in Mexico, seventy-four localities were reported in Sonora and Sinaloa. Figure 5.1 presents a revised map reflecting our current knowledge of the distribution of the desert tortoise in Mexico. Based on patterns of variation in mitochondrial DNA in tortoise populations, Lamb and McLuckie (ch. 4 of this volume) recognized several major phylogeographic lineages: tortoises in the Mohave and Sonoran Deserts and those in the tropical areas south of about Guaymas in Sonora and Sinaloa.

Baja California

The natural occurrence of the desert tortoise in Baja California is not likely. Ottley and Velazques (1989) described a new species of desert tortoise *(Xerobates lepidocephalus)* from near La Paz, in the Cape Region in Baja California Sur. *Xerobates* represents the *G. agassizii* lineage that has been viewed as a genus or a subgenus of *Gopherus* (Lowe 1990). However, Crumly and Grismer (1994) interpreted the rarity of material (one live tortoise and a shell) and the lack of morphological differences from the desert tortoise as indicating that the animals were probably transported to the Cape Region from Sonora. José Luis León de la Luz (pers. comm. 1999) has not encountered tortoises in twenty years of botanical exploration in the Cape Region. The only fossil *Gopherus* reported from the Baja Peninsula was from near Bahía de Los Angeles in Baja California (Crumly and Grismer 1994). These indeterminate fossil remains could be from a desert tortoise or a Bolson tor-

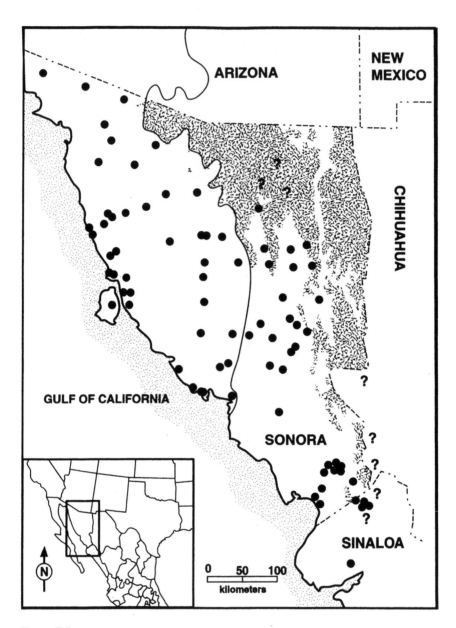

Figure 5.1. Map of the distribution of the desert tortoise in Mexico (modified from Fritts and Jennings 1994). The boundary of the Sonoran Desert (solid vertical line) follows Brown and Lowe (1978) and Turner (1982). The stippled areas in Sonora are zones of upland nondesert vegetation, including grassland, oak woodland, and pine-oak forest.

toise (*Gopherus flavomarginatus*; McCord, ch. 3 of this volume). We think the possibility of a desert tortoise population in Baja California remote because none have been reported from a considerable number of recent biological surveys in the region (e.g., Grismer 1994), and tortoises or their sign are usually conspicuous in the field.

Sonora

The first reports of desert tortoises from Mexico were by Van Denburgh (1922) and Bogert and Oliver (1945) from near Alamos, Sonora, but these investigators did not describe the habitat. Fritts and Jennings (1994) reported that tortoises in Sonora live in the Sonoran Desert (Arizona Upland, Central Gulf Coast, Plains of Sonora, and Lower Colorado River Valley subdivisions). They appear to be most abundant at elevations of 300–500 m (or somewhat higher) in rock outcrops in desertscrub and foothills thornscrub and are rare or absent in flatter areas away from mountains. Tortoises are absent in the higher elevations of the Sierra Madre Occidental in eastern Sonora. The animals are not found in the hyperarid desert areas of the lower Colorado River Valley in northwestern Sonora, except in the Mexican Biosphere Reserve (El Pinacate y Gran Desierto de Altar; Búrquez M. and Martínez Y. 1997; Carlos Castillo, pers. comm. 1999). The desert tortoise should be searched for in the Sierra del Rosario and other isolated desert mountains in this region.

The eastern limits of the range in Sonora are defined by the transition from tropical deciduous forest to oak woodland at 800- to 1,000-m elevation. The northeastern limits are poorly known. The ranges of many foothills thornscrub dominants including desert feather tree *(tepeguaje, Lysiloma watsonii)*, organpipe cactus *(pitahaya, Stenocereus thurberi)*, *papelío (Jatropha cordata)*, torotes *(Bursera fagaroides, B. laxiflora)*, and tree ocotillo *(Fouquieria macdougalii)* extend well beyond the Sonoran Desert in northeastern Sonora (Turner et al. 1995). We expect desert tortoise populations to be found well northeast of the localities presented in Fritts and Jennings (1994). The distribution and ecology of desert tortoises in the adjacent Chihuahuan Desert biotic province portion of northeastern Sonora and adjacent Arizona warrants study.

Fritts and Jennings (1994) reported a perceived distributional gap between Alamos and Guaymas. The area south of Guaymas is an open coastal thornscrub with some Sonoran desertscrub species present. In March 1993, Thomas R. Van Devender discovered a tortoise shell near a beach house in Camahuiroa, a coastal Mayo Indian village 60 km south of Navojoa. Samuel L. Friedman (pers. comm. 1994) observed an adult crossing a nearby road. These indicate that tortoises are present in at least low densities in coastal thornscrub at less than 10-m elevation in southern Sonora. Vicente Tajia (pers. comm. 1998), a resident of Teachive, an inland Mayo Indian village 35 km south-southeast of Navojoa, reported that tortoises were common in foothills thornscrub at 120-m elevation on the isolated volcanic Cerro Terucuchi and were occasionally found in flatter coastal thornscrub areas from Teachive to Las Bocas on the coast. All of the area between Camahuiroa and Las Bocas along the coast and inland to beyond Teachive are part of the Comunidad Indígena de los Mayos, a political unit equivalent to Native American reservations in the United States. Another important consideration with the distribution gap is that Mexico Highway 15 passes through poor tortoise habitat in the seasonally flooded lowlands and agricultural areas that is mostly on the Comunidad Indígena de los Yaquis. Access to the Yaqui lands is highly restricted; vast areas remain unsurveyed for tortoises. Tortoise populations will likely be discovered in future surveys in foothills thornscrub on a series of basaltic hills and mountains on the coastal plain from the Yaqui's Sierra Bacatete south to the Sinaloan border.

Tortoises also live between sea level and 300 m, but these populations have received little attention. In central Sonora, the Seri Indians have a rich cultural and ecological knowledge of desert tortoises near Desemboque and Punta Chueca on the coast of the Gulf of California and on Isla Tiburón (Nabhan, ch. 15 of this volume). Brent E. Martin has observed tortoises and their sign on desert hills by the coast near Bahía de Kino and at Estero Sargento (fig. 5.2). In 1988, a freshly dead adult male was found washed up on the beach in Coloradito Canyon in the Sierra Bacha just south of Puerto Libertad (Brent E. Martin, pers. obs.).

Treviño et al. (1994) surveyed nineteen tortoise transects in central Sonora. The transects were in Plains of Sonora subdivision desertscrub in the valleys and in foothills thornscrub on the emergent hills. They found high frequencies of live tortoises (52.6 percent) and scat (73.6 percent of the transects). The highest counts of tortoises were at the highest-elevation site (1,000 m). Treviño et al. detected an average of 2.0 tortoise signs (e.g., tortoise, scat, burrow) per person-hour of survey, which suggests that walking surveys can detect tortoise sign fairly quickly (i.e., about one sign per half-hour in the field), although the team required 8.6 person-hours to locate each live tortoise.

Chihuahua

Although the desert tortoise has not been previously reported in Chihuahua, tropical deciduous forest habitats similar to those inhabited by tortoises near Alamos, Sonora, extend eastward in the Río Mayo and Río Fuerte drainages well beyond the border of Sonora. At Choquincahui on Arroyo El Cobre and Rancho San Pedro at the Cajón of the Río Cuchujaqui east of Alamos, tortoises have been observed in tropical deciduous forest only 12–15 km west of the Chihuahuan border (Brent E. Martin, pers. obs.; George M. Ferguson, pers. comm. 2000). According to former Chihuahuans now residing in Alamos, *tortugas* live in Chihuahua near Arechuibo (Lalo Estrada and Emiliano Granillo, pers. comm. 1998). Considering that Arechuibo is at 1600-m elevation, well above tropical deciduous forest, and that the box turtle *(Terrapene nelsoni)* is also called "tortuga," sightings, shells, or photographs are needed to better document the species' presence in Chihuahua.

Sinaloa and Southern Limits

The first report of desert tortoises in Sinaloa was by Loomis and Geest (1964), who found a female near El Fuerte, about 65 km south of Alamos, Sonora. Hardy and McDiarmid (1969) reported that the desert tortoise in Sinaloa is restricted to low elevations in tropical thorn woodland (an archaic term now called thornscrub or tropical deciduous forest). Fritts and Jennings (1994) show only five locality records in northernmost Sinaloa. Recently, tortoises were seen crossing the Choix–El Fuerte road on 21 August 1993 (Randy D. Babb, pers. comm. 1999; fig. 5.3), 27 June 1996 (David A. Yetman, pers. comm. 1996), and 3 July 1999 (Brent E. Martin, pers. obs.).

In 1993, Horacio Vargas Velazquez, a student at the Universidad de Occidente Unidad Los Mochis (Sinaloa), did his *servicio social profesional*

Figure 5.2. Sonoran desert tortoises live along the coast in western Sonora. Left:
Central Gulf Coast desertscrub in the Sierra Bacha south of Puerto Libertad.
The cacti are teddybear cholla *(choya güera, Opuntia bigelovii)* and old man cactus

(senita, Lophocereus schottii). (Photograph by C. David Bertelsen) Right: Tortoises
are found in sheltersites in coastal bluffs overlooking the Gulf of California.
(Photograph by Thomas Van Devender)

Figure 5.3. A Sinaloan desert tortoise observed on 21 August 1993 about 12.7 km
west of the junction of the road to Santa Ana on Choix–El Fuerte road, Sinaloa.
The vegetation in the area is transitional between thornscrub and tropical deciduous forest. Note the bleached scutes or exposed bone on the shell, which is probably the result of bacterial or fungal infection. (Photograph by Randy D. Babb)

(a required program of bachelor's students) by studying a tortoise population in the Valle del Carrizo under the guidance of Charles H. Lowe (Vargas V. 1994). The habitat is mostly rockless rolling hills, and the vegetation
is transitional between foothills thornscrub and tropical deciduous forest.
From September through December, Vargas encountered ten tortoises (six
female, four male) estimated to be eight to twenty-five years in age by scute
ring counts. Four radio-telemetered animals showed maximum movements
of 800 m for males and less than 400 m for females during the four-month
period. The animals were not active on cloudy or cold days with temperature
below 18°C.

Germano et al. (1994) placed a question mark on their map of desert
tortoise range in Sinaloa to indicate that seemingly suitable tortoise habitat exists farther south than tortoises have been found in Sinaloa. There is
greater annual rainfall at the same elevations in central and southern Sonora

Figure 5.4. *Aeromonas hydrophila* infection on the skin of a captive desert tortoise in Tucson, Arizona, in April 2001 as a result of wet winter conditions. (Photograph by James L. Jarchow)

than in the southwestern United States (Búrquez M. et al. 1999). Although many studies have focused on how the desert tortoise survives the aridity of the Mohave Desert, excess rainfall may have a profound influence on its ecology and life history at the southern limits of its distribution.

James L. Jarchow (pers. comm. 1999) suggested that the tortoise's southern limit might be due to an increase in infection related to increasing rainfall. During the exceptionally wet winters of 1997–1998 and 2000–2001, many captive tortoises in Arizona were diagnosed with septicemia, an infectious disease carried through the blood stream (Jarchow et al., ch. 12 of this volume). In winter during years of heavy winter rainfall, wild tortoises have been observed on the surface. The extended hot, wet, humid conditions during the tropical summer rainy season in Sinaloa would accelerate the growth of disease-causing bacteria or other pathogens in soil and especially in the confined shelters used by tortoises. Desert tortoises are highly susceptible to infections caused by the soil bacterium *Aeromonas hydrophila*. The bacterium first attacks the tortoise as a skin infection (figs. 5.3, 5.4), but can spread quickly through the blood stream, causing serious damage to various body systems and even death.

The warm, moist, shady microclimates of the tropical deciduous forest understory (Van Devender et al. 2000) also support a diverse fungal flora, especially wood rotters and slime molds. In the rainy season in these closed-canopy forests, it would be difficult for desert tortoises to bask in sunlight. A small tortoise (about 150-mm carapace length) observed approximately 13 km west of the road to Santa Ana between Choix and El Fuerte, Sinaloa, on 21 August 1993 appeared to have fungus damage on the lateral marginals (Randy D. Babb, pers. comm. 1999; fig. 5.3). The potential health problems of these soil pathogens and fungi warrant further study.

Thus, pathogens or fungi could be involved in setting the southern and southeastern limits of desert tortoises' distribution, preventing them from living in similar tropical deciduous forest habitats that extend southward through Sinaloa to Nayarit and Jalisco (Van Devender et al. 2000). Optimal conditions (elevated precipitation and productivity during the warm season, shady microhabitat, and relatively warm temperatures and high humidity year-round) for the growth of pathogens and fungi are found in tropical deciduous forest understory.

Tortoise populations in Sinaloa are in relatively open foothills thornscrub, which has less summer rainfall than tropical deciduous forest to the northeast near Alamos, Sonora. If the desert tortoise is found farther south in Sinaloa, the populations will likely be in foothills thornscrub on isolated hills near the coast, because large areas of coastal thornscrub have been converted to agriculture. However, the original distribution and abundances of tortoises below about 200-m elevation on the coastal plain in southern Sonora and northern Sinaloa will be difficult to assess because of extensive human impacts (e.g., large areas of cultivated land) on these areas.

Tropical Deciduous Forest

Lowe (1990) reported that the desert tortoise is primarily an inhabitant of thorn forest, which is an archaic equivalent to tropical deciduous forest and perhaps some denser coastal thornscrub. Because tortoise habitat in tropical deciduous forest is poorly known, we present descriptions of areas near Alamos, Sonora, where we (Brent E. Martin and Thomas R.

Van Devender) have observed tortoises (see also Averill-Murray et al., fig. 6.5, this vol.). The forest canopy at about 12 m height is mostly formed by *mauto (Lysiloma divaricatum)*, although various other trees including *amapa (Tabebuia impetiginosa)*, *brasil (Haematoxylum brasiletto)*, kapok *(pochote, Ceiba acuminata)*, torotes *(Bursera grandifolia, B. penicillata, B. stenophylla)*, and tree morning glory *(palo santo, Ipomoea arborescens)* are common. Arborescent cacti, especially *etcho (Pachycereus pecten-aboriginum)* and organpipe cactus are important, but unlike in thornscrub and desertscrub, the trees tower above the cacti.

The climate of the Alamos area is tropical, and freezing temperatures are rare. Rainfall at 400- to 600-m elevation is about 620–700 mm/year. The primary rainy season is summer, from July to September, but tropical storms or hurricanes may bring heavy rains from late September to November. There is a secondary rainy season in the winter months of December and January or February, the southernmost expression of the Pacific frontal storms that bring winter rains to the Mohave and Sonoran Deserts. A long dry season begins in late January or February and extends to the beginning of the summer rains in early July. Within about ten days of the first heavy rainfall, the forest canopy leafs out. Sunlight is mostly intercepted by the canopy, and the understory becomes shady, humid, warm, and highly productive in growth of plants and fungi. The peak activity period of tortoises in tropical deciduous forest is presumably during the summer rainy season, as it is in the Sonoran Desert of Arizona. Another possibility is that tortoise activity might be concomitant with the maximum productivity of light-limited columnar cacti in tropical deciduous forest, which occurs after the rains decrease in September and October and the canopy opens up (Park Nobel, pers. comm. 1998). In the fall, the forest is partially deciduous, allowing more light to reach the ground. The forest does not become fully deciduous until the end of the winter rains. During the tropical dry season, when sunlight passes through the canopy, conditions on the ground are hot and dry—essentially similar to those in the Sonoran Desert.

We do not know if tortoises are active in the spring or remain sedentary until the rains begin. However, there are other important environmental differences between tortoise habitat near Alamos and in the two northern deserts. There are few, if any, spring annuals in thornscrub or tropical deciduous forest communities. Although May and June are the hottest and driest months, humidity remains relatively high. During the dry season, trees with showy magenta, purple, yellow, or white flowers brighten the forest, with a new species flowering about every six weeks. John F. Wiens (pers. comm. 1999) noted that his captive tortoises in Tucson relish the fallen flowers of desert willow *(Chilopsis linearis)*. Thus, tropical deciduous forest tortoises might eat the fallen flowers of amapa (another Bignoniaceae), tree morning glory, or other plants.

Sinaloan tortoises are occasionally found crossing the road between Alamos and Navojoa in tropical deciduous forest on the northern slopes of the Sierra de Alamos (fig. 5.5). In Cañón de las Piedras on the northeastern side of the Sierra de Alamos, tortoise are regularly seen in a granite boulder habitat at 550-m elevation on steep slopes not too different from those in Arizona habitats, except that the area is in tropical deciduous forest with a closed canopy (Peter A. Holm and Elizabeth B. Wirt, pers. comm. 1998).

Live tortoises and shells have also been found at 260- to 280-m elevation near the Río Cuchujaqui (fig. 5.5), a tropical river that originates near Chínipas in southwestern Chihuahua, flows east to the Sierra de Alamos in Sonora, and then southward back into Sinaloa to join the larger Río Fuerte. In many parts of its course, the Cuchujaqui forms a 5- to 15-m wide flood-prone channel in the bottom of a 30-m deep, 800-m wide canyon incised into mudstone bedrock. Tropical deciduous forest grows on the bluffs above and ledges in the canyon, with Mexican bald cypress *(sabino, Taxodium distichum* var. *mexicanum)* and occasional Bonpland willow *(sauce, Salix bonplandiana)* along the stream channel. A live tortoise was found in the forest on top of a bluff. One complete shell found at the edge of the stream in a steep narrow side canyon had possibly died from a 10-m fall from the dense forest above and into the canyon. Another fragmentary shell in Arroyo El Mentidero just above its junction with the Río Cuchujaqui was in a more open and disturbed habitat on one side of the 500-m wide canyon (Brent E. Martin and Thomas R. Van Devender, pers. obs.). Tortoises also have been seen in foothills thornscrub and tropical deciduous forest from Tecoripa to Tepoca (about 28°30' N), but the extent of their presence in tropical vegetation is not well known. Tropical deciduous forest in southern Sonora, northern Sinaloa, and adjacent Chihuahua are relatively intact and should be surveyed for tortoises. We need to know if they live throughout this habitat or only in the Alamos area.

Figure 5.5. Sinaloan tortoise habitat in tropical deciduous forest at 260-m elevation along the Río Cuchujaqui southeast of Alamos, Sonora, during the spring dry season. During the summer monsoon heavy rains fall from July to September; dense green foliage closes the tree canopy and shades the understory. (Photograph by Thomas Van Devender)

Persons familiar with the desert tortoise only in the United States might be surprised to find that tortoises in tropical deciduous forest share the habitat with such tropical amphibians and reptiles as the Mazatlán toad *(Bufo mazatlanensis)*, a box turtle *(Terrapene nelsoni)*, painted wood turtle *(Rhinoclemmys pulcherrima)*, an anole *(Anolis nebuloides)*, beaded lizard *(escorpión, Heloderma horridum)*, spiny-tailed iguana *(Ctenosaura hemilopha)*, boa constrictor *(corua, Boa constrictor)*, a coral snake *(Micrurus diastema)*, indigo snake *(babatuco, Drymarchon corais)*, and parrot snake *(güirotillo, Leptophis diplotropis)*. The habitat is also used by jaguar *(tigre, Felis onca)*, which is a predator on desert tortoises in tropical communities in Sonora (Carlos A. Lòpez G., pers. comm. 1999).

Sheltersites

Use of sheltersites by desert tortoises in the United States varies from self-constructed deep burrows (usually 1- to 3-m long, occasionally to 10 m) in the Mohave Desert to hiding under rocky overhangs, using shallower burrows, packrat (*Neotoma* spp.) houses, and shrubs in the Sonoran Desert and more tropical areas to the south (Burge 1978, 1979; Germano et al. 1994; Averill-Murray et al., ch. 7 of this volume). Burrows in south-central Arizona in the northeastern Sonoran Desert varied from 0.25- to 1.25-m deep and were constructed in a variety of substrates, including silt, gravel, and hardened lake sediments (Bailey et al. 1995; Averill-Murray et al., ch. 7 of this volume).

Little is known about use of sheltersites in Mexico, but most descriptions agree with those in the United States. Auffenberg (1969) reported that tortoises in Sonora occasionally dig shallow hollows into the bases of arroyo walls but are otherwise completely "nomadic" throughout most of the year with a very large activity range. Germano et al. (1994) stated that desert tortoises commonly dig short burrows in Sonoran desertscrub and tropical deciduous forest. Reyes O. and Bury (1982) found that tortoises on Isla Tiburón often construct pallets or shallow burrows 1- to 2-m long into arroyo walls or at the base of large shrubs or small trees. As in the Sonoran Desert in Arizona, tortoises also simply hide under vegetation. Interestingly, the Seri Indian name for desert tortoise is *ziix hehet cöqiij* ("live-thing branchy bushes what-it sits-under"; Felger et al. 1981; Nabhan, ch. 15 of this volume).

In northern Sinaloa, tortoises use packrat houses, dry cacti, or burrows dug in friable soil by the tortoises themselves, nine-banded armadillos *(Dasypus novemcinctus)*, or other animals for shelter (Vargas V. 1994). Treviño et al. (1994) reported twenty-one observations of tortoises in sheltersites in central Sonora. The average lengths were 53.6 cm (range = 30–81 cm, $n = 5$) for soil burrows (excluding one 12-cm long burrow), 29.3 cm (26–34 cm, $n = 4$) for pallets, and 81 cm (18–190 cm, $n = 11$) for caves. The shelters were found in volcanic ($n = 16$) and sedimentary ($n = 4$) rocks, and underneath a mesquite tree (*Prosopis* sp., $n = 1$).

Reproduction

There are no published data on the reproductive biology of the desert tortoise in Mexico. Eggs from a captive female (about 270-mm carapace length) kept outdoors in the yard of Arón I. Espinoza R. (pers. comm. 1999) in Hermosillo, Sonora, hatched on 6 November 1999. In Tucson most tortoises hatch in early October, although they occasionally hatch as late as November (James L. Jarchow, pers. comm. 1999). Sometimes eggs even hatch in the nest, but young emerge in the spring. Reproductive features of desert tortoises vary in different areas (Germano 1994). The minimum age of first reproduction (mean values) is thirteen years in the western Mohave Desert, sixteen to twenty-two years in the Sonoran Desert, and fourteen years in tropical Sonoran habitats. Based on data from the Mohave Desert, the minimum size at first reproduction for female desert tortoises is 176- to 189-mm carapace length. Female Sonoran tortoises in Arizona begin to lay eggs at 220-mm carapace length and about nineteen to twenty years in age (Averill-Murray et al., ch. 7 of this volume; Germano et al., ch. 11 of this volume).

Discussion and Conservation

Drought has been implicated as contributing to population declines of Mohave desert tortoises, but likely causes loss of individuals for only one or a few recruitment years for long-lived organisms like tortoises (Germano 1994; Bury and Corn 1995). The only documented declines in Sonoran tortoise populations were in the Maricopa and Sand Tank Mountains of Arizona, where losses are attributed to drought or shell disease (Averill-Murray et al., ch. 6 of volume). Successful recruitment may have a cyclic pattern (e.g., when regional precipitation is high for two years in a row) or be a chaotic phenomenon (e.g., when unpredictable summer thunderstorms drench one area). In the biseasonal precipitation regimes of the Sonoran Desert and tropical communities to the south, droughts rarely extend beyond one season, and reliable monsoonal rains and forage may provide water and food in the hottest part of the summer. In coastal and tropical areas, elevated humidity moderates extreme summer temperature, reduces evaporation, and supports more-succulent vegetation.

On Isla Tiburón, density estimates of tortoises were estimated as 29–87 per square kilometer (75–225 per square mile; Reyes O. and Bury 1982). These densities are relatively high and comparable to abundance levels of tortoises in habitats in the Mohave and Sonoran Deserts. Although there are many coyotes *(Canis latrans)* on Isla Tiburón, this native predator does not appear to significantly impact the tortoise population (Reyes O. and Bury 1982). Nor has periodic harvest for food by Seri and mestizo (a Mexican of mixed Spanish and Indian descent) fishermen (Felger et al. 1981) impacted the population. Isla Tiburón was established as a wildlife reserve in 1963 and included in a 1978 decree protecting the islands in the central Gulf of California (Búrquez M. and Martínez Y. 1997). Thus, the tortoise population on Isla Tiburón is in a large, natural reserve that is presently uninhabited and has no domesticated or feral livestock (Nabhan, ch. 15 of this volume).

In earlier times, Mexican traders carried live desert tortoises as a source of fresh meat (Felger et al. 1981). Tortoises were also used as food by native Indians (Bury and Corn 1995; Nabhan, ch. 15 of this volume). Fritts and Jennings (1994) report that some tortoise populations have declined in rural areas with extensive human activity. Capture and use of tortoises by humans appears to be mostly opportunistic and is limited in scope because it is difficult and time-consuming to search for tortoises in their steep, rocky slope habitats. In Teachive, Sonora, tortoises are eaten sporadically as a *birria* (meat stew) by some villagers but are not kept as pets (Vicente Tajia, pers. comm. 1998). It is common knowledge that during *las aguas* (the summer rains from July to September) is the time to look for tortoises when they come out of the rocks.

In twelve interviews at six ranches in central Sonora, Treviño et al. (1994) found that most people (67 percent) agreed that the number of tortoises had declined. All but one person stated that they eat or had eaten desert tortoise on some occasion. Human use of tortoises for food varied widely from occasional take at most of the ranches to complete disinterest in tortoises at one site. At times *tortugueros* ("tortoise hunters") collected tortoises in sufficient numbers to sell for food in the smaller towns (Mario Treviño, pers. comm. 1993), including Benjamín Hill (Socorro Guerrero R., pers. comm. 1998). However, human impact on tortoises in rural Sonora is part of a larger demographic picture. The number of people living on ranches and *ejidos* (communally owned lands used for agriculture and grazing) has declined, while city populations have swelled. Today the younger, more-urban generations are accustomed to obtaining their food in supermarkets and probably would be repulsed by the inhumane treatment involved in processing tortoises for food (Martha L. Reina G., pers. comm. 1998). The family of Rosa Mendoza in Benjamín Hill has always had tortoises in the house, formerly for food, now for pets. On a recent visit, her mother, Misha, said, *"¡Que bonita la tortuga! Traigánme una hacha. Vamos a matarla para hacer una sopa con arroz"* ("Oh, what a beautiful tortoise! Fetch me an axe. We will kill it to make a soup with rice").

Today, in larger cities such as Hermosillo, captive tortoises are increasingly common in yards. These animals were mostly obtained by city dwellers who found them in rural areas. Although baby tortoises were formerly sold in pet stores in Hermosillo, their sale is now prohibited because they are a federally protected species. Occasionally, hatchlings of captives in Hermosillo are still sold by private individuals (Carmen Reina G., pers. comm. 1999). Presently, the Instituto de Medio Ambiente y el Desarrollo Sustenable del Estado de Sonora provides publications that discourage the keeping of tortoises as pets (Meléndez T. 1998). A wealthy businessman has established a tortoise sanctuary, releasing all tortoises brought to him on his ranch near Cerro Colorado on the north side of Hermosillo. Similarly, well-intentioned college students from Hermosillo have been releasing substantial numbers of captive tortoises on Isla Tiburón because it is a protected natural area (Francisco Molina F., pers. comm. 1999). Unfortunately, public awareness against releasing captive or displaced animals into wild populations based on concerns about disease, carrying capacity, and population integrity that are prevalent in the United States has not been well established in Sonora.

The collection and distribution of desert tortoises as pets in Mexico was never organized into state, interstate, national, or international trade networks as in the United States. Only a single desert tortoise from Mexico has ever been intercepted at the United States border by U.S. Customs (Roy C. Averill-Murray, pers. comm. 1998). This is not surprising considering that tortoises are also found in the southwestern United States, are available free through adoption programs, and cannot legally be sold. In contrast, other reptiles that are highly desired in the pet trade but do not live north of the international border, such as beaded lizards and San Esteban chuckwallas *(Sauromalus varius)*, have been intercepted more frequently at the U.S. border.

The clearing of Sonoran desertscrub and foothills thornscrub to plant pastures of buffelgrass *(Pennisetum ciliare)* presents unique regional problems for desert tortoises in central and southern Sonora (Esque et al., ch. 13 of this volume). Buffelgrass is a stout, almost shrubby, African grass that responds quickly to rainfall, accumulating vast amounts of fuel. Buffelgrass burns to the ground but resprouts quickly from its massive underground roots. However, unlike in desert grasslands and higher woodlands, fire has never been an important ecological process in Sonoran desertscrub, thornscrub, and tropical deciduous forest. As a result, fire is now catastrophic in these communities, killing most of the dominant trees, shrubs, and cacti. Thus, recurrent fires are converting vast areas of tortoise habitat to tracts that resemble African savannah.

The tortoise population near El Batamote, a popular roadside stop 47 km north of Hermosillo on Mexico Highway 15, provides an excellent example. The vegetation is Plains of Sonora desertscrub on the flats and foothills thornscrub on the slopes. Dominant plants in the granite boulders include *chuparosa (Justicia californica)*, desert ironwood *(palo fierro, Olneya tesota)*, papelío, torotes, *samota (Coursetia glandulosa)*, and tree morning glory. In 1978, Thomas R. Van Devender found a pile of butchered and burned shells from about six tortoises and a skeleton of a boa constrictor at the site. Later, both Fritts and Jennings (1994) and Treviño et al. (1994) surveyed tortoises at this site. It was also a study area for Parish Indian mallow *(Abutilon parishii)*, a herbaceous perennial that was formerly a candidate species with the U.S. Fish and Wildlife Service (Van Devender et al. 1994). Since about 1992, buffelgrass and fire have been rapidly spreading from a cleared, planted pasture upslope into the granite boulder habitat, decimating the foothills thornscrub (Esque et al., ch. 13 of this vol., fig. 13.3). The fires are so hot that large torotes are reduced to ash pits and the varnished weathering surfaces of the boulders are exfoliating. Tortoise mortality in burned areas surely occurs (see Esque et al., ch. 13 of this volume), but we lack research of its effects in Mexico. Although suitable tortoise habitat still exists in a large area on the nearby hills, buffelgrass is a self-perpetuating exotic plant that will likely dominate the entire area.

More surveys of tortoises are urgently needed in Mexico, especially in thornscrub and tropical deciduous forest, to assess distribution, population sizes, and habitat features. These should be conducted away from ranches, towns, and cultivated lands. We have little data on reproduction (e.g., clutch size, season of egg-laying) or home range dynamics in Mexico. The only diet study is from tropical deciduous forest on the Sierra de Alamos in southern Sonora (Van Devender et al., ch. 8 of this volume). This basic knowledge is critically needed not only to build an effective plan for the protection of tortoises in Mexico but to better understand the species in the United States. We encourage the continued cooperation of biologists from Mexico and the United States to study and conserve this species on both sides of the border.

As in the United States today, humans severely impact tortoise populations in some areas in Mexico. Although a major demographic shift of people from the ranches and ejidos to the towns and cities may have reduced contact with tortoises, especially in areas of rugged or remote terrain, human populations and destruction of natural communities will likely increase. Interviews are needed with people in larger cities, small towns, and ranches to learn more about changing cultural attitudes and uses of tortoises as food or pets.

We think the future of the desert tortoise can be best assured in the United States and Mexico by increased public awareness through creative educational programs stressing the ecological importance of this large, charismatic animal. Discouraging the collection of tortoises for food or pets so others can see them in their habitats and the need for habitat protection should be emphasized (Treviño et al. 1994). We also suggest that the tortoise be promoted in Mexico as a symbol for the conservation of all desert wildlife.

ACKNOWLEDGMENTS

We thank Stephanie A. Meyer for her observations of tortoises near Alamos, Sonora, and for interviewing former Chihuahuan residents. Elizabeth B. Wirt and Peter A. Holm shared their observation of tortoises on the Sierra de Alamos and provided a copy of the servicio social profesional report by Horacio Vargas Velazquez on Sinaloan tortoises. Randy Babb provided photographs and observations of tortoises in southern Sonoran and northern Sinaloa. We thank Ana Lilia Reina G. for providing her notes on the folklore, uses of, and attitudes toward tortoises in four generations of Reinas in El Llano, Benjamín Hill, and Hermosillo, Sonora. Carlos Castillo of SEMARNAP and Cristina Meléndez T. of IMADES in Hermosillo commented on the manuscript.

LITERATURE CITED

Auffenberg, W. 1969. *Tortoise behavior and survival.* Rand McNally, Chicago.

Auffenberg, W., and R. Franz. 1978. *Gopherus agassizii.* Pp. 212.1–212.2 *in* William J. Riemer, ed. *Catalogue of American amphibians and reptiles.* American Society of Ichthyologists and Herpetologists, Bethesda, Md.

Bailey, S. J., C. R. Schwalbe, and C. H. Lowe. 1995. Hibernaculum use by a population of desert tortoises *(Gopherus agassizii)* in the Sonoran Desert. *Journal of Herpetology* 29:361–69.

Barrett, S. L. 1990. Home range and habitat of the desert tortoise *(Xerobates agassizii)* in the Picacho Mountains of Arizona. *Herpetologica* 46:202–6.

Berry, K. H. 1989. *Gopherus agassizii,* desert tortoise. Pp. 5–7 *in* I. R. Swingland and M. W. Klemens, eds. *The conservation biology of tortoises.* Occasional Papers no. 5, IUCN Species Survival Committee, Gland, Switzerland.

Bogert, C. M., and J. A. Oliver. 1945. A preliminary analysis of the herpetofauna of Sonora. *Bulletin of the American Museum of Natural History* 83:301–415.

Brown, D. E., and C. H. Lowe. 1978. *Biotic communities of the Southwest.* Map. General Technical Report no. RM-41, U.S. Dept. of Agriculture Forest Service, Denver.

Burge, B. L. 1978. Physical characteristics and patterns of utilization of cover sites used by *Gopherus agassizii* in southern Nevada. *Proceedings of the Desert Tortoise Council Symposium* 1978:80–111.

———. 1979. A survey of the present distribution of the desert tortoise, *Gopherus agassizii,* in Arizona. *Proceedings of the Desert Tortoise Council Symposium* 1979:27–74.

Búrquez M., A., and A. Martínez Y. 1997. Conservation and landscape transformation in Sonora, Mexico. *Journal of the Southwest* 39:371–98.

Búrquez M., A., A. Martínez Y., R. S. Felger, and D. Yetman. 1999. Vegetation and habitat diversity at the southern edge of the Sonoran Desert. Pp. 36–67 *in* R. H. Robichaux, ed. *Ecology of Sonoran Desert plants and plant communities.* University of Arizona Press, Tucson.

Bury, R. B., and P. S. Corn. 1995. Have desert tortoises undergone a long-term decline in abundance? *Wildlife Society Bulletin* 23:41–47.

Bury, R. B., and D. J. Germano, eds. 1994. *Biology of North American tortoises.* Fish and Wildlife Research no. 13, U.S. Dept. of the Interior National Biological Survey, Washington, D.C.

Bury, R. B., R. A. Luckenbach, and L. R. Muñoz. 1978. Observations on *Gopherus agassizii* from Isla Tiburón, Sonora, Mexico. *Proceedings of the Desert Tortoise Council Symposium* 1978:69–79.

Bury, R. B., T. C. Esque, L. A. Defalco, and P. A. Medica. 1994. Distribution, habitat use, and protection of the desert tortoise in the eastern Mojave Desert. Pp. 57–72 *in* R. B. Bury and D. J. Germano, eds. *Biology of North American tortoises*. Fish and Wildlife Research no. 13, U.S. Dept. of the Interior National Biological Survey, Washington, D.C.

Crumly, C. R., and L. L. Grismer. 1994. Validity of the tortoise *Xerobates lepidocephalus* Ottley and Velazques in Baja California. Pp. 33–37 *in* R. B. Bury and D. J. Germano, eds. *Biology of North American tortoises*. Fish and Wildlife Research no. 13, U.S. Dept. of the Interior National Biological Survey, Washington, D.C.

Felger, R. S., M. B. Moser, and E. W. Moser. 1981. The desert tortoise in Seri Indian culture. *Proceedings of the Desert Tortoise Council Symposium* 1981:113–20.

Fritts, T. H., and R. D. Jennings. 1994. Distribution, habitat use, and status of the desert tortoise in Mexico. Pp. 49–56 *in* R. B. Bury and D. J. Germano, eds. *Biology of North American tortoises*. Fish and Wildlife Research no. 13, U.S. Dept. of the Interior National Biological Survey, Washington, D.C.

Germano, D. J. 1994. Comparative life histories of North American tortoises. Pp. 175–85 *in* R. B. Bury and D. J. Germano, eds. *Biology of North American tortoises*. Fish and Wildlife Research no. 13, U.S. Dept. of the Interior National Biological Survey, Washington, D.C.

Germano, D. J., and R. B. Bury. 1994. Research on North American tortoises: A critique with suggestions for the future. Pp. 187–204 *in* R. B. Bury and D. J. Germano, eds. *Biology of North American tortoises*. Fish and Wildlife Research no. 13, U.S. Dept. of the Interior National Biological Survey, Washington, D.C.

Germano, D. J., R. B. Bury, T. C. Esque, T. H. Fritts, and P. A. Medica. 1994. Range and habitats of the desert tortoise. Pp. 73–84 *in* R. B. Bury and D. J. Germano, eds. *Biology of North American tortoises*. Fish and Wildlife Research no. 13, U.S. Dept. of the Interior National Biological Service, Washington, D.C.

Grismer, L. L. 1994. The origin and evolution of the peninsular herpetofauna of Baja California, Mexico. *Herpetological Natural History* 2:51–106.

Hardy, L. M., and R. W. McDiarmid. 1969. The amphibians and reptiles of Sinaloa, Mexico. *University of Kansas Museum of Natural History Publication* 18:39–252.

Loomis, R. B., and J. C. Geest. 1964. The desert tortoise *Gopherus agassizii* in Sinaloa, Mexico. *Herpetologica* 20:203.

Lowe, C. H. 1990. Are we killing the desert tortoise with love, science, and management? Pp. 84–102 *in* K. R. Beaman, F. Chaporaso, S. McKeown, and M. D. Graff, eds. *Proceedings of the first international symposium on turtles and tortoises: Conservation and captive husbandry*. Chapman University, Chapman, Calif.

Meléndez T., C. 1998. Tortuga del desierto o del monte *(Gopherus agassizii)*. P. 27 *in Instituto de Medio Ambiente y el Desarrollo Sustenable del Estado de Sonora Entorno*. Conservación e Investigación para el Desarrollo Sustenable, Hermosillo, Sonora.

Murray, R. C., C. R. Schwalbe, S. J. Bailey, S. P. Cuneo, and S. D. Hart. 1996. Reproduction in a population of the desert tortoise, *Gopherus agassizii*, in the Sonoran Desert. *Herpetological Natural History* 4:83–88.

Ottley, J. R., and V. M. Velazques. 1989. An extant, indigenous tortoise population in Baja California Sur, Mexico, with the description of a new species of *Xerobates* (Testudines: Testudinidae). *Great Basin Naturalist* 49:496–502.

Patterson, R. 1982. The distribution of the desert tortoise *(Gopherus agassizii)*. Pp. 51–55 *in* R. B. Bury, ed. *North American tortoises: Conservation and ecology.* Wildlife Research Report no. 12, U.S. Fish and Wildlife Service, Washington, D.C.

Reyes O., S., and R. B. Bury. 1982. Ecology and status of the desert tortoise *(Gopherus agassizii)* on Tiburón Island, Sonora. Pp. 39–49 *in* R. B. Bury, ed. *North American tortoises: Conservation and ecology.* Wildlife Research Report no. 12, U.S. Fish and Wildlife Service, Washington, D.C.

Secretaría de Desarrollo Social. 1994. Poder ejecutivo. Diario Oficial de la Federación. Tomo 488 no. 10, Mexico, D.F., lunes 16 de mayo de 1994.

Smith, H. M., and R. B. Smith. 1979. *Synopsis of the herpetofauna of Mexico.* Vol. VI, *Guide to Mexican turtles, bibliographic addendum III.* John Johnson, North Bennington, Vt.

Treviño, M. A., M. E. Haro, S. L. Barrett, and C. R. Schwalbe. 1994. Preliminary desert tortoise surveys in central Sonora, México. *Proceedings of the Desert Tortoise Council Symposia* 1987–1991:379–88.

Turner, R. M. 1982. Sonoran desertscrub. *Desert Plants* 4:181–221.

Turner, R. M., J. E. Bowers, and T. L. Burgess. 1995. *Sonoran Desert plants: An ecological atlas.* University of Arizona Press, Tucson.

Van Denburgh, J. 1922. The reptiles of western North America. Vol. I, Snakes and turtles. *Occasional Papers of the California Academy of Science* 10:617–1024.

Van Devender, T. R., C. D. Bertelsen, and J. F. Wiens. 1994. *Abutilon parishii* S. Watson. Status Report, U.S. Fish and Wildlife Service, Phoenix.

Van Devender, T. R., A. C. Sanders, R. K. Wilson, and S. A. Meyer. 2000. Flora and vegetation of the Río Cuchujaqui, a tropical deciduous forest near Alamos, Sonora. Pp. 36–101 *in* R. H. Robichaux and D. Yetman, eds. *The tropical deciduous forest of the Alamos: Biodiversity of a threatened ecosystem.* University of Arizona Press, Tucson.

Vargas V., H. 1994. *Estudio de la densidad poblacional de la tortuga del desierto* (Gopherus agassizii) *en la zona norte del Estado de Sinaloa.* Informe final de Servicio Social Profesional, Universidad Occidente Unidad Los Mochis, Sinaloa.

CHAPTER 6

Population Ecology of the Sonoran Desert Tortoise in Arizona

ROY C. AVERILL-MURRAY, A. PETER WOODMAN,
AND JEFFREY M. HOWLAND

Desert tortoise populations have been studied in Arizona since the mid-1970s, but few specific populations have been studied for more than a few years. Much has been learned about population ecology of desert tortoises during the past twenty-five years, but many unanswered questions remain. It is difficult to identify clear ecological patterns from one- or two-year studies of an animal that may live fifty years or more. Long-term studies provide valuable information about population attributes that are impossible to perceive over a shorter time period (Gibbons 1990). Although we can only speculate about several aspects of how desert tortoise populations change through time and relate to their environment, the scope of studies conducted to date allows us to draw general conclusions about populations and make preliminary inferences about factors affecting them. Many of the data summarized and discussed in this chapter originate from unpublished theses and agency reports (table 6.1) resulting from statewide monitoring efforts (extensive mark-recapture studies) and specific ecological studies (intensive radio-telemetry studies).

The works referenced in this chapter have been conducted across the range of the desert tortoise in the Sonoran Desert of Arizona (fig. 6.1). Study sites include a wide range of biotic communities within or extending from the Sonoran Desert (Brown 1982; see Van Devender, ch. 1 of this volume)—from the Arizona Upland subdivision (West Silver Bell Mountains) and combined elements of the Arizona Upland and lower Colorado River subdivisions (Eagletail Mountains) to desert grassland (Tortolita Mountains). Other populations have been studied in ecotonal areas consisting of Sonoran desertscrub with elements of Mohave desertscrub and juniper woodland (Hualapai [Mountain] Foothills), interior chaparral (Little Shipp

TABLE 6.1. Desert tortoise populations studied in the Sonoran Desert, Arizona. Estimated density of adults, scaled to 1 mi² (95% confidence limits). Observed tortoise numbers: F, female; M, male; U, unsexed; X, carcasses. Type: m, monitoring plot (mark-recapture); t, radio-telemetry study. Citations are listed numerically at the end of the table.

Locality	Abbreviation	Year	Density	F:M:U:X	Type	Citation
ARRASTRA MTS.	am	1987	20 (15–25)	9:6:3:16	m	27
		1997	24 (18–30)ᵃ	8:5:1:2	m	36
BONANZA WASH	bw	1992	—	6:8:3:13	m	31
		1997	27 (16–38)ᵃ	4:6:3:2	m	36
EAGLETAIL MTS.	et	1987	—	22:12:8:8	m	19
		1990	31 (26–36)	21:8:3:1	m	21
		1991	30 (28–32)ᵃ	16:9:7:5	m	7
		1992	29 (27–31)ᵃ	12:10:5:1	m	31
		1993	30 (26–34)ᵃ	13:10:14:3	m	32
		1994	30 (28–32)ᵃ	17:11:19:9	m	33
		1998	39 (35–43)ᵃ	17:14:8:5	m	37
EASTERN BAJADAᵇ	eb	1990	—	12:21:12:5	m	23
		1993	67 (51–83)ᵃ	14:29:3:10	m	32
		1993–1994	—	5:8:0:?	t	11
				14:25:12:14ᶜ		
		1997	61 (50–72)ᵃ	23:20:2:6	m	36
GRANITE HILLS	gh	1990	68 (24–112)	16:16:15:8	m	21
		1991	63 (50–76)ᵃ	30:19:21:4	m	7
		1992	60 (56–64)ᵃ	23:22:30:2	m	31
		1993	90 (78–102)ᵃ	31:24:40:2	m	32
		1994	69 (66–72)ᵃ	31:29:49:3	m	33
		1998	60 (59–61)ᵃ	20:16:20:13	m	37
		1998	—	22:33:5:8	m	30
HARCUVAR MTS.	hm	1991–1992	—	6:15:0:?	t	25
		1993	72 (65–79)ᵃ	15:29:2:5	m	32
		1997	77 (67–87)ᵃ	23:27:4:6	m	36
HARQUAHALA MTS.	hq	1988	—	9:8:4:4	m	8
		1994	15 (13–17)ᵃ	10:7:2:0	m	33
HUALAPAI FOOTHILLS	hf	1991	—	13:19:5:8	m	7
		1996	52 (44–60)ᵃ	13:21:2:6	m	35

TABLE 6.1. Continued

Locality	Abbreviation	Year	Density	F:M:U:X	Type	Citation
LITTLE SHIPP						
WASH	ls	1980	—	2:2:2:?	t	16
			—	18:16:17:?	m	16
		1990	85 (71–100)	42:26:16:9	m	21
		1991	79 (75–83)[a]	37:30:15:2	m	7
		1991–1992	—	6:4:0:?	t	25
		1992	107 (97–117)[a]	42:34:12:2	m	31
		1993	107 (100–114)[a]	47:36:20:9	m	32
		1994	97 (91–103)[a]	34:27:16:3	m	33
		1998	98 (90–106)[a]	30:18:10:9	m	37
MARICOPA MTS.	mm	1987	146 (69–223)	24:33:1:65	m	27
		1990	—	6:7:4:54	m	21
		1993–1994	—	14:0:0:?	t	28
MAZATZAL MTNS	mz	1991–93	—	10:1:0:?	t	13,15
		1992	150 (83–218)	19:27:5:8	m	12,14
		1995	114 (91–137)	24:25:17:3	m	14
		1996–1999	—	30:27:14:4[c]	t	1,9
NEW WATER MTS.	nw	1988	—	8:7:1:2	m	20
		1999	32 (30–35)	9:8:5:3	m	38
ORGAN PIPE CACTUS NATIONAL MONUMENT	orpi					
Ajo Mt. Drive		1996	75 (21–225)	11:12:6:8	m	29
Quitobaquito Hills		1997	34 (18–60)	16:6:3:1	m	29
Twin Peaks		1996	28 (8–73)	9:6:0:0	m	29
PICACHO MTS.	pm	1982–1983	—	9:5:0:?	t	4,26
RINCON MTS.	rm	1988	—	4:2:2:?	t	18
SAN PEDRO VALLEY	sp	1988	—	9:10:1:?	m	17
		1990–1992	—	4:4:0:?	t	2,3
		1991	—	18:16:9:11	m	7
		1995	125 (103–147)[a]	36:48:6:9	m	34
SAND TANK MTS.	st	1992	—	19:15:0:31[d]	m	6
		1994	—	2:5:6:32	m	5
SANTAN MTS.	sn	1990	—	3:4:1:?	m	22
		1991	—	16:10:3:3	m	24

TABLE 6.1. Continued

Locality	Abbreviation	Year	Density	F:M:U:X	Type	Citation
TORTOLITA MTS.	tt	1980–1989	—	8:8:2:?	m	10
		1990–1992	—	3:4:0:?	t	10
TORTILLA MTS.	tl	1992	—	29:20:3:12	m	31
		1996	97 (82–112)[a]	34:26:12:9	m	35
TUCSON MTS.	tm	1988	—	2:0:1:?	t	18
WEST SILVER BELL MTS.	ws	1991	—	39:20:5:11	m	7
		1995	134 (112–156)[a]	40:35:16:8	m	34
WICKENBURG MTS.	wm	1991	—	5:10:0:2	m	7

[a] Density was calculated using tortoises marked from previous and current surveys; therefore, estimates are not independent between surveys.
[b] Mohave Desert population included in the Sonoran Desert in Endangered Species Act decisions by the U.S. Fish and Wildlife Service (1990).
[c] Total numbers of tortoises observed.
[d] Combined data from two 4-km² plots within 2 km of each other.
Citations: 1, Averill-Murray and Klug (2000); 2, Bailey (1992); 3, Bailey et al. (1995); 4, Barrett (1990); 5, Dames and Moore, Tucson (1994); 6, Geo-Marine, Inc. (1994); 7, Hart et al. (1992); 8, Holm (1989); 9, Klug and Averill-Murray (1999); 10, Martin (1995); 11, McLuckie et al. (1996); 12, Murray (1993); 13, Murray and Schwalbe (1993); 14, Murray and Schwalbe (1997); 15, Murray et al. (1995, 1996); 16, Schneider (1981); 17, Schnell and Drobka (1988); 18, Shaw and Goldsmith (1988); 19, Shields and Woodman (1987); 20, Shields and Woodman (1988); 21, Shields et al. (1990); 22, SWCA, Inc. (1990a); 23, SWCA, Inc. (1990b); 24, SWCA, Inc. (1992); 25, Trachy and Dickinson (1993); 26, Vaughan (1984); 27, Wirt (1988); 28, Wirt and Holm (1997); 29, Wirt et al. (1999); 30, Woodman and Shields (1988); 31, Woodman et al. (1993); 32, Woodman et al. (1994); 33, Woodman et al. (1995); 34, Woodman et al. (1996); 35, Woodman et al. (1997); 36, Woodman et al. (1998); 37, Woodman et al. (1999); 38, Woodman et al. (2000).

Wash), and desert grassland (San Pedro Valley). We will also provide comparative information from populations in the Mohave Desert in Arizona and elsewhere. One population on the eastern *bajada* of the Black Mountains (hereafter, referred to as Eastern Bajada; fig. 6.1) is included in the Sonoran population south and east of the Colorado River for purposes of the Endangered Species Act (U.S. Fish and Wildlife Service 1990) even though it is in the Mohave Desert (McLuckie et al. 1999).

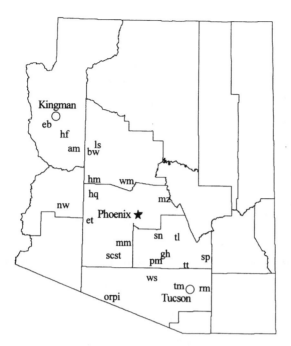

Figure 6.1. Desert tortoise populations studied in the Sonoran Desert, Arizona. Site codes are defined in table 6.1.

Population Characteristics

Abundance and Sex Ratios

Local desert tortoise population densities vary widely within the Sonoran Desert, ranging from fifteen to more than one hundred adults per square mile (2.6 km^2; table 6.1). Three caveats should be considered regarding these estimates. First, various methods have been used to calculate tortoise abundance (number in a general area or population) and density (number per specific unit area). Second, most of the plots contain some areas used by few to no tortoises, which affects the accuracy of density estimates within occupied habitat. Third, except for the Mazatzal Mountain population, density was derived simply by dividing estimated abundance by plot area. This method of density estimation disregards the fact that tortoise habitat usually extends beyond the study plot boundaries and tortoises living near the plot boundaries range both inside and outside the plot. Dividing abundance by plot area assumes that all the tortoises actually live *within* the plot boundaries and usually produces overestimates of density; that is, tortoises are actually distributed over a larger area than estimated in this way. Regardless, these estimates do illustrate the magnitude of geographic variation in tortoise density in Arizona. Later, we discuss specific factors contributing to the wide range of local tortoise densities in the Sonoran Desert.

The range of tortoise densities observed on Sonoran Desert monitoring plots falls within the range historically observed in the Mohave Desert. Since 1977, 13–219 adult tortoises have been found on plots (which generally measure 1 mi^2) in the eastern Mohave Desert, whereas 19–402 have been found on plots in the western Mohave (data summarized by Corn 1994). However, in some cases the low ends of these ranges may reflect previous declines in Mohave Desert populations (U.S. Fish and Wildlife Service 1994).

Most high tortoise densities observed in the Mohave Desert have been within intermountain valleys, where friable soils allow the construction of deep burrows (Luckenbach 1982; Germano et al. 1994). At least in the past, such populations may have been expansive, spanning from one valley to the next (Luckenbach 1982). In the Sonoran Desert, tortoises are found at the highest densities on steep, rocky hills and desert mountain slopes; tortoises are generally absent from the intermountain valleys (Germano et al. 1994). As a result, local tortoise populations appear to be smaller and more isolated in the Sonoran Desert than historical populations in the Mohave. Tortoise occupation of valley-floor habitats in the Mohave Desert may be a relatively recent trait relative to their evolutionary history; the desert tortoise's tropical ancestors lived in a warmer climate, where burrowing was less important for avoiding temperature extremes (Van Devender, ch. 2 of this volume).

With only a few exceptions, sex ratios typically are balanced with approximately equal numbers of males and females (table 6.1). About twice as many males as females were found on the Harcuvar Mountain plot in 1993. In contrast, females have outnumbered males by similar margins on the West Silver Bell (1991) and Eagletail Mountain plots. Cumulative sex ratios (over all years surveyed) in the Sonoran Desert do not differ statistically from 1:1 for any plot. Mohave Desert populations also typically have 1:1 sex ratios (Goodlett et al. 1996, 1997). Although males outnumbered females by about

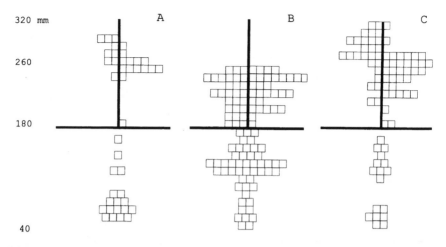

Figure 6.2. Representative size distributions (frequency histograms) of individuals from three Sonoran desert tortoise populations in 1994: (A) Eagletail Mountains; (B) Granite Hills; and (C) Little Shipp Wash. Each row represents a 10-mm increment in midline carapace length (MCL). Males are on the left side of the vertical lines, females on the right; individuals less than 180 mm MCL (below the horizontal lines) are of undetermined sex.

2:1 in 1990 and 1993 surveys of the Eastern Bajada plot, combined data from all surveys resulted in approximately equal numbers of males and females.

For visual comparison of populations, abundance of individuals and sex ratios may be illustrated together in size distribution histograms (fig. 6.2). Size distributions observed from desert tortoise populations across Arizona share some similar characteristics, and regional variation in other characteristics is also evident. A distinct gap is typically present in the distributions at the smaller adult and larger juvenile sizes (around 180 mm midline carapace length [CL]; fig. 6.2). In fact, of the fifteen monitoring plots listed in table 6.1 on which at least twenty tortoises were marked, only the Granite Hills population in south-central Arizona lacks this characteristic gap. In contrast, only six of sixteen populations in California exhibit a comparable gap in size distributions (figs. 5-1 through 5-30 in Berry and Nicholson 1984). No tortoises measuring 180–199 mm CL were found in a 1996 survey of the Beaver Dam Slope population in the Mohave Desert, located in extreme northwestern Arizona; however, eight tortoises measuring 150–

179 mm were found (Goodlett et al. 1996). This population has experienced high turnover and has a high incidence of disease symptoms (Goodlett et al. 1996). It remains to be seen if the subadult tortoises will fill in the adult size distribution.

Small individuals are underrepresented in size distributions because they are more difficult to find in the structurally complex habitat in the Sonoran Desert (Van Devender, ch. 1 of this volume). As figures 6.2A and 6.2B illustrate, however, relatively large numbers of juveniles may be found in some populations during some years, usually characterized and preceded by above-average rainfall and forage availability. Even in these cases, each juvenile is recaptured fewer times on average, if at all, than the adults, making density estimation of juveniles impossible. Finally, differences in tortoise growth characteristics at different populations result in population-specific maximum sizes and sexual size dimorphism (figs. 6.2 and 6.3; see Growth below).

Growth

Desert tortoises grow most rapidly early in life and reach 36–47 percent of their maximum CL before growth begins to slow (fig. 6.3; Murray and Klug 1996). Rapid early growth contributes to high juvenile survivorship (described below). Maximum sizes, however, differ between sexes and among populations. For example, Hart (1996) found that individuals in populations north of the Gila River tend to reach larger sizes than individuals in populations south of the river. Additionally, Murray and Klug (1996) found that males reach larger average maximum sizes than females at Little Shipp Wash (299 and 267 mm CL, respectively) and the Eagletail Mountains (288 and 268 mm). In fact, males reach larger average sizes than females at all thirteen monitoring plots surveyed to date north of the Gila River. On the Granite Hills plot, the sexes reach about the same size (males = 244 mm, females = 243 mm), which is significantly smaller than in the Eagletail Mountain and Little Shipp Wash populations (figs. 6.2, 6.3). Females reach the same or larger average sizes at three plots south of the Gila River (Sand Tank Mountains, San Pedro Valley, West Silver Bell Mountains), but males are larger than females in the Maricopa Mountains. The underlying reasons for these patterns are currently unknown.

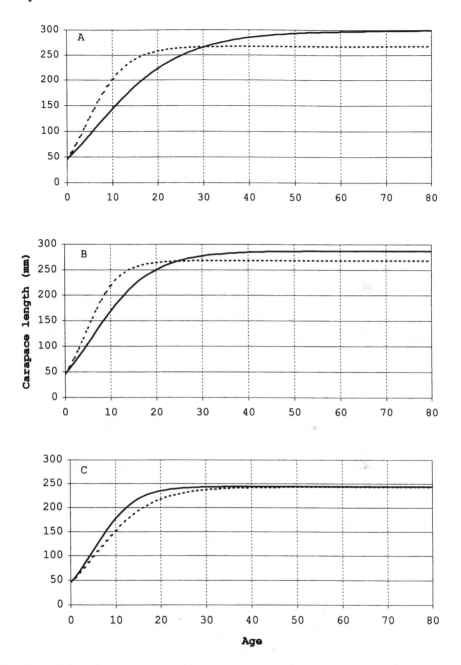

Figure 6.3. Growth curves for female (dashed) and male (solid) desert tortoises on the (A) Little Shipp Wash, (B) Eagletail Mountains, and (C) Granite Hills plots.

TABLE 6.2. Upper size estimates for desert tortoises across their range. Estimates are mean upper decile carapace lengths, unless otherwise indicated. Sample size is given in parentheses.

Location	Males	Females
WESTERN MOHAVE DESERT	283 (24)[a]	246 (15)[a]
EASTERN MOHAVE DESERT	260 (54)[a]	233 (34)[a]
NORTHERN SONORAN DESERT		
Little Shipp Wash	311 (27)[b]	290 (35)[b]
	299 (54)[c]	267 (83)[c]
Eagletail Mts.	301 (11)[b]	287 (17)[b]
	287 (23)[c]	268 (34)[c]
SOUTHERN SONORAN DESERT		
Granite Hills	254 (29)[b]	255 (31)[b]
	244 (69)[c]	243 (77)[c]
SINALOA	282 (22)[a]	265 (10)[a]

[a] Germano (1994b).
[b] Data from 1994 survey (Woodman et al. 1995).
[c] Growth curve estimates using data from surveys through 1994 (Murray and Klug 1996).

The largest desert tortoise on record from the Sonoran Desert of Arizona is a 322-mm CL female found on the Harcuvar Mountain plot in 1997. The large size of this female tortoise is anomalous, given the trend for males to reach larger sizes than females in northern populations. Interestingly, the largest wild tortoise found in the Mohave Desert is also a female, from the Lucerne Valley, California, with a CL of 378 mm (A. Peter Woodman, pers. obs.).

Growth of tortoises in the Mohave Desert is generally similar to that in the Sonoran Desert. Annual growth declines as CL increases (Germano 1994a; Karl 1998), and maximum size varies geographically, with tortoises in the western Mohave reaching larger sizes than those in the eastern Mohave (Germano 1994a). Geographic variation in size does not grade smoothly across the entire range of the desert tortoise, however (table 6.2). The largest tortoises seem to be found at both ends (western Mohave Desert and Sinaloa, Mexico) and the middle of the distribution (northern Sonoran Desert), with smaller tortoises distributed in between (eastern Mohave and southern

Sonoran Deserts). On average, males reach larger sizes than females across the Mohave Desert (Minden and Keller 1981; Berry and Nicholson 1984; Goodlett et al. 1996, 1997; Karl 1998). Even though growth varies each year with rainfall and forage availability within populations (Medica et al. 1975; Karl 1998), factors controlling growth relative to different populations (and species) of tortoises are unknown (Germano 1994a).

Reproduction

Female tortoises lay an average of about five eggs, but between one and twelve eggs in a clutch have been recorded in the Sonoran Desert (Murray et al. 1996; Wirt and Holm 1997; Klug and Averill-Murray 1999; Averill-Murray and Klug 2000; fig. 6.4). Clutch size is not related to body size—some of the smallest reproductive tortoises have laid some of the largest clutches of eggs and vice versa. Thus far, the smallest recorded size at which Sonoran desert tortoises have laid eggs in the wild was 220 mm CL, in the Mazatzal (Murray et al. 1996) and Maricopa (Wirt and Holm 1997) Mountains. Ages of these tortoises were not measured directly, but comparison of growth curves (fig. 6.3) indicates that females reach this size in ten to twenty years. More precise estimates of age at maturity are not possible because insufficient data on growth (which varies geographically, as shown above) exist for these populations.

Sonoran desert tortoises lay a maximum of one clutch in a year, at least in Arizona, but many females may not reproduce each year (Murray et al. 1996; Wirt and Holm 1997; Klug and Averill-Murray 1999; Averill-Murray and Klug 2000). Preliminary data indicate that recent rainfall influences mean clutch size and the proportion of females reproducing each year (Averill-Murray and Klug 2000). In dry years, smaller tortoises are particularly less likely to lay eggs than larger ones (table 6.3). Only two of six female tortoises studied in the Maricopa Mountains in 1994 laid eggs after almost ten years of drought, whereas all seven female tortoises under observation laid eggs at a nearby site, which was evidently less influenced by drought (Wirt and Holm 1997). Egg-laying in the Sonoran Desert generally occurs near the onset of the summer rainy season and has been observed from early June to early August, but the average egg-laying date each year does not appear to be directly related to recent rainfall (table 6.3).

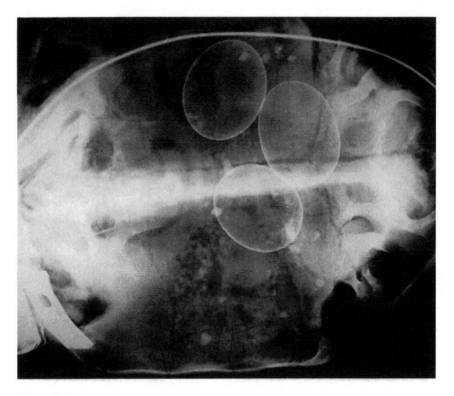

Figure 6.4. Radiograph of a female tortoise from Sugarloaf Mountain, northeast of Phoenix, Maricopa County, in July 1993. Three eggs are visible.

Rainfall also greatly influences tortoise reproduction in the Mohave Desert, where tortoises may lay as many as three clutches in a year. Winter rainfall and subsequent spring annual plant production can affect clutch frequency and annual egg production (Turner et al. 1984, 1986; Henen 1994; Karl 1998). Average annual reproductive output varies from about five to nine eggs per female per year depending on the environment, and most females usually lay at least a few eggs each year (Turner et al. 1986; Karl 1998; Mueller et al. 1998; Wallis et al. 1999). In especially productive years, however, reproductive output reaches asymptotic levels and may be constrained by other factors, such as body size and maternal nutrient reserves (Wallis et al. 1999). The extent that reserves are expended one year for reproduction can also affect the energy available for egg production the next year (Henen 1997; Mueller et al. 1998). Because they have smaller nutrient reserves than

TABLE 6.3. Rainfall and egg production for desert tortoises in the Mazatzal Mountains. From Murray et al. (1996), Klug and Averill-Murray (1999), and Averill-Murray and Klug (2000). MCL, midline carapace length.

	1993	1997	1998
PRECEDING RAINFALL (CM)[a]	67.9	18.9	41.0
PERCENT REPRODUCING *(n)*[b]	80% (10)	36% (11)	72% (18)
MEAN CLUTCH SIZE (SD)	5.7 (2.43)	3.8 (1.26)	5.7 (1.49)
MEAN REPRODUCTIVE MCL (SD)	247 (12.9)	253 (11.5)	248 (16.0)
MEAN OVIPOSITION DATE (SD, RANGE)	27 June (14 days, 9 June–25 July)	1 July (7 days, 20 June–4 July)	13 July (9 days, 12 June–3 August)

[a] Total rainfall during 10 months prior to eggs detectable on radiographs (July–April).
[b] MCL ≥ 220 mm; no tortoise laid more than one clutch.

larger females, smaller tortoises are particularly susceptible to laying fewer clutches and laying them later (Wallis et al. 1999).

Female tortoises as small as 176 mm CL have laid eggs in the Mohave Desert (Germano 1994b). Minimum age at first reproduction may be as low as nine years (Germano 1994a), but estimates range up to twenty-six years (Medica et al. 1975; Turner et al. 1987; Germano 1994a; Karl 1998; Mueller et al. 1998). In addition, tortoises lay their eggs earlier in the Mohave Desert than in the Sonoran, from late April through mid-July (Turner et al. 1986; Karl 1998; Wallis et al. 1999).

Mean summer rainfall in the Arizona Upland subdivision of the Sonoran Desert is more reliable and about 85 mm (380 percent) greater than in the Mohave Desert (Turner and Brown 1982; Germano 1994a), and the western Mohave receives even less summer rain than the eastern Mohave (Turner 1982; Wallis et al. 1999). Geographic variation in rainfall in the Mohave Desert also correlates with variation in reproductive traits. Eastern Mohave females lay eggs at smaller body sizes, lay proportionally smaller eggs, lay more eggs, and lay eggs earlier than western Mohave females (Wallis et al. 1999). Beginning reproduction at smaller body sizes and laying proportionally smaller eggs may enable eastern Mohave females to produce more offspring, and laying earlier may enable hatchlings to capitalize on potential summer rains and forage before winter (Wallis et al. 1999). Likewise, Sonoran tortoises (at least in the Arizona Upland) may be investing their entire reproductive output in a single clutch prior to more reliable rainfall (Murray et al. 1996). Additional research is needed to determine whether reproductive output varies geographically in the Sonoran Desert, especially given the variation in body size (see above), and to help illuminate rangewide patterns in desert tortoise reproductive ecology and life history.

Survivorship

Desert tortoises may exceed thirty-five years of age in the Sonoran Desert (Germano 1992) and have been observed to live at least fifty years in the eastern Mohave Desert (Hardy 1976). As would be expected in a long-lived animal, adult desert tortoises have exceptionally high survivorship. In an intensive study of three Arizona populations, annual survival estimates were 94 percent or higher for adult tortoises (table 6.4; Howland and Klug 1996). Confidence intervals generally indicate these estimates are not significantly different from the 98 percent level estimated necessary for long-term persistence of tortoise populations in the Mohave Desert (U.S. Fish and Wildlife Service 1994). Although less precise, estimates of mean annual survivorship for juvenile tortoises (i.e., all tortoises of less than 180 mm CL) are only slightly lower than those for adults, ranging from 84 to 93 percent (table 6.4). However, the data supporting these estimates are biased toward larger juveniles (fig. 6.5). Survivorship of hatchlings and one- to two-year-olds is probably much lower, and rates of nest survivorship are also unknown. Because of their secretive nature and cryptic appearance, studies specifically addressing the ecology of these small tortoises have yet to be conducted in the Sonoran Desert.

Life-history traits of turtles, such as delayed sexual maturity, iteroparity (reproducing more than once during the lifetime), and high adult survival rates, require high survival for adults as well as relatively high survival for juveniles to maintain viable populations (Congdon et al. 1993). In fact, viability of tortoise populations in the western Mohave Desert is most sensitive to survival of large adult females (Doak et al. 1994). Although specific reproductive traits may differ between tortoises from the Sonoran and Mohave

TABLE 6.4. Mean survivorship of desert tortoises at the Eagletail Mountains, Granite Hills, and Little Shipp Wash plots, computed with Program Jolly (Model A; Pollock et al. 1990). Juveniles less than 180 mm midline carapace length; adults greater than or equal to 180 mm midline carapace length.

Plot	Size Class	Survivorship	95% Confidence Interval
EAGLETAIL MTS.	Juvenile	0.84	0.55–1.00
	Adult	0.97	0.93–1.00
GRANITE HILLS	Juvenile	0.93	0.75–1.00
	Adult	0.95	0.91–0.99
LITTLE SHIPP WASH	Juvenile	0.85	0.26–1.00
	Adult	0.94	0.90–0.97

Deserts, the same general pattern of survivorship undoubtedly contributes to population persistence in the Sonoran Desert.

Potentially higher survivorship of juvenile tortoises in the Sonoran Desert may actually make population persistence less tenuous than in the Mohave. Although hatchlings have enough energy reserves from embryonic yolks to survive through late summer and winter (see discussion in Averill-Murray et al., ch. 7 of this volume), they potentially can accumulate additional energy by foraging. Sonoran females lay a single clutch prior to the summer rainy season, and the eggs hatch at the end of the summer rainy season when annuals and herbaceous perennial forage are available in most years. In the Sonoran Desert a few hatchlings have been observed foraging in late September (Holm 1989; Hart et al. 1992; Woodman et al. 1994). In years with late summer and fall rains, food plants may remain green until the first frost in late November or early December. The importance of late summer and fall foraging in Sonoran tortoise hatchlings needs to be evaluated but may be important in enhanced survivorship. In the western Mohave Desert, where females produce multiple clutches, the situation is very different. Hatchlings rarely, if ever, have green annuals available and wait until the next spring to feed. The primary function of hatchling emergence in August and September may be dispersal (up to 1.5 km away from the nest site), not foraging (David J. Morafka, pers. comm. 1999). However, as noted earlier (see Reproduction), eastern Mohave females appear to lay their eggs early

Figure 6.5. A juvenile desert tortoise living in tropical deciduous forest in Arroyo Uvalamita (Anolis Canyon) northwest of Tepoca, Municipio de Onavas, Sonora, in August 1998. (Photograph by Cecil Schwalbe)

enough for hatchlings to take advantage of late summer forage resulting from unpredictable summer rains (Wallis et al. 1999).

The only documented exception to high survivorship in Sonoran tortoise populations was in the Maricopa Mountains, where the tortoise population suffered a major decline in the mid to late 1980s. Wirt (1988) found fifty-seven live adult tortoises and sixty-five carcasses on the 1-mi^2 Maricopa plot in 1987. Only three years later, Shields et al. (1990) found only seventeen live tortoises and fifty-four additional carcasses on the plot, including at least fifteen carcasses of tortoises that had been marked alive in 1987. The reasons for this decline are unclear, but a major drought affecting the Maricopa Mountains from 1984 to 1992 may have contributed to increased mortality (Wirt and Holm 1997). A relatively high proportion of tortoises with a shell disease was also observed at this plot (A. Peter Woodman, pers. obs.); the disease has been correlated with a decline in at least one Mohave Desert population (Berry 1997). This condition, however, has been observed in varying proportions in virtually all tortoise populations studied in Arizona to date, with no apparent detrimental effects to those populations or individuals (Dickinson et al., ch. 10 of this volume). Relatively high numbers of carcasses compared to live tortoises have also been found on the Arrastra Mountain, Bonanza Wash, and Sand Tank Mountain plots. These cases could represent accumulated mortality over a number of years, especially for the Arrastra Mountain and Bonanza Wash plots, at which few carcasses have been found in subsequent surveys; a previous short-term decline; or a longer-term decline in progress.

Specific causes of mortality are usually impossible to determine. Disease has contributed to widespread mortality of tortoises in the Mohave Desert (U.S. Fish and Wildlife Service 1994), but no population-level effects have been determined in the Sonoran Desert (see Dickinson et al., ch. 10 of this volume; Howland and Rorabaugh, ch. 14 of this volume). Predation affects all tortoise populations to varying degrees. For example, of thirteen adult mortalities estimated to have occurred on the Little Shipp Wash plot between 1986 and 1993, most were attributed to mountain lion *(Felis concolor)* predation, including seven of eight carcasses found in 1993. Mountain lions are one of the few, if not only, natural predators capable of breaking through an adult tortoise's shell, but other carnivores, including coyote (*Canis latrans*; Hohman and Ohmart 1980), kit fox (*Vulpes macrotis*; Coombs 1977), bobcat (*Felis rufus*; Woodbury and Hardy 1948), gray fox *(Urocyon cinereoargenteus)*, and badger *(Taxidea taxus)*, may prey on hatchlings, juveniles, or eggs or kill adults by chewing exposed limbs. Feral dogs have been implicated in tortoise mortality on the Eastern Bajada and Bonanza Wash plots.

Other potential predators of smaller tortoises include golden eagle (*Aquila chrysaetos*; Luckenbach 1982) and other raptors, common raven *(Corvus corax)*, and greater roadrunner *(Geococcyx californianus)*. Although increased predation on hatchling and juvenile tortoises by ravens near urban areas and along power lines crossing the desert has contributed to the decline of Mohave tortoise populations (U.S. Fish and Wildlife Service 1994), predation by aerial predators has not resulted in any noticeable population effects in the Sonoran Desert, probably because of the relative complexity of Sonoran tortoise habitat. Although not documented, some snakes, including coachwhip *(Masticophis flagellum)*, gopher snake *(Pituophis melanoleucus)*, and kingsnake *(Lampropeltis getula)*, may also eat tortoise eggs or juveniles. Finally, Gila monsters *(Heloderma suspectum)* are known to eat tortoise eggs (Barrett and Humphrey 1986).

Population Regulation

As previously discussed, desert tortoise abundance and density vary widely across the Sonoran Desert. One hypothesis to explain this variation suggests that a relatively fixed number of sheltersites available to the population ultimately limits its size. Since 1992 a component of monitoring plot surveys has been to individually mark and number active tortoise burrows as well as the tortoises themselves. No correlation ($r^2 = 0.05$) exists between the number of burrows marked on a plot and tortoise density, but only about half the sites have had burrows marked in more than one survey (fig. 6.6). Not all burrows are found in a single survey, and only those with tortoises actually inside or nearby are marked. A significant correlation does exist between the number of burrows and tortoise density on those plots on which more intensive effort has been made to find, confirm, and number tortoise burrows ($r^2 = 0.76$; fig. 6.6).

Tortoises use burrows to regulate heat and moisture. Females require adequate soil development to excavate nests, which are usually located inside the burrow. In most Sonoran populations, tortoise burrows are relatively permanent sheltersites often found below large rocks or boulders (Averill-Murray et al., ch. 7 of this volume). Researchers rarely find newly excavated burrows, because existing rock crevices and patches with suitable friable soil that tortoises can excavate already have burrows. Only some burrows have soil deep enough for nesting.

Closer examination of intensively studied sites augments the available evidence that sheltersite abundance limits tortoise abundance. On the Eagletail Mountain plot, for example, individual tortoises have often been found in the same burrow repeatedly within an annual survey. Some individuals have even been found in the same burrows during six consecutive annual surveys—an exceptional degree of sheltersite fidelity relative to other plots. A volcanic dike running across the Eagletail Mountain plot has many large rocks but relatively little soil development, resulting in few quality sheltersites and a small population. In comparison, the Little Shipp Wash

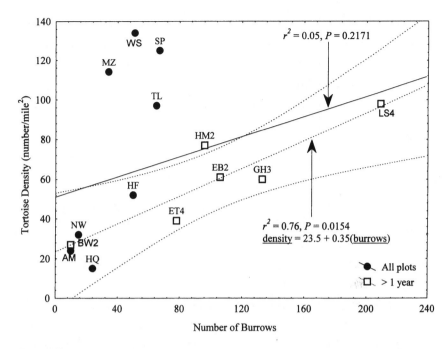

Figure 6.6. Correlation of the number of desert tortoise burrows to population density in the Sonoran Desert of Arizona. Codes refer to the study sites listed in table 6.1. Burrow numbers were obtained from monitoring plot surveys in a single year (closed circles) or multiple years (open squares).

and Mazatzal Mountain study sites with large populations are in areas where rapidly eroding granitic bedrock forms extensive soils and abundant suitable sheltersites (see fig. 6.6).

Given natural limitations on sheltersite availability, especially those with burrows suitable for nesting, sheltersite usage is complex. Individual tortoises use multiple burrows, including those used by other tortoises. However, Murray and Klug (1996) found that over a two- to three-year period, female tortoises were found in other female's burrows (i.e., those in which another female had previously been found) significantly less often than in burrows used by males. Other short-term studies have shown the same pattern in both the Mohave (Burge 1977) and Sonoran Deserts (Bailey et al. 1995). At one northeastern Mohave Desert site, female tortoises shared burrows with females less often than males with males or males with females

(Bulova 1994). During the nesting season at this site, female tortoises avoid burrows containing another female's feces, suggesting that chemical cues may indicate whether another female has already nested in a particular burrow (Bulova 1997). Avoiding such burrows might be advantageous if females risk breaking their own or previously laid eggs. Broken eggs may result in bacterial infection of embryos in adjacent eggs or an increased chance of detection by egg predators.

Another hypothesis may also help explain tortoise population regulation. As described previously, variation in rainfall may influence reproductive success, juvenile survivorship, and recruitment. About ten years of relatively low juvenile survivorship will gradually deplete new generations, so few tortoises in a given cohort (a group of similar-aged tortoises) reach sexual maturity to be recruited into the adult population. Wirt and Holm (1997) reported a low proportion of Sonoran tortoises reproducing in the Maricopa Mountains following an extended drought, and Turner et al. (1986) and Karl (1998) found that clutch frequency of Mohave tortoises was correlated with winter rainfall and spring forage in California populations. A series of dry years may limit reproductive output or result in high mortality among smaller tortoises, resulting in smaller surviving cohorts from those years. These missing cohorts may be reflected as gaps in size distributions in the smaller adult sizes (figs. 6.2A, 6.2C). However, successive years with precipitation producing adequate forage may result in high reproductive output and juvenile survivorship (Morafka 1994). Although such years may be those represented by bulges in juvenile size distributions (figs. 6.2A, 6.2B), we have not studied those populations long enough to see increases in adult densities.

How individual populations in the Sonoran Desert interrelate is even less understood than their local dynamics. Tortoises in the Sonoran Desert exist naturally in disjunct local populations, generally in low, desert mountain foothills (Germano et al. 1994; Van Devender, ch. 1 of this volume). Although observations of Sonoran tortoises dispersing far away from rocky ridge habitats are rare, populations, at least theoretically, may depend on occasional cross-valley immigration for genetic interchange and long-term survival. Local tortoise populations receiving high precipitation for two or three years may increase in size, thus increasing the probability of individuals at or approaching sexual maturity dispersing across the valleys (e.g., Morafka 1994). Such tortoises of both sexes have been observed to make relatively long-distance movements (up to 3 km over a period of several weeks) away from their normal observed activity centers. These animals crossed areas of atypical tortoise habitat, including an approximately 1-km-wide alluvial fan and steep, boulder-free slopes occupied by few to no resident tortoises (Roy C. Averill-Murray, pers. obs.).

Most local tortoise populations in the Sonoran Desert appear stable at present, but they are increasingly fragmented by urban and agricultural development (Howland and Rorabaugh, ch. 14 of this volume). Given the fact that individual tortoises live for decades, potential impacts of population isolation may not be seen for many years. The degree to which local populations depend on interchange with other populations for long-term persistence is unknown, as are the effects of dismissing intermountain valleys as "unsuitable habitat" in Sonoran desert tortoise conservation efforts.

ACKNOWLEDGMENTS

This chapter is the result of the hard work and dedication of many biologists who have studied desert tortoises over the past two decades and whose work is cited below. The U.S. Fish and Wildlife Service, U.S. Bureau of Land Management, and Arizona Game and Fish Department (Heritage Fund and Nongame Checkoff) have contributed significant funding to tortoise monitoring efforts as well as to the completion of this chapter. Comments and other assistance from Linda Allison, Laurie Averill-Murray, Terry Johnson, Chris Klug, Earl McCoy, Phil Medica, Henry Mushinsky, Tom Van Devender, Martin Whiting, and an anonymous reviewer improved the manuscript.

LITERATURE CITED

Averill-Murray, R. C., and C. M. Klug. 2000. Reproduction in Sonoran desert tortoises: 1999 progress report. *Proceedings of the Desert Tortoise Council Symposium* 1999:31–34.

Bailey, S. J. 1992. Hibernacula use and home range of the desert tortoise *(Gopherus agassizii)* in the San Pedro Valley, Arizona. M.S. thesis, University of Arizona, Tucson.

Bailey, S. J., C. R. Schwalbe, and C. H. Lowe. 1995. Hibernaculum use by a population of desert tortoises *(Gopherus agassizii)* in the Sonoran Desert. *Journal of Herpetology* 29:361–69.

Barrett, S. L. 1990. Home range and habitat use of the desert tortoise *(Xerobates agassizii)* in the Picacho Mountains, Pinal County, Arizona. *Herpetologica* 46:202–6.

Barrett, S. L., and J. H. Humphrey. 1986. Agonistic interactions between *Gopherus agassizii* (Testudinidae) and *Heloderma suspectum* (Helodermatidae). *Southwestern Naturalist* 31:261–63.

Berry, K. H. 1997. Demographic consequences of disease in two desert tortoise populations in California, USA. Pp. 91–99 *in* J. Van Abbema, ed. *Proceedings: Conservation, restoration, and management of tortoises and turtles—An international conference.* New York Turtle and Tortoise Society, New York.

Berry, K. H., and L. L. Nicholson. 1984. Attributes of populations at twenty-seven sites in California. Ch. 5 *in* K. H. Berry, ed. *The status of the desert tortoise* (Gopherus agassizii) *in the United States.* Report to U.S. Fish and Wildlife Service from the Desert Tortoise Council on Order no. 11310-81.

Brown, D. E., ed. 1982. Biotic communities of the American Southwest: United States and Mexico. *Desert Plants* 4:1–342.

Bulova, S. J. 1994. Patterns of burrow use by desert tortoises: Gender differences and seasonal trends. *Herpetological Monographs* 8:133–43.

———. 1997. Conspecific chemical cues influence burrow choice by desert tortoises *(Gopherus agassizii)*. *Copeia* 1997:802–10.

Burge, B. L. 1977. Daily and seasonal behavior, and areas utilized by the desert tortoise *Gopherus agassizii* in southern Nevada. *Proceedings of the Desert Tortoise Council Symposium* 1977:59.

Congdon, J. D., A. E. Dunham, and R. C. Van Loben Sels. 1993. Delayed sexual maturity and demographics of Blanding's turtles *(Emydoidea blandingii)*: Implications for conservation and management of long-lived organisms. *Conservation Biology* 7:826–33.

Coombs, E. M. 1977. *Implications of behavior and physiology on the desert tortoise* (Gopherus agassizii) *concerning their declining populations in southwestern Utah, with inferences on related desert ectotherms.* Report to U.S. Bureau of Land Management, St. George, Utah.

Corn, P. S. 1994. Recent trends of desert tortoise populations in the Mojave Desert. Pp. 85–93 *in* R. B. Bury and D. J. Germano, eds. *Biology of North American tortoises.* Fish and Wildlife Research no. 13, U.S. Dept. of the Interior National Biological Survey, Washington, D.C.

Dames and Moore, Tucson. 1994. *Luke Air Force legacy studies: Desert tortoise surveys.* Report to U.S. Air Force, 56th CES/CEVN, Luke Air Force Base, Contract no. F02604-93-D0020.

Doak, D., P. Kareiva, and B. Klepetka. 1994. Modeling population viability for the desert tortoise in the western Mojave Desert. *Ecological Applications* 4:446–60.

Geo-Marine, Inc. 1994. *Sonoran desert tortoise inventories for Luke Air Force Base, Arizona.* Report to U.S. Army Corps of Engineers, Fort Worth, Tex.

Germano, D. J. 1992. Longevity and age-size relationships of populations of desert tortoises. *Copeia* 1992:367–74.

———. 1994a. Growth and age at maturity of North American tortoises in relation to regional climates. *Canadian Journal of Zoology* 72:918–31.

———. 1994b. Comparative life histories of North American tortoises. Pp. 175–85 *in* R. B. Bury and D. J. Germano, eds. *Biology of North American tortoises.* Fish and Wildlife Research no. 13, U.S. Dept. of the Interior National Biological Survey, Washington, D.C.

Germano, D., R. B. Bury, T. C. Esque, T. H. Fritts, and P. A. Medica. 1994. Range and habitats of the desert tortoise. Pp. 73–84 *in* R. B. Bury and D. J. Germano, eds. *Biology of North American tortoises*. Fish and Wildlife Research no. 13, U.S. Dept. of the Interior National Biological Survey, Washington, D.C.

Gibbons, J. W. 1990. Recommendations for future research on freshwater turtles: What are the questions? Pp. 311–17 *in* J. W. Gibbons, ed. *Life history and ecology of the slider turtle*. Smithsonian Institution Press, Washington, D.C.

Goodlett, G., P. Woodman, M. Walker, and S. Hart. 1996. *Desert tortoise population survey at Beaver Dam Slope exclosure desert tortoise study plot: Spring 1996*. Report to Arizona Game and Fish Dept., Phoenix.

Goodlett, G., M. Walker, and P. Woodman. 1997. *Desert tortoise population survey at Virgin Slope desert tortoise study plot: Spring 1997*. Report to Arizona Game and Fish Dept., Phoenix.

Hardy, R. 1976. The Utah population: A look in the 1970s. *Proceedings of the Desert Tortoise Council Symposium* 1976:84–88.

Hart, S. 1996. Demographics of three desert tortoise populations in the Sonoran Desert. *Proceedings of the Desert Tortoise Council Symposium* 1993:27–36.

Hart, S., P. Woodman, S. Bailey, S. Boland, P. Frank, G. Goodlett, D. Silverman, D. Taylor, M. Walker, and P. Wood. 1992. *Desert tortoise population studies at seven sites and a mortality survey at one site in the Sonoran Desert, Arizona*. Report to Arizona Game and Fish Dept. and U.S. Bureau of Land Management, Phoenix.

Henen, B. T. 1994. Seasonal and annual energy and water budgets of female desert tortoises *(Gopherus agassizii)* at Goffs, California. Ph.D. diss., University of California, Los Angeles.

———. 1997. Seasonal and annual energy budgets of female desert tortoises. *Ecology* 78:283–96.

Hohman, J. P., and R. D. Ohmart. 1980. *Ecology of the desert tortoise on the Beaver Dam Slope, Arizona*. Report to U.S. Bureau of Land Management, St. George, Utah.

Holm, P. A. 1989. *Desert tortoise monitoring baseline study: Harquahala Mountains*. Report to U.S. Bureau of Land Management, Phoenix.

Howland, J. M., and C. M. Klug. 1996. Results of five consecutive years of population monitoring at three Sonoran desert tortoise plots. *Proceedings of the Desert Tortoise Council Symposium* 1995:74–87.

Karl, A. E. 1998. Reproductive strategies, growth patterns, and survivorship of a long-lived herbivore inhabiting a temporally variable environment. Ph.D. diss., University of California, Davis.

Klug, C. M., and R. C. Averill-Murray. 1999. Reproduction in Sonoran desert tortoises: A progress report. *Proceedings of the Desert Tortoise Council Symposium* 1997–1998:59–62.

Luckenbach, R. A. 1982. Ecology and management of the desert tortoise *(Gopherus agassizii)* in California. Pp. 1–37 *in* R. B. Bury, ed. *North American tortoise conservation and ecology*. Wildlife Research Report no. 12, U.S. Fish and Wildlife Service, Washington, D.C.

Martin, B. E. 1995. Ecology of the desert tortoise *(Gopherus agassizii)* in a desert grassland community in southern Arizona. M.S. thesis, University of Arizona, Tucson.

McLuckie, A. M., C. R. Schwalbe, and T. Lamb. 1996. *Genetics, morphology, and ecology of the* desert tortoise (Gopherus agassizii) *in the Black Mountains, Mohave County, Arizona.* Report to U.S. Bureau of Land Management, Phoenix; Transwestern Pipeline Company, Houston; and Arizona Game and Fish Dept., Phoenix.

McLuckie, A. M., T. Lamb, C. R. Schwalbe, and R. D. McCord. 1999. Genetic and morphometric assessment of an unusual tortoise *(Gopherus agassizii)* population in the Black Mountains of Arizona. *Journal of Herpetology* 33:36–44.

Medica, P. A., R. B. Bury, and F. B. Turner. 1975. Growth of the desert tortoise *(Gopherus agassizii)* in Nevada. *Copeia* 1975:639–43.

Minden, R. L., and S. M. Keller. 1981. *Population analysis of the desert tortoise* (Gopherus agassizii) *on the Beaver Dam Slope, Washington County, Utah.* Publication no. 81-14, Utah State Dept. of Natural Resources and Energy, Salt Lake City.

Morafka, D. J. 1994. Neonates: Missing links in the life histories of North American tortoises. Pp. 161–73 *in* R. B. Bury and D. J. Germano, eds. *Biology of North American tortoises.* Fish and Wildlife Research no. 13, U.S. Dept. of the Interior National Biological Survey, Washington, D.C.

Mueller, J. M., K. R. Sharp, K. K. Zander, D. L. Rakestraw, K. R. Rautenstrauch, and P. E. Lederle. 1998. Size-specific fecundity of the desert tortoise *(Gopherus agassizii). Journal of Herpetology* 32:313–19.

Murray, R. C. 1993. Mark-recapture methods for monitoring Sonoran populations of the desert tortoise *(Gopherus agassizii).* M.S. thesis, University of Arizona, Tucson.

Murray, R. C., and C. M. Klug. 1996. Preliminary data analysis from three desert tortoise long-term monitoring plots in Arizona: Sheltersite use and growth. *Proceedings of the Desert Tortoise Council Symposium* 1996:10–17.

Murray, R. C., and C. R. Schwalbe. 1993. *The desert tortoise on national forest lands in Arizona.* Report to U.S. Dept. of Agriculture Forest Service, Coronado, Prescott, and Tonto National Forests, Ariz.

———. 1997. *Second survey of the Four Peaks desert tortoise monitoring plot: Testing abundance estimation procedures.* Report to Arizona Game and Fish Dept., Phoenix.

Murray, R. C., C. R. Schwalbe, S. J. Bailey, S. P. Cuneo, and S. D. Hart. 1995. *Desert tortoise* (Gopherus agassizii) *reproduction in the Mazatzal Mountains, Maricopa County, Arizona.* Report to Arizona Game and Fish Dept. and U.S. Dept. of Agriculture Tonto National Forest, Phoenix.

———. 1996. Reproduction in a population of the desert tortoise, *Gopherus agassizii,* in the Sonoran Desert. *Herpetological Natural History* 4:83–88.

Pollock, K. H., J. D. Nichols, C. Brownie, and J. E. Hines. 1990. *Statistical inference for capture-recapture experiments.* Wildlife Monographs no. 107, The Wildlife Society, Bethesda, Md.

Schneider, P. B. 1981. *A population analysis of the desert tortoise,* Gopherus agassizii, *in Arizona.* Report to U.S. Bureau of Land Management, Phoenix.

Schnell, J., and D. Drobka. 1988. *The 1988 desert tortoise survey: San Pedro population.* Report to U.S. Bureau of Land Management, Safford, Ariz.

Shaw, W. W., and A. Goldsmith. 1988. *Final report on desert tortoise ecology 1988.* Report to Southwest Parks and Monuments Association, Tucson, Ariz.

Shields, T., and P. Woodman. 1987. *A study of the desert tortoise in the Eagletail Mountains, Maricopa County, Arizona.* Report to U.S. Bureau of Land Management, Phoenix.

———. 1988. *Some aspects of the ecology of the desert tortoise at the New Water study plot, Arizona*. Report to U.S. Bureau of Land Management, Phoenix.

Shields, T., S. Hart, J. Howland, N. Ladehoff, T. Johnson, K. Kime, D. Noel, B. Palmer, D. Roddy, and C. Staab. 1990. *Desert tortoise population studies at four plots in the Sonoran Desert, Arizona*. Report to Arizona Game and Fish Dept. and U.S. Fish and Wildlife Service, Albuquerque.

SWCA, Inc. 1990a. *1990 Santan Regional Park desert tortoise study plot, Pinal County, Arizona*. Report to Greiner Engineering, Inc., Tucson.

———. 1990b. *Final report for 1990 East Bajada desert tortoise study plot, Mohave County, Arizona*. Report to U.S. Bureau of Land Management, Kingman, Ariz.

———. 1992. *1991 Santan Regional Park desert tortoise study plot, Pinal County, Arizona*. Report to Greiner Engineering, Inc., Tucson.

Trachy, S., and V. M. Dickinson. 1993. *Use areas and sheltersite characteristics of Sonoran desert tortoises*. Report to Arizona Game and Fish Dept., Phoenix.

Turner, F. B., P. A. Medica, and C. L. Lyons. 1984. Reproduction and survival of the desert tortoise *(Scaptochelys agassizii)* in Ivanpah Valley, California. *Copeia* 1984:811–20.

Turner, F. B., P. Hayden, B. L. Burge, and J. B. Roberson. 1986. Egg production by the desert tortoise *(Gopherus agassizii)* in California. *Herpetologica* 42:93–104.

Turner, F. B., P. A. Medica, and R. B. Bury. 1987. Age-size relationships of desert tortoises *(Gopherus agassizii)* in southern Nevada. *Copeia* 1987:974–79.

Turner, R. M. 1982. Mohave desertscrub. *Desert Plants* 4:157–68.

Turner, R. M., and D. E. Brown. 1982. Sonoran desertscrub. *Desert Plants* 4:181–221.

U.S. Fish and Wildlife Service. 1990. Endangered and threatened wildlife and plants: Determination of threatened status for the Mojave population of the desert tortoise. *Federal Register* 55:12178–91.

———. 1994. *Desert tortoise (Mojave population) recovery plan*. U.S. Fish and Wildlife Service, Portland, Ore.

Vaughan, S. L. 1984. Home range and habitat use of the desert tortoise *(Gopherus agassizii)* in the Picacho Mountains, Pinal County, Arizona. M.S. thesis, Arizona State University, Tempe.

Wallis, I. R., B. T. Henen, and K. A. Nagy. 1999. Egg size and annual egg production by female desert tortoises *(Gopherus agassizii)*: The importance of food abundance, body size, and date of egg shelling. *Journal of Herpetology* 33:394–408.

Wirt, E. B. 1988. *Two desert tortoise populations in Arizona*. Report to U.S. Bureau of Land Management, Phoenix.

Wirt, E. B., and P. A. Holm. 1997. *Climatic effects on survival and reproduction of the desert tortoise* (Gopherus agassizii) *in the Maricopa Mountains, Arizona*. Report to Arizona Game and Fish Dept., Phoenix.

Wirt, E. B., P. A. Holm, and R. H. Robichaux. 1999. *Survey and monitoring of the desert tortoise* (Gopherus agassizii) *at Organ Pipe Cactus National Monument*. Report to National Park Service, Ajo, Ariz.

Woodbury, A. M., and R. Hardy. 1948. Studies of the desert tortoise, *Gopherus agassizii*. *Ecological Monographs* 18:145–200.

Woodman, P., and T. Shields. 1988. *Some aspects of the ecology of the desert tortoise in the Harcuvar study plot, Arizona*. Report to U.S. Bureau of Land Management, Phoenix.

Woodman, P., S. Boland, P. Frank, G. Goodlett, S. Hart, D. Silverman, T. Shields, and P. Wood. 1993. *Desert tortoise population surveys at five sites in the Sonoran Desert, Arizona.* Report to Arizona Game and Fish Dept. and U.S. Bureau of Land Management, Phoenix.

Woodman, P., S. Hart, S. Boland, P. Frank, D. Silverman, G. Goodlett, P. Gould, D. Taylor, M. Vaughn, and P. Wood. 1994. *Desert tortoise population surveys at five sites in the Sonoran Desert of Arizona, 1993.* Report to Arizona Game and Fish Dept. and U.S. Bureau of Land Management, Phoenix.

Woodman, P., S. Hart, P. Frank, S. Boland, G. Goodlett, D. Silverman, D. Taylor, M. Vaughn, and M. Walker. 1995. *Desert tortoise population surveys at four sites in the Sonoran Desert of Arizona, 1994.* Report to Arizona Game and Fish Dept. and U.S. Bureau of Land Management, Phoenix.

Woodman, P., S. Hart, S. Bailey, and P. Frank. 1996. *Desert tortoise population surveys at two sites in the Sonoran Desert of Arizona, 1995.* Report to Arizona Game and Fish Dept., Phoenix.

Woodman, P., P. Frank, G. Goodlett, and M. Walker. 1997. *Desert tortoise population surveys at two sites in the Sonoran Desert of Arizona, 1996.* Report to Arizona Game and Fish Dept., Phoenix.

Woodman, P., P. Frank, S. Hart, G. Goodlett, M. Walker, D. Roddy, and S. Bailey. 1998. *Desert tortoise population surveys at two sites in the Sonoran Desert of Arizona, 1997.* Report to Arizona Game and Fish Dept., Phoenix.

Woodman, P., P. Frank, and D. Silva. 1999. *Desert tortoise population surveys at three sites in the Sonoran Desert of Arizona, 1998.* Report to Arizona Game and Fish Dept., Phoenix.

Woodman, P., D. Silva, and M. Walker. 2000. *Desert tortoise population survey at the New Water Mountains, Arizona, 1999.* Arizona Game and Fish Dept., Phoenix.

CHAPTER 7

Activity and Behavior of the Sonoran Desert Tortoise in Arizona

ROY C. AVERILL-MURRAY, BRENT E. MARTIN,
SCOTT JAY BAILEY, AND ELIZABETH B. WIRT

Most early accounts of desert tortoise habitats and behavior were obtained from animals in the Mohave Desert (Miller 1932; Grant 1936, 1946; Bogert 1937; Woodbury and Hardy 1940, 1948). The desert tortoise was generally depicted as an inhabitant of arid sandy valleys who is most active during spring, foraging on fresh spring foliage produced by winter rainfall and retreating into burrows for extended periods during hot, dry summers and cold winters (see Carr 1952; Stebbins 1954; Ernst and Barbour 1972). By 1980 important habitat differences between the Mohave and Sonoran Deserts were being recognized (Burge 1979, 1980), but relatively little attention was being directed toward the Sonoran Desert (Hohman et al. 1980; Grover and DeFalco 1995).

The first ecological studies of desert tortoises in the Sonoran Desert quickly verified the differences in activity between the Mohave and Sonoran populations. Vaughan (1984) confirmed that tortoises in the Picacho Mountains of Arizona were more active in August and September than in other months and were only found on rocky hillsides. In the Maricopa Mountains of Arizona, tortoises proved to be much more difficult to locate during spring than summer (Wirt 1988). Sweeping generalizations simply cannot capture the ecological intricacies of a species as wide-ranging as the desert tortoise. In this chapter, we describe activity and behavior of the desert tortoise in the Sonoran Desert based on recent studies conducted in Arizona (see Averill-Murray et al., table 6.1 of this volume).

Sheltersites

Desert tortoises are closely associated with sheltersites in both the Mohave and Sonoran Deserts (Lowe 1964; Burge 1978). For example, Nagy

Figure 7.1. A female Sonoran desert tortoise equipped with a radio-transmitter in front of her sheltersite at Sugarloaf Mountain, northeast of Phoenix, Maricopa County, Arizona. (Photograph by Roy Averill-Murray)

and Medica (1986) estimated that tortoises in the Mohave Desert of southern Nevada spend approximately 98 percent of their lifetimes inside burrows. The proportions of surface activity versus subsurface inactivity have not been similarly quantified for tortoises in the Sonoran Desert, but our work indicates that Sonoran tortoises do spend a comparable amount of time underground. Such limited activity combined with a cryptic appearance (a tortoise on a rocky hillside mimics a stone very well) results in a false perception of rareness to otherwise keen wildlife observers.

Sonoran tortoises use many types of shelters (Lowe 1990; fig. 7.1). In this chapter we use the term *burrow* to specifically refer to a subsurface cavity formed by erosion and/or excavated by a tortoise or another animal (Burge 1978). Sonoran tortoises typically use burrows below rocks and boulders but occasionally under shrubs or in open ground as well. On lower *bajadas*, tortoises also use burrows eroded or excavated into the hard calcium carbonate (caliche) soils along incised arroyo (dry stream) banks. We use the term *shelter* more generally, because tortoises also use the mounds of cholla (*Opuntia* spp.) stems, sticks, and other debris in white-throated packrat *(Neotoma albigula)* houses; large crevices in bedrock; or simply the shade of large rocks or shrubs. Animals, including snakes, lizards, rodents, javelinas *(Tayassu tajacu)*, birds, and insects and other invertebrates, also seek refuge from the elements in shelters used by tortoises (see Luckenbach 1982; Grover and DeFalco 1995).

Tortoises use multiple shelters during the year. For example, tortoises in the Tortolita Mountains in desert grassland used as many as twenty-seven (average twenty-one) shelters in a two-year period (Martin 1995). Two or more tortoises may use the same shelter, either at the same or different times, although females tend to avoid sharing shelters with other females during the activity season (Bailey 1992; Martin 1995; Murray and Klug 1996; Averill-Murray et al., ch. 6 of this volume). These observations are comparable to those in the Mohave Desert, where tortoises used up to twenty shelters during a one-year period in southern Nevada (Burge 1978) and an average of twelve burrows per year (ranging from four to twenty-three) at Yucca Mountain, Nevada (TRW Environmental Safety Systems, Inc. 1997). The number of shelters available to tortoises in the Sonoran Desert depends on the complexity and geological structure of the terrain, and local population sizes are highly correlated with shelter availability (Averill-Murray et al., ch. 6 of this volume).

Seasonal Activity and Behavior

As we examine the typical annual activity patterns of Sonoran tortoises, we will frame our discussion by three seasonal periods based on tortoise activity relative to environmental conditions. The Sonoran Desert spring (March through June) is characterized by temperatures increasing from mild to hot, decreasing rainfall, and variable tortoise activity. The hot, dry period from mid-May through June in late spring, also referred to as the "arid foresummer," is a time of minimal biological activity (Van Devender, ch. 1 of this volume). Summer (July through October) is hot and generally includes peak rainfall and tortoise activity. An autumn season is not included here because the transition from summer to winter is more subtle than in other biomes. Temperatures generally begin decreasing in October, and this month is usually characterized by at least moderate tortoise activity. Tortoises hibernate during the cool months (November through February) of the winter rainy season. To provide a broader overview of the species, we will also compare and contrast Sonoran Desert observations with those from the Mohave Desert.

Spring Activity

Spring, with its mild temperatures, low humidity, showy wildflowers, and conspicuous wildlife, is a pleasant time to be in the Sonoran Desert. In general, however, the desert tortoise is more difficult to find in the spring relative to summer (Wirt 1988), regardless of the weather. The scarcity of tortoises in spring is at least partly due to the fact that male tortoises typically terminate hibernation about one month later (average late April) than females (late March) and remain less active (Vaughan 1984; Bailey 1992; Martin 1995). For example, in radio-equipped tortoises in the Tortolita Mountains, females were more active and used more home range area than males in spring. However, both sexes were less active, used less area within their home ranges, and participated in less frequent social interactions during spring compared to summer. These seasonal behavioral patterns corroborated ten years of mark-recapture study, when only 14 percent of all desert tortoise observations, primarily of females, occurred during spring (Martin 1995). Likewise, a study of female tortoises in the Mazatzal Mountains revealed that females are roughly half as active during spring as in summer, although they seem to be feeding relatively more often when active in the spring, with about equal numbers of foraging observations as in summer (fig. 7.2). This spring forage may be important for developing eggs to be laid in late spring or early summer (see Averill-Murray et al., ch. 6 of this volume).

Greater activity by female than male tortoises during the spring and early summer has also been observed at other Sonoran Desert sites and in the Mohave Desert (Bulova 1994; Bailey et al. 1995; McLuckie et al. 1996; TRW Environmental Safety Systems, Inc. 1997). In parts of the Mohave desert tortoise activity in summer and fall may be greater than previously reported (Bulova 1994; Zimmerman et al. 1994). Even so, tortoise activity peaks during the spring in the Mohave Desert (Luckenbach 1982). In the drier western Mohave, the highest proportion of tortoise activity occurs from late April to

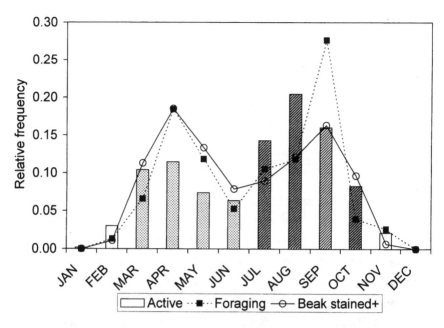

Figure 7.2. Relative monthly activity, foraging observations, and evidence of foraging for female desert tortoises in the Mazatzal Mountains, Arizona, 1991–1993 and 1996–1999. Open bars, winter; stippled bars, spring; hatched bars, summer. Active, tortoises outside a shelter; foraging, tortoises observed eating; beak stained+, feeding observations plus observations of tortoises with plant parts in their beaks or stained with vegetative material. The seasonal relative frequencies of these categories are active: 0.05 in winter, 0.36 in spring, 0.59 in summer; foraging: 0.04 in winter, 0.42 in spring, 0.54 in summer; beak stained+: 0.02 in winter, 0.51 in spring, 0.47 in summer.

early May (Marlow 1985). Summer activity becomes more common eastward as the amount and dependability of erratic summer precipitation increases. In the Black Mountains in the eastern Mohave Desert, near Kingman, Arizona, tortoises were most active in April and May, with a secondary peak during September and October (McLuckie et al. 1996).

Summer Activity

Whereas most humans sensibly avoid outdoor activity in the Sonoran Desert during the torrid summer months, the characteristic summer rainfall rouses desert tortoises from their late spring and foresummer lethargy. Thus the peak seasonal activities of humans and tortoises in the Sonoran Desert are diametrically opposed.

The summer rainy season is often referred to as the "summer monsoon season." The predictable arrival of summer rainfall, usually in July, initiates new growth in perennial plants and the germination of annuals. Tortoises drink freely and rehydrate from puddles left by summer thunderstorms. We have observed tortoises eating damp soil (later passed as mud scat) and extracting water from the surface of the ground through their nose. Tortoises apparently can sense oncoming rain and move to natural depressions (Roy C. Averill-Murray, pers. obs.) or construct water catchments prior to the onset of rainfall (Mohave Desert; Medica et al. 1980). Establishment of a positive water balance allows them to begin assimilating energy from food plants (Van Devender et al., ch. 8 of this volume). There is a lag of about ten days after the summer rains begin before perennials have substantial new growth or summer annuals germinate and accumulate much biomass (T. R. Van Devender, pers. comm. 1999). In the meantime, rehydrated tortoises eat dried grasses and dead spring annuals (Roy C. Averill-Murray, pers obs.). In dry summers when light, irregular rains are enough to rehydrate tortoises but not stimulate plant growth, consumption of dried plant foods becomes even more important, as has been observed in the Mohave Desert (Nagy and Medica 1986).

In the Sonoran Desert, social behavior also peaks during the summer monsoon season. Male-male combat and courtship involve a similar repertoire, including much posturing, head-bobbing, chasing, and ramming (Ruby and Niblick 1994). Combat can be especially violent, with one male ramming the other with his enlarged gular shields and using them as a lever to flip his opponent (Grant 1936; Ruby and Niblick 1994; fig. 7.3). Sometimes, the victor continues this ramming after having flipped his opponent, eventually returning him to an upright position (Vaughan 1984). The loser usually takes this opportunity to escape.

During courtship, the male rams the female and bites at her exposed limbs, repeatedly moving into position to mount her (Householder 1950; Black 1976; Ruby and Niblick 1994; fig. 7.4). Mounting is aided by a concavity in the male's plastron that allows the male's shell to fit over the curved carapace of the female. In sandy habitats, the circling and positioning during courtship and mating leave a cluster of tracks (Woodbury and

Figure 7.3. Male-male combat in Sonoran desert tortoises observed on 1 October 1995 in the Tucson Mountains, Pima County, Arizona. The winner was older and lighter than his opponent. Overturned vanquished males normally right themselves without injury. (Photograph by Roger Repp)

Hardy 1948), known as a courtship ring, but these tracks are rarely left in the rocky habitat of the Sonoran Desert. Because the females can store sperm, one summer's matings produce the next summer's clutch of eggs, and some evidence suggests that individual clutches may be the products of females mating with multiple males (Palmer et al. 1998). Females usually lay eggs near the beginning of the summer rainy season, primarily during June and July (Averill-Murray et al., ch. 6 of this volume), often inside burrows having adequate soil development to allow excavation of a nest cavity. Although tortoises mate both during the spring and summer in the Mohave Desert (Black 1976; Rostal et al. 1994; Goodlett et al. 1996), females develop shelled eggs following spring emergence before mating activities (Rostal et al. 1994). Spring mating may not fertilize eggs but simply stimulate growth and maturation of reproductive organs and gametes in females (e.g., Gist et al. 1990).

Gila monsters *(Heloderma suspectum)*, which sometime share the

Figure 7.4. Mating Sonoran desert tortoises observed at dusk on 28 September 2000 in the Tucson Mountains, Pima County, Arizona. (Photograph by Jim Jarchow)

same burrows with tortoises, are known to eat desert tortoise eggs. Female tortoises aggressively defend their nests against Gila monsters (Barrett and Humphrey 1986). In one instance, an intruding Gila monster was expelled from the nest site, although not before it had consumed an egg (Barrett and Humphrey 1986). We have seen females position themselves at their burrow entrances in response to humans, blocking the entrance to intruders while limiting their own exposure. Females that lay eggs often remain close to their nest sites throughout most of the summer, whereas those that are not reproductive use multiple shelters (Murray et al. 1996). We have also found mammalian carnivore tracks at excavated tortoise nests, but nest defense behavior has not been reported against mammals (other than humans). Nest defense in the wild has only been suggested for one other species of turtle, the yellow mud turtle (*Kinosternon flavescens*; Iverson 1990).

Incubation lasts about three months, with eggs usually hatching in September and October. Some eggs, however, apparently overwinter in the nest in the Mohave and possibly the Sonoran Desert, with hatchlings emerging in the spring (Luckenbach 1982). Hatchlings measure 43–48 mm carapace length when they leave the nest and are relatively soft and vulnerable. Like many other turtles, sex of hatchling tortoises is determined by incubation temperature in the nest. In the northeastern Mohave Desert nest temperatures exceeding 32°C produced females, whereas lower temperatures produced males (Spotila et al. 1994). This threshold temperature may vary across the range of the desert tortoise, as it does in other widely ranging turtle species (Bull et al. 1982).

Several hypotheses have been proposed to explain temperature-dependent sex determination in turtles, including (1) differential fitness between sexes incubated at different temperatures, (2) production of unisexual clutches to prevent inbreeding, (3) adaptation for biased adult sex ratios in the population, and (4) phylogenetic inertia preventing evolution of genetic sex determination due to low genetic variation (Ewert and Nelson 1991). None of these hypotheses, however, are clearly supported with much evidence (Ewert and Nelson 1991; Burke 1993; Ewert et al. 1994). The typically balanced adult sex ratios observed in most desert tortoise populations (Averill-Murray et al., ch. 6 of this volume) particularly argues against hypothesis 3.

Gibbons and Nelson (1978) suggested that, in species laying a single clutch annually, natural selection should favor a combination of egg-laying season and developmental times that would result in the highest probability of hatching at the time most advantageous for emergence. Hatchlings emerging immediately upon completion of embryonic development have the potential to initiate feeding and growth toward maturity (Gibbons and Nelson 1978). Because female tortoises lay a single clutch of eggs just prior to the relatively predictable monsoonal rains and production of new summer forage in the Sonoran Desert (Averill-Murray et al., ch. 6 of this volume), hatchlings who emerge in the presence of green plants carrying protein-rich fruits have an opportunity to forage for a few weeks to a month, depending on additional rainfall, before hibernating (Murray et al. 1996).

Delayed emergence after overwintering in the nest is predicted for species in which egg-laying occurs over an extended time (i.e., in multiple clutches) or for which food sources for juveniles are likely to be unfavorable throughout a major portion of the hatching period (Gibbons and Nelson

1978). Tortoises in the Mohave Desert produce multiple clutches in the spring and summer (Turner et al. 1984, 1986; Karl 1998; Mueller et al. 1998; Wallis et al. 1999), and fresh forage is not reliably available to hatchlings until the following spring. Therefore, it would seem to be advantageous for Mohave (especially western) hatchlings to wait to emerge until spring when they have a better chance to obtain food.

Unfortunately, hatchling tortoises are very difficult to study; only twelve have been observed on Sonoran Desert summer study plots since 1987 (Averill-Murray et al., ch. 6 of this volume). The little available information from both deserts does not neatly fit Gibbons and Nelson's (1978) predictions for hatchling emergence. In the eastern Mohave Desert, females lay their eggs in May and June, incubation generally lasts 84 to 97 days, and hatchlings emerge in September and early October (Roberson et al. 1985). In the western Mohave Desert, females lay eggs in late May and late June, incubation lasts 85 to 110 days, and eggs hatch from late August through late September (David J. Morafka and Jeff Lovich, pers. comm. 1999). As previously noted, fresh annuals are not usually available for hatchlings until the spring, and Luckenbach (1982) suggested that captive hatchlings in California even ignored food between emergence and the onset of winter dormancy. Hatchling desert tortoises appear to have sufficient energy from their mothers stored as yolk to allow emergence, dispersal, growth, and hibernation for four to six months (James L. Jarchow, pers. comm. 1998; David J. Morafka, pers. comm. 1999). Under favorable feeding conditions produced by unpredictable summer rainfall, female tortoises laying multiple clutches in the Mohave Desert occasionally allow some hatchlings to emerge, grow to larger sizes, and benefit from increased survival through hibernation (Rostal et al. 1994; Wallis et al. 1999).

Few data are available on hatchling emergence and behavior in the Sonoran Desert. Hatchlings emerged asynchronously from a nest in the Mazatzal Mountains beginning by at least 15 October 1998, with the last hatchling observed leaving the nest on 29 October (Averill-Murray and Klug 1999). In 1999 at the same site, estimated oviposition dates were extremely variable, ranging from 9 July to 30 August (Roy C. Averill-Murray, pers. obs.). Only hatchlings from the earliest clutches could possibly emerge in time to take advantage of any available forage. Of the twelve hatchlings observed on summer monitoring plots, two were foraging, one on fallen ocotillo

(Fouquieria splendens) leaves in September 1988 (Holm 1989) and the other on javelina scat in October 1991 (Hart et al. 1992).

Spring-Summer Thermoregulation and Physiological Ecology

During May and June in the arid foresummer, many parts of the Sonoran Desert are rainless and surface temperatures often exceed 50°C. Desert tortoises cannot tolerate body temperatures above 39.5–43.0°C (Brattstrom 1965), so tortoises must seek shelter to avoid lethal thermal extremes. Surveys of four tortoise populations in Arizona during summer 1994 revealed that tortoises were active at temperatures from 20.0 to 45.0°C (recorded 1 cm aboveground; Woodman et al. 1995). Desert tortoises can tolerate shell-surface temperatures above their lethal limit for short durations only if their deep body temperatures remain within the preferred range (McGinnis and Voigt 1971).

The ability of tortoises to thermoregulate behaviorally (physically move to favorable temperatures) depends on the availability, depth, and slope aspect of shelters. Tortoises at the Picacho Mountains used deeper burrows in summer than in other seasons (Barrett 1990). Deep burrows provide cooler, more stable temperatures than shallow burrows or vegetative cover, and shelters on northerly aspects are cooler than those of similar depth on southerly slopes (Elizabeth B. Wirt, unpubl. data). Packrat houses, in which desert tortoises often seek shelter, offer considerable protection from the extremes and fluctuations of daily ambient temperatures and provide higher relative humidity (Brown 1968). During the hottest times of the year, tortoises are generally active in the morning, inactive during midday, active again in late afternoon, and may remain outside of any shelter overnight (Hart et al. 1992; Woodman et al. 1995). However, tortoises occasionally may be active during midday on cloudy days, especially after rains have cooled the air.

In the Mohave Desert, variation in body temperature was most often a response to microhabitat selection, primarily use of burrows (Zimmerman et al. 1994). Tortoises remain cooler in burrows during midday in summer, but often sleep under bushes at night (Woodbury and Hardy 1948; Burge 1977; Luckenbach 1982; Nagy and Medica 1986; Zimmerman et al. 1994).

Throughout the year, however, the pattern of burrow use relative to burrow depth is opposite that in the Sonoran Desert. Tortoises in the Mohave Desert use deeper burrows during winter than summer (Woodbury and Hardy 1948; Burge 1978; TRW Environmental Safety Systems, Inc. 1997) to avoid much colder winter temperatures, due to northern latitudes or cold-air drainage in valley floors.

The physiological ecology of desert tortoises has only been studied in the Mohave Desert, but we briefly summarize it here. Tortoises drink rainwater and obtain water from forage in spring. Female tortoises in the Mohave Desert have used this forage to achieve positive energy balance in the spring (Henen 1997), but other studies have found that tortoises were unable to obtain enough energy to balance expenditures (Minnich 1977; Nagy and Medica 1986; Peterson 1996a). Negative energy balance even while feeding during spring could be due to high water content in the forage outweighing dry matter (energy) intake (Nagy and Medica 1986). Tortoises might also consume food at less than their maximum rate because of harmful components in the diet, such as excess potassium, which tortoises cannot excrete without losing water (Nagy and Medica 1986; Peterson 1996a).

As drought conditions prevail during the Mohave Desert summer, tortoises endure negative water balance (Nagy and Medica 1986). They generally cease feeding, spend more time inactive in their burrows, and retain large volumes of water in their bladders. Water stored in the bladder dilutes excess dietary salts (e.g., potassium) and metabolic wastes (Minnich 1977; Nagy and Medica 1986) but may also be resorbed into the bloodstream (Peterson 1996a, 1996b). Concentrating their urine also allows tortoises to feed during the *initial* stages of drought, even after food plants dehydrate (Nagy and Medica 1986). The arrival of summer rainfall is important because drinking rainwater allows tortoises to flush their systems of excess salts and rehydrate their bladders; they can then obtain energy for growth and maintenance from still-dry forage and fresh vegetation as it becomes available (Minnich 1977; Nagy and Medica 1986; Peterson 1996a, 1996b; Nagy et al. 1997).

Desert tortoise energetics are regionally variable, however, depending on rainfall (Henen et al. 1998). For example, summer rainfall is nearly nonexistent in the western Mohave Desert, erratic but more common in the eastern Mohave and western Sonoran Deserts, and comprises half or more of the annual precipitation in the eastern Sonoran Desert (Turner and Brown 1982). Variation in the amount and seasonality of rainfall and resulting vegetation constrains tortoises' abilities to obtain energy for maintenance, growth, and reproduction. Additional study in the Sonoran Desert is necessary to determine the degree and specific mechanisms by which rainfall affects life-history traits of desert tortoises.

Winter and Hibernation

As late summer temperatures begin to cool and day length decreases, Sonoran tortoises contract their movements toward their winter shelters, and daily activity shifts from mornings and evenings to midday. Tortoises generally begin hibernation during October and November, but onset of hibernation varies widely throughout the Sonoran Desert and has been observed from late September through mid-December (Vaughan 1984; Bailey et al. 1995; Martin 1995). Some tortoises begin hibernation when afternoon temperatures still reach 28–30°C, when others in the population are still very active. Temperature probably plays a more important role than day length in the onset of hibernation at a particular location. For example, radio-equipped tortoises in the San Pedro Valley began hibernation over a narrow time period (late October to November) in 1990, characterized by early and frequent near-freezing temperatures, but over a wide period (October to December) during the more moderate fall of 1991 (Bailey et al. 1995; Scott J. Bailey, unpubl. data). Mohave tortoises at Yucca Mountain also entered hibernation earlier during cooler years (Rautenstrauch et al. 1998).

The onset of hibernation varies greatly among individuals within each sex, but the average date that male (5 November) and female (12 November) tortoises entered hibernation did not differ significantly for combined data from three populations in southern Arizona (Picacho Mountains, San Pedro Valley, and Tortolita Mountains; Martin 1995). Similarly, onset of hibernation in the Black Mountains (males, 25 November; females, 17 November; in 1993; McLuckie et al. 1996) and Yucca Mountain (males, 22 October; females, 19 October; 1991–1994; Rautenstrauch et al. 1998) in the Mohave Desert did not differ significantly between males and females. Interestingly, an immature tortoise at Yucca Mountain entered a burrow on 18 August 1993 and did not reemerge until the following spring.

In both the Sonoran and Mohave Deserts, most tortoises hibernate more than 100 days per year, and many have been observed to hibernate more than 200 days (Burge 1977; Vaughan 1984; Nagy and Medica 1986; Bailey et al. 1995; Martin 1995). A male tortoise in the Tortolita Mountains was inactive for 315 and 292 days during consecutive years (Martin 1995). Most tortoises remain inside their winter shelter (hibernaculum) throughout the winter (Burge 1977; Nagy and Medica 1986; Bailey et al. 1995; Martin 1995), but they may be active on warm winter days. Individuals have been observed to bask, move, and even forage during midwinter in both the Mohave and Sonoran Deserts (Woodbury and Hardy 1948; Roy C. Averill-Murray and Brent E. Martin, pers. obs.; Mercy Vaughn and Roger Repp, pers. comm. 1998), although the significance of winter behavior is not clear. Tortoises at Yucca Mountain were found outside burrows in less than 1 percent of more than four thousand observations between 16 November and 28 February 1991–1995 (Rautenstrauch et al. 1998). Most of the tortoises outside had been disturbed by predators or humans. Winter basking is especially evident in ill or injured tortoises and is possibly initiated as an infection-fighting immune response (Scott J. Bailey, pers. obs.). As in other vertebrate classes, the thermal set-points of many reptiles have been shown to increase when pyrogenic bacteria are injected into the body (Kluger 1979). In wet winters, basking tortoises may also simply be leaving damp hibernacula to dry out, which inhibits the growth of fungus on the shell (James L. Jarchow, pers. comm. 1999)

Desert tortoises hibernate in similar types of shelters (often in the same particular sites) as they use during the active season, including soil burrows, rock shelters, conglomerate or caliche caves, packrat houses, and shallow scrapes below shrubs. However, tortoises in southern Arizona avoided certain types of shelters during winter that were frequently used during the active season (Bailey et al. 1995; Martin 1995). For example, tortoises in the Tortolita Mountains avoided soil burrows and vegetation during hibernation, whereas tortoises in the San Pedro Valley avoided terrace gravel burrows. In both the Sonoran and Mohave Deserts, some tortoises occupy the same hibernaculum winter after winter (Woodbury and Hardy 1948; Auffenberg 1969; Martin 1995). Hibernaculum fidelity and avoidance of shelter types in winter may be a response to specific thermal buffering capabilities of specific shelters and shelter types from lethal minimum temperatures (less than −4.4°C, the supercooling limit of desert tortoises; Lowe et al. 1971), especially if the hibernacula that provide adequate protection from minimum temperatures compose a relatively low proportion of all available shelters.

Female tortoises tend to overwinter in shallower hibernacula than males (Lowe 1990; Bailey et al. 1995; Martin 1995). Combined data from the Picacho Mountains, San Pedro Valley, and Tortolita Mountains indicate that males used hibernacula greater than 0.5 m deep significantly more often than females (Martin 1995). The average length of hibernacula used by four female tortoises during two winters in the San Pedro Valley was 24.4 cm, and that of five males averaged 118.3 cm (Bailey et al. 1995). In two studies, males often hibernated deep enough within shelters that they were not visible from the entrance. In contrast, females were most often visible just inside the entrance, and in many cases a portion of their carapaces were not covered by the hibernaculum roof (Bailey et al. 1995; Martin 1995). The trend of male tortoises hibernating in deeper burrows than females has also been observed in the Mohave Desert (Burge 1978; Burge et al. 1985; McLuckie et al. 1996), where hibernaculum depth is greater overall for both sexes (Bailey et al. 1995).

Hibernacula used by females were both warmer and colder than those used by males in the San Pedro Valley (Bailey et al. 1995). Temperatures within the deeper hibernacula used by males remained relatively stable, fluctuating an average of 8.8°C. In contrast, temperatures within the shallow hibernacula used by females fluctuated over a significantly wider range of temperatures, an average of 20.2°C. Minimum temperatures of hibernacula used by females were consistently lower (average 4.3°C) than those of hibernacula used by males (average 9.3°C) during two winters of study. Maximum temperatures measured in hibernacula used by females (average 24.5°C) were consistently warmer than in those used by males (average 18.2°C).

Thermal buffering, the difference between the temperature measured within a hibernaculum and the temperature measured simultaneously on the surface outside the shelter, is also significantly greater for hibernacula of males than those of females, at least in some populations (Lowe 1990; Bailey et al. 1995). In the San Pedro Valley, temperatures measured in the shallow hibernacula of females were more similar to ambient temperatures than were the temperatures measured in the deeper hibernacula of males (Bailey et al. 1995). Minimum hibernaculum temperatures averaged 12.2°C warmer than minimum surface ambient temperatures for males, but for females averaged only 5.7°C warmer than minimum surface ambient temperatures. Maximum hibernaculum temperatures were 24.1°C cooler than maximum surface ambient temperatures for males, but were only 11.0°C cooler for females.

Although onset of hibernation among male and female tortoises occurs over a similar period, termination dates differ considerably (Bailey 1992; Bailey et al. 1995; Martin 1995). The average date that male tortoises terminated hibernation (28 April) was significantly later than that of females (22 March) at three sites in southern Arizona (Martin 1995). Similarly, males exited their hibernacula significantly later (26 March) than females (14 March) near the Black Mountains in the Mohave Desert (McLuckie et al. 1996). Rautenstrauch et al. (1998) also noticed a trend for males to exit hibernation later than females in Nevada, but this was statistically significant in only one of four years.

Hibernation characteristics may influence tortoise reproductive physiology. For example, annual testicular cycles of male Hermann's tortoises *(Testudo hermanni)* are affected by temperature (Kuchling 1982). By selecting shallow hibernacula, which warm faster than deeper ones in the spring, female desert tortoises may be better able to actively respond to warming temperatures sooner, take advantage of spring forage, and increase their energy intake for egg production. However, no significant differences existed between thermal characteristics of male and female tortoise hibernacula near the Black Mountains (McLuckie et al. 1996). The average hibernaculum temperature range for females was 14.0°C (8.7–22.7°C) and was 12.2°C (7.8–20.0°C) for males. Likewise, thermal buffering did not differ for male and female tortoise hibernacula at this site (McLuckie et al. 1996).

Home Range

Several short-term (approximately two-year) telemetry studies provide some insight into seasonal and annual home ranges of desert tortoises. In these studies, males and females had mean range sizes of 9.2–25.8 and 2.6–23.3 ha, respectively (table 7.1). Reflecting seasonal activity discussed

TABLE 7.1. Home range area estimates for adult male and female desert tortoises in the Sonoran Desert. Mean range areas (ha) are listed, followed by the range of estimates for each sex and site; sample sizes are given in parentheses. Range areas were computed with the minimum convex polygon method and corrected for sample size bias (Jennrich and Turner 1969).

Study Site	Males	Females	Combined Mean ± SD
HARCUVAR MTS.[a]	9.2 (9)	4.7 (4)	7.8 ± 6.70
	1.0–22.3	2.7–7.5	
LITTLE SHIPP WASH[a]	21.7 (4)	23.3 (6)	22.6 ± 15.63
	15.2–31.5	3.4–51.5	
MAZATZAL MTS.[b]	—	12.8 (10)	12.8 ± 14.15
		2.3–50.7	
PICACHO MTS.[c]	25.8 (5)	15.3 (9)	19.1 ± 17.34
	5.6–53.4	2.8–48.5	
SAN PEDRO VALLEY[d]	11.0 (4)	2.6 (4)	6.7 ± 6.88
	3.9–22.2	1.1–6.0	
TORTOLITA MTS.[e]	16.3 (3)	13.1 (3)	16.1 ± 9.39
	3.3–25.2	8.1–21.6	
MEAN ± SD	15.8 ± 12.28	13.3 ± 14.72	14.3 ± 13.72

[a] Trachy and Dickinson (1993).
[b] Murray et al. (1995).
[c] Barrett (1990).
[d] Bailey (1992).
[e] Martin (1995).

earlier, tortoises in the Tortolita Mountains used a greater proportion of their overall range areas during summer than in winter and spring (Martin 1995).

Individual home ranges overlap both within and between the sexes. Each study found no significant differences in home range areas between males and females, but combined analysis reveals modest but significant differences between sexes (female home ranges smaller) and among sites (table 7.2). Estimates from six sites in the Mohave Desert also suggest that female ranges are about half as large as males' (Berry 1986). The relatively smaller home ranges in the Harcuvar Mountains and San Pedro Valley (table 7.2) are difficult to explain in terms of substrate, topography, vegetation, or climate.

TABLE 7.2. Two-factor analysis of variance of desert tortoise home ranges (top) and Tukey's multiple-comparisons test for unequal sample sizes for differences between sites (bottom; *P*-values reported). Data exclude the Mazatzal Mountains site, which lacks data for males, and were log-transformed to correct for heterogeneous variances.

Effect	df	MS	*F*	*P*
SITE	4	0.7381	5.0250	0.0022
SEX	1	0.6647	4.5252	0.0395
SITE × SEX	4	0.1127	0.7676	0.5525
ERROR	41	0.1469		

Site	Little Shipp Wash	Picacho Mts.	San Pedro Valley	Tortolita Mts.
HARCUVAR MTS.	0.0258	0.0539	0.9874	0.3760
LITTLE SHIPP WASH		0.9662	0.0169	0.9746
PICACHO MTS.			0.0700	1.0000
SAN PEDRO VALLEY				0.1826

Several factors must be considered when comparing differences between home range sizes of male and female tortoises in the Sonoran Desert. Male tortoises are typically less active than females during the spring, but during summer (mating season) males may range over larger distances than the more sedentary nesting females. Also, seasonal and annual variation in individual activity areas has been observed. For example, a female tortoise occupied the same 0.8-ha area in the Mazatzal Mountains during both 1992 and 1993, with the exception of June 1992 when she moved to a burrow about 0.75 km away, where she was found in the same burrow every week and presumable nested. She returned to her original activity area in September 1992 (Murray et al. 1996). This female made a similar movement in 1998 from the same activity area to a different nesting location approximately 300 m toward her June–September 1992 burrow. Another female occupied a total area of 50.7 ha in 1992 and 1993, but her two annual activity areas (about 14 ha each) did not overlap (Murray et al. 1996). Several male tortoises at other sites hibernated long distances from their primary or peak-summer-activity areas (Vaughan 1984; Martin 1995). Martin (1995) found that tortoises in the Tortolita Mountains first encountered in a mark-recapture study and later equipped with radio-transmitters, used portions of their home range areas for more than a decade. For such long-lived organisms as desert tortoises, long-term studies are needed to better understand seasonal, annual, and life-long uses of home range areas.

Finally, little is known about activity of juvenile or immature tortoises. We have observed a few immature or young adult tortoises (of both sexes) make relatively long-distance movements away from their usual activity areas. Additional study may reveal whether these observations represent movements between smaller activity areas within a larger range or if emigration occurs as tortoises approach sexual maturity (see Averill-Murray et al., ch. 6 of this volume).

Evolution of Differing Activity Peaks in the Sonoran and Mohave Deserts

The fact that peak tortoise activity is in the summer in the Sonoran Desert and in the spring in the Mohave Desert is not surprising if one considers the climates of the two regions. Precipitation changes along major gradients within the tortoise's range. Total annual rainfall decreases at lower elevations from central Arizona westward into the lower Colorado River Valley and from Central Sonora to the Gulf of California (Turner and Brown 1982). Overlaid on this is a general shift from summer rainfall to winter rainfall both from the Chihuahuan Desert in Texas and New Mexico (summer dominated) to the Sonoran Desert in Arizona (winter and summer) to the Mohave Desert in California (winter rainfall). There is also a major transition from the summer monsoonal rainfall regimes of the New World tropics as far north as southern Sonora northwestward to the Pacific frontal winter rainfall in the Mediterranean climates of California. Summer temperatures are uniformly hot throughout the tortoise's range. The situation is aggravated by nearly rainless conditions in the Mohave Desert and the low erratic rainfall of the lower Colorado River subdivision of the Sonoran Desert in southwestern Arizona and southeastern California (Turner and Brown 1982). Winter temperatures are coldest in the Mohave Desert, which has regular freezing temperatures. Southward, the intensity and duration of freezing temperatures decreases until the winters are mostly frost-free in thornscrub and tropical deciduous forest of tropical southern Sonora and northern Sinaloa, Mexico (Lowe 1990; Bury et al., ch. 5 of this volume).

Most of the desert tortoise's distribution is within Sonoran desert-scrub and thornscrub in biseasonal to summer-maximum rainfall regimes. The behavioral and physiological adaptations to aridity and heat found in modern desert tortoises were presumably inherited from ancestors living in thornscrub and tropical deciduous forest (Van Devender, ch. 2 of this volume). A shift to spring peak activity in the Mohave Desert probably evolved in direct response to the development of Mediterranean winter-rainfall climates during the formation of the Mohave Desert, as well as the enhancement of winter rainfall during Pleistocene glacial climates (Van Devender, ch. 2 of this volume).

ACKNOWLEDGMENTS

We acknowledge the support of the University of Arizona, including the Arizona Cooperative Fish and Wildlife Research Unit, the Graduate Student Research Fund, and the Native American Training Program in the School of Renewable Natural Resources; the Arizona Game and Fish Department (Heritage Fund); the U.S. Forest Service; and the U.S. Fish and Wildlife Service (Partnerships for Wildlife and Section 6 funding). Charles H. Lowe, Cecil R. Schwalbe, Paul Krausman, and many other individuals provided guidance and support. This chapter was greatly improved by the comments of Linda Allison, Earl McCoy, Jeff Howland, Terry Johnson, Phil Medica, Henry Mushinsky, Tom Van Devender, Martin Whiting, and an anonymous reviewer.

LITERATURE CITED

Auffenberg, W. 1969. *Tortoise behavior and survival*. Rand McNally, Chicago.

Averill-Murray, R. C., and C. M. Klug. 1999. Reproduction in desert tortoises in the Sonoran Desert. *Sonoran Herpetologist* 12:70–73.

Bailey, S. J. 1992. Hibernacula use and home range of the desert tortoise *(Gopherus agassizii)* in the San Pedro Valley, Arizona. M.S. thesis, University of Arizona, Tucson.

Bailey, S. J., C. R. Schwalbe, and C. H. Lowe. 1995. Hibernaculum use by a population of desert tortoises *(Gopherus agassizii)* in the Sonoran Desert. *Journal of Herpetology* 29:361–69.

Barrett, S. L. 1990. Home range and habitat of the desert tortoise *(Xerobates agassizii)* in the Picacho Mountains of Arizona. *Herpetologica* 46:202–6.

Barrett, S. L., and J. A. Humphrey. 1986. Agonistic interactions between *Gopherus agassizii* (Testudinidae) and *Heloderma suspectum* (Helodermatidae). *Southwestern Naturalist* 31:261–63.

Berry, K. H. 1986. Desert tortoise *(Gopherus agassizii)* relocation: Implications of social behavior and movement. *Herpetologica* 42:113–25.

Black, J. H. 1976. Observations on courtship behavior of the desert tortoise. *Great Basin Naturalist* 36:467–70.

Bogert, C. M. 1937. Note on the growth rate of the desert tortoise, *Gopherus agassizii*. *Copeia* 1937:191–92.

Brattstrom, B. H. 1965. Body temperature in reptiles. *American Midland Naturalist* 73:376–85.

Brown, J. H. 1968. *Adaptation to environmental temperature in two species of woodrats*, Neotoma cinerea *and* N. albigula. Miscellaneous Publication no. 135, University of Michigan Museum of Zoology, Ann Arbor.

Bull, J. J., R. C. Vogt, and C. J. McCoy. 1982. Sex determining temperatures in turtles: A geographic comparison. *Evolution* 36:326–32.

Bulova, S. J. 1994. Patterns of burrow use by desert tortoises: Gender differences and seasonal trends. *Herpetological Monographs* 8:133–43.

Burge, B. L. 1977. Daily and seasonal behavior, and areas utilized by the desert tortoise *Gopherus agassizii* in southern Nevada. *Proceedings of the Desert Tortoise Council Symposium* 1977:59–94.

———. 1978. Physical characteristics and patterns of utilization of cover sites used by *Gopherus agassizii* in southern Nevada. *Proceedings of the Desert Tortoise Council Symposium* 1978:80–111.

———. 1979. A survey of the present distribution of the desert tortoise, *Gopherus agassizii*, in Arizona. *Proceedings of the Desert Tortoise Council Symposium* 1979:27–74.

———. 1980. Survey of the present distribution of the desert tortoise, *Gopherus agassizii*, in Arizona: Additional data, 1979. *Proceedings of the Desert Tortoise Council Symposium* 1980:36–60.

Burge, B. L., G. R. Stewart, J. E. Roberson, K. Kirtland, R. J. Baxter, and D. C. Pearson. 1985. Excavation of winter burrows and relocation of desert tortoises *(Gopherus agassizii)* at the Twenty-nine Palms Marine Corps Air Ground Combat Center. *Proceedings of the Desert Tortoise Council Symposium* 1985:32–39.

Burke, R. L. 1993. Adaptive value of sex determination mode and hatchling sex ratio bias in reptiles. *Copeia* 1993:854–59.

Carr, A. 1952. *Handbook of turtles of the United States, Canada, and Baja California*. Comstock Publishing Associates, Ithaca, N.Y.

Ernst, C. H., and R. W. Barbour. 1972. *Turtles of the United States*. University of Kentucky Press, Lexington.

Ewert, M. A., and C. E. Nelson. 1991. Sex determination in turtles: Diverse patterns and some possible adaptive values. *Copeia* 1991:50–69.

Ewert, M. A., D. R. Jackson, and C. E. Nelson. 1994. Patterns of temperature-dependent sex determination in turtles. *Journal of Experimental Zoology* 270:3–15.

Gibbons, J. W., and D. H. Nelson. 1978. The evolutionary significance of delayed emergence from the nest by turtles. *Evolution* 32:297–303.

Gist, D. H., J. A. Michaelson, and J. M. Jones. 1990. Autumn mating in the painted turtle *(Chrysemys picta)*. *Herpetologica* 46:331–36.

Goodlett, G., P. Woodman, M. Walker, and S. Hart. 1996. *Desert tortoise population survey at Beaver Dam Slope exclosure desert tortoise study plot: Spring 1996.* Report to Arizona Game and Fish Dept., Phoenix.

Grant, C. 1936. The southwestern desert tortoise, *Gopherus agassizii. Zoologica* 21:225–29.

———. 1946. Data and field notes on the desert tortoise. *Transactions of the San Diego Society of Natural History* 10:399–402.

Grover, M. C., and L. A. DeFalco. 1995. *Desert tortoise* (Gopherus agassizii): *Status-of-knowledge outline with references.* General Technical Report no. INT-GTR-316, U.S. Dept. of Agriculture, Intermountain Research Station, Ogden, Utah.

Hart, S., P. Woodman, S. Bailey, S. Boland, P. Frank, G. Goodlett, D. Silverman, D. Taylor, M. Walker, and P. Wood. 1992. *Desert tortoise population studies at seven sites and a mortality survey at one site in the Sonoran Desert, Arizona.* Report to Arizona Game and Fish Dept. and U.S. Bureau of Land Management, Phoenix.

Henen, B. T. 1997. Seasonal and annual energy budgets of female desert tortoises *(Gopherus agassizii). Ecology* 78:283–96.

Henen, B. T., C. C. Peterson, I. R. Wallis, K. H. Berry, and K. A. Nagy. 1998. Effects of climatic variation on field metabolism and water relations of desert tortoises. *Oecologia* 117:365–73.

Hohman, J. P., R. D. Ohmart, and J. Schwartzmann. 1980. *An annotated bibliography of the desert tortoise* (Gopherus agassizii). Special Publication no. 1, Desert Tortoise Council, Long Beach, Calif.

Holm, P. A. 1989. *Desert tortoise monitoring baseline study: Harquahala Mountains.* Report to U.S. Bureau of Land Management, Phoenix.

Householder, V. H. 1950. Courtship and coition of the desert tortoise. *Herpetologica* 6:11.

Iverson, J. B. 1990. Nesting and parental care in the mud turtle, *Kinosternon flavescens. Canadian Journal of Zoology* 68:230–33.

Jennrich, R. I., and F. B. Turner. 1969. Measurement of non-circular home range. *Journal of Theoretical Biology* 22:227–37.

Karl, A. E. 1998. Reproductive strategies, growth patterns, and survivorship of a long-lived herbivore inhabiting a temporally variable environment. Ph.D. diss., University of California, Davis.

Kluger, M. J. 1979. *Fever.* Princeton University Press, Princeton, N.J.

Kuchling, G. 1982. Effects of temperature and photoperiod on spermatogenesis in the tortoise, *Testudo hermanni. Amphibia-Reptilia* 2:329–41.

Lowe, C. H. 1964. *The vertebrates of Arizona.* University of Arizona Press, Tucson.

———. 1990. Are we killing the desert tortoise with love, science, and management? *In* K. R. Beaman, F. Caporaso, S. McKeown, and M. D. Graff, eds. *Proceedings of the first international symposium on turtles and tortoises: Conservation and captive husbandry.* Chapman University, Chapman, Calif.

Lowe, C. H., P. J. Lardner, and E. A. Halpern. 1971. Supercooling in reptiles and other vertebrates. *Comparative Biochemical Physiology* 39A:125–35.

Luckenbach, R. A. 1982. Ecology and management of the desert tortoise *(Gopherus agassizii)* in California. Pp. 1–37 *in* R. B. Bury, ed. *North American tortoise conservation and ecology*. Wildlife Research Report no. 12, U.S. Fish and Wildlife Service, Washington, D.C.

Marlow, R. W. 1985. Is time a limiting resource for *Gopherus agassizii? Proceedings of the Desert Tortoise Council Symposium* 1982:133.

Martin, B. E. 1995. Ecology of the desert tortoise *(Gopherus agassizii)* in a desert-grassland community in southern Arizona. M.S. thesis, University of Arizona, Tucson.

McGinnis, S. M., and W. G. Voigt. 1971. Thermoregulation in the desert tortoise, *Gopherus agassizii. Comparative Biochemical Physiology* 40:119–26.

McLuckie, A. M., C. R. Schwalbe, and T. Lamb. 1996. *Genetics, morphology, and ecology of the desert tortoise* (Gopherus agassizii) *in the Black Mountains, Mohave County, Arizona*. Report to U.S. Bureau of Land Management, Transwestern Pipeline Company, and Arizona Game and Fish Dept.

Medica, P. A., R. B. Bury, and R. A. Luckenbach. 1980. Drinking and construction of water catchments by the desert tortoise, *Gopherus agassizii*, in the Mojave Desert. *Herpetologica* 36:301–4.

Miller, L. 1932. Notes on the desert tortoise *(Testudo agassizii)*. *Transactions of the San Diego Society of Natural History* 7:189–208.

Minnich, J. E. 1977. Adaptive responses in the water and electrolyte budgets of native and captive desert tortoises, *Gopherus agassizii*, to chronic drought. *Proceedings of the Desert Tortoise Council Symposium* 1977:102–29.

Mueller, J. M., K. R. Sharp, K. K. Zander, D. L. Rakestraw, K. R. Rautenstrauch, and P. E. Lederle. 1998. Size-specific fecundity of the desert tortoise *(Gopherus agassizii)*. *Journal of Herpetology* 32:313–19.

Murray, R. C., and C. M. Klug. 1996. Preliminary data analysis from three desert tortoise long-term monitoring plots in Arizona: Sheltersite use and growth. *Proceedings of the Desert Tortoise Council Symposium* 1996:10–17.

Murray, R. C., C. R. Schwalbe, S. J. Bailey, S. P. Cuneo, and S. D. Hart. 1995. *Desert tortoise* (Gopherus agassizii) *reproduction in the Mazatzal Mountains, Maricopa County, Arizona*. Report to Arizona Game and Fish Dept. and Tonto National Forest, Phoenix.

———. 1996. Reproduction in a population of the desert tortoise, *Gopherus agassizii*, in the Sonoran Desert. *Herpetological Natural History* 4:83–88.

Nagy, K. A., and P. A. Medica. 1986. Physiological ecology of desert tortoises in southern Nevada. *Herpetologica* 42:73–92.

Nagy, K. A., D. J. Morafka, and R. A. Yates. 1997. Young desert tortoise survival: Energy, water, and food requirements in the field. *Chelonian Conservation and Biology* 2:396–404.

Palmer, K. S., D. C. Rostal, J. S. Grumbles, and M. Mulvey. 1998. Long-term sperm storage in the desert tortoise *(Gopherus agassizii)*. *Copeia* 1998:702–5.

Peterson, C. C. 1996a. Ecological energetics of the desert tortoise *(Gopherus agassizii)*: Effects of rainfall and drought. *Ecology* 77:1831–44.

———. 1996b. Anhomeostasis: Seasonal water and solute relations in two populations of the desert tortoise *(Gopherus agassizii)* during chronic drought. *Physiological Zoology* 69:1324–58.

Rautenstrauch, K. R., A.L.H. Rager, and D. L. Rakestraw. 1998. Winter behavior of desert tortoises in south-central Nevada. *Journal of Wildlife Management* 62:98–104.

Roberson, J. B., B. L. Burge, and P. Hayden. 1985. Nesting observations of free-living desert tortoises *(Gopherus agassizii)* and hatching success of eggs protected from predators. *Proceedings of the Desert Tortoise Council Symposium* 1985:91–99.

Rostal, D. C., V. A. Lance, J. S. Grumbles, and A. C. Alberts. 1994. Seasonal reproductive cycle of the desert tortoise *(Gopherus agassizii)* in the eastern Mojave Desert. *Herpetological Monographs* 8:72–82.

Ruby, D. E., and H. A. Niblick. 1994. A behavioral inventory of the desert tortoise: Development of an ethogram. *Herpetological Monographs* 8:88–102.

Spotila, J. R., L. C. Zimmerman, C. A. Binckley, J. S. Grumbles, D. C. Rostal, A. List Jr., E. C. Beyer, K. M. Phillips, and S. J. Kemp. 1994. Effects of incubation conditions on sex determination, hatching success, and growth of hatchling desert tortoises, *Gopherus agassizii. Herpetological Monographs* 8:103–16.

Stebbins, R. C. 1954. *Amphibians and reptiles of western North America.* McGraw-Hill, New York.

Trachy, S., and V. M. Dickinson. 1993. *Use areas and sheltersite characteristics of Sonoran desert tortoises.* Report to Arizona Game and Fish Dept., Phoenix.

TRW Environmental Safety Systems, Inc. 1997. *Patterns of burrow use by desert tortoises at Yucca Mountain, Nevada.* Report to U.S. Dept. of Energy, Las Vegas.

Turner, F. B., P. A. Medica, and C. L. Lyons. 1984. Reproduction and survival of the desert tortoise *(Scaptochelys agassizii)* in Ivanpah Valley, California. *Copeia* 1984:811–20.

Turner, F. B., P. Hayden, B. L. Burge, and J. B. Roberson. 1986. Egg production by the desert tortoise *(Gopherus agassizii)* in California. *Herpetologica* 42:93–104.

Turner, R. M., and D. E. Brown. 1982. Sonoran desertscrub. *Desert Plants* 4:181–221.

Vaughan, S. L. 1984. Home range and habitat use of the desert tortoise *(Gopherus agassizii)* in the Picacho Mountains, Pinal County, Arizona. M.S. thesis, Arizona State University, Tempe.

Wallis, I. R., B. T. Henen, and K. A. Nagy. 1999. Egg size and annual egg production by female desert tortoises *(Gopherus agassizii)*: The importance of food abundance, body size, and date of egg shelling. *Journal of Herpetology* 33:394–408.

Wirt, E. B. 1988. *Two desert tortoise populations in Arizona.* Report to U.S. Bureau of Land Management, Phoenix.

Woodbury, A. M., and R. Hardy. 1940. The dens and behavior of the desert tortoise. *Science* 92:529.

———. 1948. Studies of the desert tortoise, *Gopherus agassizii. Ecological Monographs* 18:145–200.

Woodman, A. P., S. Hart, P. Frank, S. Boland, G. Goodlett, D. Silverman, D. Taylor, M. Vaughn, and M. Walker. 1995. *Desert tortoise population surveys at four sites in the Sonoran Desert of Arizona, 1994.* Report to Arizona Game and Fish Dept., Phoenix, and U.S. Bureau of Land Management, Arizona State Office, Phoenix.

Zimmerman, L. C., M. P. O'Connor, S. J. Bulova, J. R. Spotila, S. J. Kemp, and C. J. Salice. 1994. Thermal ecology of the desert tortoises in the eastern Mohave Desert: Season patterns of operative and body temperatures, and microhabitat utilization. *Herpetological Monographs* 8:45–59.

CHAPTER 8

Grasses, Mallows, Desert Vine, and More

Diet of the Desert Tortoise in Arizona and Sonora

THOMAS R. VAN DEVENDER, ROY C. AVERILL-
MURRAY, TODD C. ESQUE, PETER A. HOLM,
VANESSA M. DICKINSON, CECIL R. SCHWALBE,
ELIZABETH B. WIRT, AND SHERYL L. BARRETT

The desert tortoise *(Gopherus agassizii)* ranges from Nevada and Utah south to northern Sinaloa (Germano et al. 1994). It is the largest (in mass) herbivorous reptile in this region and lives in a variety of habitats including winter-rainfall desertscrub in the Mohave Desert in Nevada, Utah, California, and Arizona; biseasonal rainfall desertscrub in the Sonoran Desert in Arizona and Sonora; and primarily summer-rainfall thornscrub and tropical deciduous forest in Sonora and Sinaloa (Lowe 1964; Turner and Brown 1982; Bury et al., ch. 5 of this volume; Van Devender, ch. 1 of this volume; fig. 8.1). The plants that tortoises eat and the nutrition that they derive from them (Germano et al., ch. 11 of this volume; Jarchow et al., ch. 12 of this volume; Oftedal, ch. 9 of this volume) are the physiological foundations that fuel tortoises' basic natural history, behavior, and reproduction. The diet of the Mohave desert tortoise has been extensively studied (Jennings 1993; Esque 1994).

In this chapter, we briefly describe the different techniques that researchers use to quantify the diets of desert tortoises. We then visit several tortoise study sites throughout the Sonoran Desert (see Averill-Murray et al., fig. 6.1 of this volume) and discuss and compare the diets of tortoises in those populations. Diet information for tortoises living in tropical communities in southern Sonora is presented for the first time here. Food availability is discussed, and a summary of the diet of the Sonoran desert tortoise is presented.

Figure 8.1. Desert tortoise feeding on prickly pear *(Opuntia engelmannii)* in the Tucson Mountains, Pima County, Arizona. (Photograph by Thomas A. Wiewandt)

Methods Used for Quantifying Diets

Observation and Bite Counts

Observations by field biologists of tortoises feeding provide basic dietary information. Woodbury and Hardy (1948) discussed the diet of tortoises based on their observations in the Mohave Desert of southwestern Utah. Appendix 8.1 gives feeding observations of Sonoran tortoises in Arizona from 1984 to 1997 based on Vaughan (1984), Johnson et al. (1990), Ogden (1993), Snider (1993), Martin (1995), Averill-Murray and Klug (2000), and Vanessa M. Dickinson (unpubl. data). Considering that each population plot survey represents thirty to sixty days of observation time, tortoise biologists made relatively few feeding observations of tortoises.

In recent years, desert tortoise diets have been the focus of bite-count studies in the western Mohave Desert in the Desert Tortoise Natural Area, California (Jennings 1993); the eastern Mohave Desert in the Ivanpah Valley, California (Avery 1995); and the Beaver Dam Slope area of southwestern Utah and northwestern Arizona (Esque 1994). In these studies, biologists observed feeding tortoises, often individuals equipped with radio-transmitters. Parameters including location, temperature, weather conditions, the animals identity, and tortoise activities were recorded. When a tortoise fed, the observer recorded the exact number of bites, the species of plant (if unknown the plant was collected for later identification), phenological condition, the parts fed on, and the distance travelled between each feeding.

Fragment Analyses

The earliest dietary account for the desert tortoise was by Ortenburger and Ortenburger (1927), who only found grasses in the stomachs of tortoises they examined from the Sonoran Desert in Pima County, Arizona. Johnson et al. (1990) summarized the plants identified in analyses of scat and fecal pellets. Although the bulk of the plant material in fecal pellets is not distinctive, many reproductive structures (florets, fruits, and seeds) and spines that pass intact through tortoise digestive tract are easily identified by comparison with reference specimens. The pellets are disaggregated by hand and sorted using a binocular microscope. Taxa in each pellet are scored on an internal relative abundance scale of 1 (rare, one or two specimens), 2 (common), and 3 (very common), and 4 (abundant, the most common). Although the identified fragments represent a tiny fraction of the total number of fragments or of the plant mass in a pellet, the method reliably documents the presence of a plant species in the diet of the desert tortoise.

Microhistological Analyses

Hansen et al. (1976) pioneered the use of microhistological analyses of tortoise scat in the New Water Mountains, La Paz County, Arizona, and Vaughan (1984) in the Picacho Mountains, Pinal County, Arizona. Microhistological analysis of plant epidermis in dung of herbivores is a standard range management technique that provides quantitative estimates of the percent dry weight of leaf and stem tissues consumed (Free et al. 1970; Ward 1970; Hansen et al. 1973). In this method, fecal samples are ground to uniform fragment size in a blender and mounted on slides for microscopic analysis. Microscopic plant fragments are identified using epidermal characters including cell, stomate, and trichome (hairs) structure and patterns compared against voucher samples of plant species found in the general study area. Reasonable amounts of epidermis are recognizable after passing through the digestive systems of cattle, desert tortoise, and other herbivores including the epidermis of highly digestible annuals and herbaceous perennials. Identifiable fragments are counted in each microscope field, thus allowing percent density, frequency, and relative density to be calculated. The last measure is highly correlated with the dry weight of plants consumed (Sparks and Malachek 1968). Although 35–40 percent of the fragments in a field cannot be identified, technicians are trained to accurately quantify hand-compounded mixtures. Microhistological analysis identifies and quantifies the less-distinctive but dominant leaf and stem tissues that are not easily identified in fragment analysis. The mean relative densities are general indicators of the biomass of plants in the diet.

Floristic and Seasonal Considerations

The biseasonal rainfall regime in the Arizona Upland subdivision of the northeastern Sonoran Desert supports a diverse flora in structurally complex vegetation. Visually dominant plants such as foothills paloverde *(Parkinsonia microphylla)*, ironwood *(Olneya tesota)*, and saguaro *(Carnegiea gigantea)* are never as common, nor do they cover as much ground, as subshrubs such as brittlebush *(Encelia farinosa)* and triangleleaf bursage *(Ambrosia deltoidea)*. The abundance and diversity of annuals available to a foraging tortoise are both spatially and temporally variable. The annual and grass floras of the winter-spring and summer monsoonal rainy seasons are very different. Winter-spring annuals and grasses are typically species with affinities to the Mohave Desert or more mesic temperate areas. Summer plants are typically species that are widely distributed in the eastern Sonoran Desert and the New World tropics to the south. Even in biseasonal or summer-dominated portions of the Sonoran Desert, the winter-spring annuals are more numerous than summer annuals. For example, half of the rainfall in the Tucson Mountains (300 to 350 mm/year) is in the summer monsoon (July through September); yet, 62 percent of the annuals are winter-spring, 33 percent summer, and 6 percent opportunistic (Rondeau et al. 1996). The percentages of winter-spring annuals are even greater in the floras in desert tortoise areas to the north and west: South Mountain (80 percent), White Tank Mountains (75 percent), and Organ Pipe Cactus National Monument (73 percent). Six-weeks threeawn *(Aristida adscensionis)* is one of the few opportunistic species that grows in both seasons.

The variety of growth forms of plants in the Arizona Upland has important implications for food ability for tortoises (see Forage Availability section). Woody plants can be classified into trees, shrubs, subshrubs, and woody parasites and vines. There are so few woody parasites (desert mistletoe, *Phoradendron californicum*) and vines (desert vine, *Janusia gracilis*) that they are usually presented with the common subshrubs. Succulents are plants that store water in their tissues; these include stem succulents (cacti) and rosette succulents (agaves, *Agave* spp.; sotol *Dasylirion wheeleri*; and yuccas, *Yucca* spp.). For this study, annual succulents such as purslane *(Portulaca oleracea)* were considered to be annuals instead of succulents.

Nonwoody plants are usually classified as herbaceous perennials or annuals. Herbaceous perennials die back to the roots due to drought, extreme hot or cold temperatures, grazing, or fire and regrow when favorable conditions return. They can be short- or long-lived depending on the kinds of roots (simple or tuberous) they have. Annuals, also called ephemerals, complete their life cycles in less than a year, typically in a single season. Annuals avoid seasons and years of unfavorable conditions by lying dormant as seeds in the soil.

In some cases, it may not be possible to know if plants are perennial or annual. There has been a general evolutionary trend from herbaceous perennials to annuals in the Sonoran and Mohave Deserts. Many plant genera including grama *(Bouteloua)*, deer vetch *(Lotus)*, *Ditaxis*, globe mallow *(Sphaeralcea)*, milkvetch *(Astragalus)*, muhly *(Muhlenbergia)*, spiderling *(Boerhavia)*, spurge *(Euphorbia)*, threeawn *(Aristida)*, and wild buckwheat *(Eriogonum)* include both perennial and annual species. In fragment and microhistological analyses of fecal pellets, it is often not possible to identify the material to species. In other species, such as fluffgrass *(Erioneuron pulchellum)*, *Marina parryi*, spurges *(Euphorbia polycarpa* and others), and windmills *(Allionia incarnata)*, the species are typically herbaceous perennials but often flower the first year and under harsh climatic conditions can function as annuals. In table 8.1 and appendices 8.1 and 8.2, plants that

TABLE 8.1. Plants identified in forty-four desert tortoise fecal pellets from tropical deciduous forest in the Sierra de Alamos in southern Sonora. An asterisk indicates an introduced species.

Species (Common Name)	Plant Parts	1	2	3	4	5
				Samples		
TREES AND SHRUBS						
Acacia cochliacantha (boatthorn acacia)	stem	x				
Bursera laxiflora (torote prieto)	bark		x			
Croton sp. *(vara)*	fruit	x				
Haematoxylum brasiletto (brasil)	leaflets	x				
Senna biflora (flor de iguana)	anther				x	
$N = 5$						
SUBSHRUB						
Acalypha subviscida	fruit	x	x	x		
$N = 1$						
WOODY VINE						
Gouania rosei (güirote de violín)	flower	x				
$N = 1$						
SUCCULENTS						
Opuntia sp. (prickly pear/cholla)	areole	x				
$N = 1$						
PERENNIAL GRASS						
Melinus repens (zacate rosado)*	florets	x	x	x	x	
$N = 1$						
ANNUAL GRASSES						
Eriochloa aristata (cup grass)	florets				x	x
Panicum trichoides (panicgrass)	florets	x	x	x	x	x
Setaria liebmannii (bristlegrass)	florets		x	x	x	
$N = 3$						
HERBACEOUS PERENNIALS						
Ayenia filiformis	fruit	x	x		x	x
Elytraria imbricata (cordoncillo)	leaves			x		x
Evolvulus alsinoides	seeds					x
Marina palmeri	flower		x			
Sida alamosana	seeds	x	x		x	x
Tetramerium nervosum (albahaca del monte)	fruit				x	
Tragia sp. (noseburn, *ortiguilla*)	fruit	x	x			
$N = 7$						

TABLE 8.1. Continued

Species (Common Name)	Plant Parts	Samples				
		1	2	3	4	5
PERENNIAL/ANNUAL HERBS						
Desmodium sp. (beggar's tick)	seed		x	x		
Salvia setosa (tapachorro)	calyx	x				
$N = 2$						
ANNUALS						
Brickellia sonorana	achenes	x				x
Euphorbia hirta (spurge, *golondrina*)	fruit	x				x
Mitracarpus hirtus	fruit				x	x
Mentzelia aspera (pegajosa)	fruit		x	x		
$N = 4$						
TOTAL = 25		14	10	8	9	9

straddle this evolutionary fence are classified as perennial/annual herbs. Although grasses are herbs, they are usually separated from the dicot herbs in perennial, annual, and perennial/annual grass categories, reflecting their general importance as a forage class. Forb is a growth form category in range management that typically refers to all nongrass herbs (annual and perennial) eaten by livestock but sometimes includes woody vines (desert vine) and shrubs (fairy duster, *Calliandra eriophylla*, and range ratany, *Krameria erecta*; e.g., Little Shipp Wash, Vanessa M. Dickinson, unpubl. data).

The growth form categories used in the following sections are not consistent; they reflect those used in the original studies and differences between the methods used to quantify diet. The terms *taxon* and *taxa* are often used for plant categories that include different taxonomic levels (e.g., *Aristida adscensionis* plus *Aristida* sp.) or taxonomic levels that could include more than one species (e.g., *Sphaeralcea* sp. could be *S. ambigua* and/or *S. coulteri*).

The Picacho Mountains Study

Fecal pellets collected during a 1982–1983 home range and habitat use study of desert tortoises in the Picacho Mountains, Pinal County, Arizona (Vaughan 1984; Barrett 1990), were analyzed microhistologically. The

study area, located at 540- to 720-m elevation, is a mixed desertscrub dominated by foothills paloverde, saguaro, and brittlebush on the rocky slopes,
whereas creosotebush *(Larrea divaricata)*, ironwood, and triangleleaf bursage are more common on lower *bajadas* and the basin floor.

The climate of the Picacho Mountains is typical of the northeastern
Sonoran Desert, with extremely high temperatures in summer and moderate
temperatures in fall, winter, and spring. Rains are biseasonal in winter-spring
and the monsoonal summer (July through September) with an average precipitation of 206 mm/year for the nearby town of Casa Grande (Sellers and
Hill 1974).

Fifty-two plant taxa were identified in forty-four fresh fecal pellets,
seven collected in April–May and thirty-seven in August–September. Only
twenty taxa comprised 95 percent of the dry weight of all samples (Vaughan
1984). Tortoises ate forbs (51.9 percent), trees and shrubs (41.1 percent),
grasses (4.2 percent), and cacti (2.8 percent). The few spring pellets (*n* = 7)
differed markedly in that forbs, mostly lupine *(Lupinus sparsiflorus)*, were
much more important in the diet (82.3 percent). The most abundant food
plants were desert vine (15.3 percent), lanceleaf ditaxis (*Ditaxis lanceolata*;
15.2 percent), and lupine (25.7 percent). The low values for grasses (3.3 percent) and mallows (globe mallow *[Sphaeralcea ambigua]*, 1.2 percent; Indian
mallow *[Abutilon]*, 1.0 percent) in the diet are notable, although eight different grasses were recorded in the diet: Arizona cottontop *(Digitaria californica)*, Arizona panicgrass *(Brachiaria arizonica)*, fluffgrass, Mediterranean
grass *(Schismus barbatus)*, purple threeawn *(Aristida purpurea)*, red brome
(Bromus rubens), spider grass *(Aristida ternipes)*, and tanglehead *(Heteropogon contortus)*. Only lupine, Mediterranean grass, and red brome are spring
annuals.

The Little Shipp Wash and Harcuvar Mountains Study Sites

Feeding activity of tortoises fitted with radio-transmitters was observed in 1991 (May–October), 1992 (March–November), and 1993 (March–
August) at two Sonoran Desert sites in western Arizona (Snider 1993; Vanessa M. Dickinson, unpubl. data). Little Shipp Wash is a granite boulder

landscape at 788- to 975-m elevation between Bagdad and Hillside in Yavapai County (Schneider 1981). The Harcuvar Mountains plot is northwest of Aguila at 790- to 1,010-m elevation (Woodman and Shields 1988). Both areas support Arizona Upland vegetation with foothills paloverde, saguaro, catclaw acacia *(Acacia greggii)*, and western honey mesquite *(Prosopis glandulosa)* at Little Shipp and foothills paloverde, saguaro, and ocotillo *(Fouquieria splendens)* at the Harcuvar site (Snider 1993). Mean annual rainfall over twenty-nine years (1961–1990) was 400 mm/year at Hillside and 243 mm/year at Aguila (Sellers and Hill 1974).

In 342 hours of observation, Little Shipp tortoises were observed to eat 4,353 bites of nineteen species of plants. These included grasses (52.9 percent), forbs (19.6 percent), and prickly pear (*Opuntia engelmannii*; 27.3 percent); tortoise scat accounted for the other 0.2 percent. The most commonly eaten food plants were prickly pear (27.4 percent), fluffgrass (18.1 percent), big galleta grass (*Pleuraphis rigida*; 16.8 percent), desert rock pea (*Lotus rigidus*; 10.0 percent), tobosa (*Pleuraphis mutica*; 5.4 percent), Indian wheat (*Plantago ovata*; 4.3 percent), and *Ayenia filiformis* (3.7 percent). Spring annuals (filaree *[Erodium cicutarium]*, Indian wheat, lacepod *[Thysanocarpus curvipes]*, lupine, and nievitas [*Cryptantha* spp.]) accounted for 7.7 percent of the overall observed bites.

In 378 hours of observation, Harcuvar tortoises were observed to eat 1,354 bites of twelve species. Plants bitten included grasses (37.5 percent) and forbs (62.5 percent). The most commonly eaten food plants were lupine (43.1 percent), owl clover (*Castilleja exserta*; 11.4 percent), big galleta, Mediterranean grass (9.5 percent), Indian wheat (8.0 percent), red brome (7.3 percent), and nievitas. Spring annuals (filaree, Indian wheat, lupine, nievitas, and owl clover) accounted for 66.6 percent of the observed bites.

The differences in observed diet between the two sites are likely related to the differences in the availability of plant species between the sites. Both sites are in mixed Arizona Upland desertscrub dominated by foothills paloverde and saguaro with some plants typical of the higher-elevation interior chaparral. However, there is greater floristic diversity at Little Shipp Wash, especially in grasses, and a greater abundance of prickly pears. The desertscrub at Little Shipp Wash merges with nearby desert grassland, whereas the Harcuvar site is surrounded by sparse creosotebush desertscrub of the Lower Colorado River Valley subdivision of the Sonoran Desert. The Little Shipp Wash results can be compared to fecal analyses from the same plot summarized below.

Scat Analyses at Multiple Sonoran Desert Sites

Van Devender and Schwalbe (1999) analyzed scat from free-ranging tortoises at seven tortoise population sites in the Sonoran Desert in Arizona (Arizona Game and Fish Department [AGFD] scat study). Primary sites that were intensively studied included Four Peaks (northeast of Phoenix in Maricopa County; elevation 790–855 m; Murray 1993; Van Devender, this vol., chap. 1, fig. 1.3), Granite Hills (northwest of Tucson, Pinal County; 600–700 m; Shields et al. 1990), and Little Shipp Wash (788–975 m; Shields et al. 1990) in prime Arizona Upland tortoise granite boulder habitat and the Eagletail Mountains (west of Phoenix, Maricopa County; 460–698 m; Shields et al. 1990; Van Devender, this vol., chap. 1, fig. 1.4) in marginal lower Colorado River Valley tortoise habitat. In addition, relatively small samples were studied from Bonanza Wash (southwest of Bagdad, Yavapai County; 960–1,080 m), and the Sand Tank (near Gila Bend, Maricopa County; 656–853 m), and Tortilla (i.e., Mineral, Pinal County; 656–853 m) Mountains.

Twenty-five samples of fecal pellets were collected in summer and fall in 1992 and 1995–1996 and spring in 1996. Whenever possible, samples consisted of twenty-five fresh pellets. These were especially difficult to find in March and April, ironically coinciding with the peak of spring annual production. Only two of fourteen fresh pellet samples were obtained in the spring, thus limiting interseasonal comparisons. Fresh scat samples were more numerous in August and September. In other samples ($n = 6$), fresh and weathered pellets were analyzed to reflect annual diets. The scat were analyzed using both fragment and microhistological analyses.

Fragment Analyses

The plant fragments in 603 individual pellets in the twenty-five samples were identified through comparisons to reference specimens. The most common identifiable fragments were seeds, fruits (including composite achenes, grass florets, and borage nutlets), leaves, and spines/thorns. The number of taxa in two hundred individual pellets in the Four Peaks samples ranged from 1 to 13 (average 5.2 taxa/pellet). There were 15 to 43 (average 29.2 species) taxa in the twenty-four fecal pellet samples from all sites; a sample from the gut of a dead tortoise from Little Shipp Wash yielded 7 taxa. Plants in various life forms were eaten, including trees and shrubs (12.0 percent), subshrubs/woody vines (11.3 percent), grasses (12.9 percent perennials, 8.3 percent annuals), succulents (cacti and sotol, 6.0 percent), herbaceous perennials (12.0 percent), and dicot annuals (45.9 percent, 36.8 percent spring). Annuals (including grasses and dicots) account for 54.1 percent.

Fifty-two taxa (39.1 percent) were eaten frequently (i.e., 20 percent or more of the twenty-five pellet samples) in the AGFD scat study, including the following growth forms: trees and shrubs (6.0 percent), subshrubs and woody vines (3.8 percent), spring annual grasses (2.3 percent), succulents (1.5 percent), herbaceous perennials (4.5 percent), and dicot annuals (21.1 percent). Only five of thirty-one (16.1 percent) annual grasses and dicots were summer species. Eighteen species were found in 50 percent or more of the samples. These were important in the diet and included a tree (foothills paloverde), shrubs (brittlebush, desert lavender *[Hyptis albida]*, and fairy duster), subshrubs (globe mallow and lanceleaf ditaxis), desert vine, spring annual grasses (red brome and six-weeks fescues *[Vulpia microstachys* and *V. octoflora]*), and dicot annuals (brittle spineflower *[Chorizanthe brevicornu]*, combbur *[Pectocarya recurvata]*, deer vetch *[Lotus salsuginosus]*, filaree, Indian wheat, nievitas *[Cryptantha barbigera]*, and sleepy catchfly *[Silene antirrhina]*). Combbur, filaree, foothills paloverde, and nievitas were in 75 percent or more of the samples, and red brome was in every sample.

The occurrence of plants in half or more of the twenty-five samples may not be significant because some species were only represented by occasional specimens in a few pellets from each sample. For example, most of the records of brittlebush were one to three achenes, although a tortoise eating a single flower head would consume many achenes. In contrast, the maximum numbers of specimens per pellet per sample were high for other taxa: combbur (8–87), Indian wheat (10–218), nievitas (1–109), red brome (14–500), six-weeks fescues (*V. microstachys*, 17–282; *V. octoflora*, 3–335), and sleepy catchfly (2–218). Species such as desert vine, lanceleaf ditaxis, and globe mallow were identified in fecal pellets by fine surface hairs (trichomes), which were so numerous that they comprised the matrix of the pellet and counting them was not meaningful.

Microhistological Analyses

After fragment analyses, the combined matrix material from each of the 25-pellet fragment analysis samples were analyzed microhistologically. Forty-one taxa were differentiated including seven categories for unidentified plants. The thirty-one taxa identified to genus included trees and shrubs (33.3 percent), grasses (33.3 percent), annuals (22.2 percent), and a succulent (2.4 percent). Thirteen taxa were common enough to account for 10 percent or more of the diet in at least one sample. These included shrubs (desert lavender and Indian mallow), subshrubs (globe mallow, *Herissantia*, and rose mallow *[Hibiscus]*), desert vine, herbaceous perennials *(Evolvulus, Sida*, and wild buckwheat *[Eriogonum wrightii])*, grasses (dropseed *[Sporobolus]*, grama, and threeawn), and an annual (Indian wheat).

The Maricopa Mountains Study

Wirt and Holm (1997) provided extensive diet information using fragment and microhistological scat analyses; they also compared results with available forage in their study of the tortoise population in the Maricopa Mountains northeast of Gila Bend in Maricopa County, Arizona. The site is at about 524- to 820-m elevation in Arizona Upland Sonoran desertscrub with foothills paloverde, saguaro, ironwood, triangleleaf bursage, and creosotebush (Shields et al. 1990). Average rainfall for the nearest weather stations to the Maricopa plot, Buckeye, Gila Bend, and Maricopa, received 179, 146, and 187 mm/year, respectively (Sellers and Hill 1974).

Fragment analyses of seventy-nine fecal pellets yielded identifications of seventy-five plant taxa including trees and shrubs (18.7 percent), subshrubs and perennial vine (12.0 percent), succulents (4.0 percent), grasses (5.3 percent), herbaceous perennials (6.7 percent), perennial/annual herbs (8.0 percent), and annuals (45.3 percent, 32.0 percent spring, 9.3 percent summer, 4.0 percent opportunistic in either season). The combined herbs including grasses accounted for 60.0 percent of the taxa. Grasses were relatively unimportant, with fluffgrass the only perennial.

Woody plants that were found most frequently in the fecal pellets included desert vine (43.0 percent), foothills paloverde (36.7 percent), California buckwheat (*Eriogonum fasciculatum*, 31.2 percent), globe mallow (30.2 percent), San Felipe dyssodia (*Adenophyllum porophylloides*, 13.9 percent), ironwood (13.9 percent), and wolfberry (*Lycium berlandieri*, 10.1 percent). *Ditaxis* sp. (13.9 percent) and smallseed sandmats (*Euphorbia micromera* and *E. polycarpa*, 11.4 percent) were common herbs that can be perennial or annual. Common spring annual foods were *Eucrypta chrysanthemifolia* (24.1 percent), peppergrass (*Lepidium lasiocarpum*, 20.3 percent), Indian wheat (17.7 percent), and wild carrot (*Daucus pusillus*, 11.4 percent). Spiderling *(Boerhavia erecta)* was a common (12.7 percent) summer annual in summer scats.

After fragment analyses, all of the fecal material from the pellets in each sample was combined into seven samples for microhistological analyses (Wirt and Holm 1997). A total of sixteen taxa were identified. Threeawn was not identified to species in the microhistological analyses but probably represents purple threeawn, the common species in the area. The most important diet plants indicated by the mean percent relative densities were desert vine (19.1 percent), buckwheat (16.1 percent), *Opuntia* sp. (15.2 percent), globe mallow (13.9 percent), foothills paloverde (6.2 percent), ironwood (6.1 percent), threeawn (5.8 percent), and locoweed (*Astragalus* sp., 4.0 percent). The combined values of desert vine, globe mallow, and grasses accounted for 39.4 percent of the diet.

Diet in the Tortolita Mountains

The presence and abundance of Sonoran Desert plants and animals, including desert tortoises, in higher-elevation communities are limited by cold winter temperatures (Lowe 1964; Van Devender et al. 1994) and wildfires (Esque et al., ch. 13 of this volume). In high-elevation desert grassland, chaparral, woodland, and forest, winters are colder than in the Arizona Upland, and fire plays an important role in shaping vegetation structure and composition. Tortoises enter the lower edges of some of these communities but are not typical residents.

Arizona Upland merges into desert grassland at about 1,100-m elevation in the northeastern Sonoran Desert in Arizona (McClaran and Van

Devender 1995; Van Devender, ch. 1 of this volume). Martin (1995) studied a population of tortoises in desert grassland at 1,060- to 1,100-m elevation in the Tortolita Mountains in Pinal County north of Tucson. The vegetation is a desert grassland mosaic, with scattered catclaw acacia, fairy duster, prickly pear, turpentine bush *(Ericameria laricifolia)*, velvet mesquite *(Prosopis velutina)*, and wild buckwheat *(E. wrightii)* in a grass and subshrub matrix. Estimated average rainfall for the study area is about 390 mm/year, extrapolated from weather stations in Oracle (480 mm/year at 1,385 m) and Tucson (283 mm/year at 710 m; Sellers and Hill 1974).

From 1990 to 1992 tortoises were observed feeding on twenty-three plant taxa. In 1990 seventy fresh scat samples were collected throughout the activity period from May to October. Fragment analyses of the scat yielded remains of fifty plant taxa including trees and shrubs (4 percent), subshrubs (10 percent), succulents (6 percent), grasses (30 percent), herbaceous perennials (6 percent), and annuals (44 percent, 26 percent spring obligates; Brent E. Martin and Thomas R. Van Devender, unpubl. data). A few spring annuals including combbur, filaree, Indian wheat *(Plantago patagonica)*, and peppergrass were eaten dried in summer and fall. Notable summer food plants included Indian chickweed *(Mollugo verticillata)*, spiderling *(B. erecta)*, spurges *(Euphorbia* spp.*)*, and verdolaga *(Portulaca oleracea;* Brent E. Martin and Thomas R. Van Devender, unpubl. data).

Microhistological analyses of the fecal pellets combined into four seasonal scat samples yielded only eleven taxa, all of which were identified in the fragment analyses (Brent E. Martin and Thomas R. Van Devender, unpubl. data). In the late spring sample, diets were dominated by annuals (77.3 percent mean relative density), with grasses less important (22.7 percent). In contrast, grasses were dominant (85.0 to 96.8 percent) from the arid foresummer through fall.

Diet in Tropical Deciduous Forest in Southern Sonora

In southern Sonora, tortoises live in tropical deciduous forest (Bury et al., this vol., chap. 5, fig. 5.5). In Las Piedras Canyon 3 km south of Alamos, the forest canopy at about 12-m high is mostly formed by *mauto (Lysiloma divaricatum)*, although various other trees including *amapa (Tabebuia im-*

petiginosa), brasil (Haematoxylum brasiletto), kapok (pochote, Ceiba acuminata), torotes (Bursera grandifolia, B. penicillata, and *B. stenophylla),* and tree morning glory *(palo santo, Ipomoea arborescens)* are common. Arborescent cacti, especially *etcho (Pachycereus pecten-aboriginum)* and organpipe cactus *(pitahaya, Stenocereus thurberi)* are important, but unlike in thornscrub and desertscrub where they are the tallest plants, many trees in tropical deciduous forest tower above the columnar cacti.

The climate of the Alamos area is tropical, and freezing temperatures are rare (Van Devender et al. 2000). Rainfall at 400- to 600-m elevation is about 620 to 700 mm/year. The primary rainy season is summer, from July through September, but tropical storms or hurricanes may bring additional heavy rains from late September through early November. There is a secondary rainy season in the winter months of December, January, and February, the southernmost expression of the Pacific frontal storms that bring winter rains to the Mohave and Sonoran Deserts. A long dry season begins in late January or February and extends to the beginning of July.

Tortoise scat samples were collected in Las Piedras Canyon from five burrows under large granite boulders on steep slopes from 1989 to 1992. Forty-four fecal pellets were disaggregated and divided into three samples, and identifiable plant fragments were removed. Percentages by weight of the 32.7 g of scat in the twenty-six pellets in samples 1, 2, and 3 were 48.7 percent for grasses, 43.4 percent for nongrasses, and 8.1 percent for sand. There were nine to eighteen plant taxa identified from the fragments in each sample, with a combined total of twenty-five taxa (table 8.1). The plants eaten represented trees and shrubs (20 percent), a subshrub (4 percent), a liana (*Gouania rosei,* 4 percent), a cactus (*Opuntia* sp., 4 percent), grasses (16 percent, 12 percent annuals), herbaceous perennials (28 percent), perennial/annual herbs (8 percent), and annuals (16 percent). The most common food plants were *Ayenia filiformis,* bristlegrass *(cola de zorra, Setaria liebmannii),* Natal grass *(zacate rosado, Melinus [Rhynchelytrum] repens),* panicgrass *(Panicum trichoides),* and *Sida alamosana.* Most (92 percent) of the plants are new diet records; only *A. filiformis* and *Opuntia* sp. have been previously recorded in tortoise diets, although none of the local species of *Opuntia (O. puberula, O. pubescens, O. thurberi,* or *O. wilcoxii)* grow in the United States. The distributions of only nine of these diet species extend into the United States.

Discussion and Summary

Comparison of Methods

Observations, bite-count studies, fragment analyses, and micro-histological analyses yielded similar patterns of plant life forms, as well as changing seasonal abundances in desert tortoise diets, but they differed in the number and abundances of taxa identified. Observations of tortoises feeding can be hampered by the difficulty of identifying the food plants in the thick vegetation of the Sonoran Desert compared to the Mohave Desert. This is especially true for annual plants after they germinate but before flowering during the summer rainy season. Dried annuals eaten later in the year can be difficult to identify as well. Very conspicuous plant foods such as prickly pear fruits, which are relished by some tortoises and leave purple stains on the tortoises face, are likely overestimated by observation. For example, at Little Shipp Wash, only nineteen species were observed eaten during 342 hours of observation compared to sixty-nine taxa identified in one hundred fecal pellets (September 1992 [twenty-five], September 1995 [fifty], April 1996 [twenty-five]). Clearly tortoises were eating many more plants than observed. Appendix 8.1 summarizes observations for tortoises feeding on ninety-five plant taxa in Arizona from 1980 to 2000, including herbs (48.4 percent, 30.1 percent annuals), grasses (27.4 percent), woody plants (23.2 percent), and a succulent (1.1 percent). These results for the Sonoran Desert are in contrast to those for Mohave Desert, where observational techniques appear to be more effective (Jennings 1993; Esque 1994; Avery 1995), likely because of flatter terrain, more open vegetation, and generally greater visibility.

Fragment analysis yielded far more taxa than the other methods, demonstrating a greater diversity of diet plants than previously realized. Appendix 8.2 presents 104 plant taxa not observed to be eaten by tortoises (appendix 8.1) in Arizona that were identified in fragment analyses including herbs (61.5 percent, 48.1 percent annuals), grasses (8.7 percent), woody plants (21.2 percent), and succulents (8.7 percent; Cordery et al. 1993; Van Devender and Schwalbe 1999; Brent E. Martin and Thomas R. Van Devender, unpubl. data; Thomas R. Van Devender, unpubl. data). These results indicate that some kinds of food plants, especially annuals and succulents, are less likely to be observed by biologists, whereas grasses and prickly pear fruits are especially conspicuous. Fragment analyses also documented the surprising frequencies and diversity of dried spring annual grasses and dicots eaten in summer and fall. Some fragments appear to have not been ingested intentionally. Others, such as the spiny burs of combbur enjoyed by tortoises are likely not dietary staples. In the Tortolita Mountains, thirteen species of annual grasses and herbs identified in the scat were not found in the floristic surveys of the site, suggesting that the tortoises were better collectors than Tom Van Devender and Brent Martin. The dietary contributions of many plants identified in fragment analyses are not clear because most of the fragments identified were small, comprising minuscule percentages of the mass of each pellet, and represented by a single or few specimens per taxa.

The number of plant taxa identified in tortoise diets in fragment analyses ranged from twenty to seventy-five (average of about fifty-one) at nine Sonoran Desert sites. The sites with the fewest fecal pellets (Bonanza Wash, twenty-two; Sand Tank Mountains, twenty-five; and Tortilla, twenty-five) had the fewest taxa (twenty, thirty-eight, and thirty-four, respectively), thus indicating the need for larger sample sizes. The relatively low number of plant food taxa in the Eagletail Mountains (forty-seven) likely reflects the limited flora of this low-elevation site.

The majority of the mass in tortoise fecal samples is grass blades and digested leaves and not the seeds and fruits typically identified in fragment analysis. Microhistological analyses provide good estimates of the total biomass of the relatively few plant foods that are the dietary staples. Only thirteen species were present at levels of 10 percent or more in any of the twenty-five samples in the AGFD scat study from seven Sonoran Desert sites. Most of the samples (and diets) were dominated by grasses or desert vine/mallows (DV/MA). Most of the diverse food plants documented in the diets in fragment analyses are missed in microhistological analyses either because their vegetative parts are less diagnostic at the cellular level (the identified fragments were removed from the samples) or they are not common enough to be discerned when the entire sample was reduced to microscopic scale. In general, six-weeks fescues and annuals were missing, red brome was much less abundant, and other grasses (dropseed, grama, *Hilaria, Pleuraphis*, or threeawn) were not more common than in fragment analyses. The number of plant taxa (eleven to twenty-nine per aggregate sample) identified in microhistological analyses ranged from 21.3 to 60.1 percent (average 44.4 percent) of the number of taxa (thirty-four to seventy-five) identified in fragment analyses of the same scat samples at eight Sonoran Desert sites.

Diet Summary

Tortoises are opportunistic herbivores that eat many different kinds of plants. They are not large animals and primarily forage within 0.5 m or less of the ground, being unable to normally climb plants to eat foliage, flowers, fruits, parasites, or epiphytes. They may eat such items when they fall to the ground but only the softer ones. For example, tortoises may eat the lower foliage or fallen flowers and leaves of ironwood but not the thick-walled legumes. Tortoises' beaks serve well to cut soft plant tissue but cannot chew food or strip bark like toothy mammals. In general, the woody portions of trees, shrubs, and subshrubs are not eaten, nor are the underground roots. With the exception of fruits and the occasional young cholla joints or prickly pear pads, spiny cacti are not eaten by desert tortoises.

Sonoran desert tortoises in Arizona Upland have been documented to eat 199 species of plants including herbs (55.3 percent, 39.7 percent annuals), grasses (17.6 percent), woody plants (22.1 percent), and succulents (5.0 percent; appendices 8.1 and 8.2). An additional twenty-three species were recorded in the preliminary scat study from tropical deciduous forest in the Sierra de Alamos (table 8.1). Considering the diversity of known food plants, tortoises undoubtedly eat many other species in areas that have not been studied, notably the Central Gulf Coast and Plains of Sonora subdivisions of the Sonoran Desert and coastal and foothills thornscrub in Sonora. In the late Pleistocene, the range of the desert tortoise extended as far east as the Carlsbad area in southeastern New Mexico, presently in the northern Chihuahuan Desert. Considering that they were living in pinyon-juniper-oak woodland (Van Devender et al. 1976), ice age desert tortoises likely ate plant foods not found in their modern habitats.

However, the present stable diet of the desert tortoises in the Arizona Upland is primarily grasses and DV/MA, plants that are also the preferred forage of livestock. In the AGFD scat study, the most important food items in general at seven Sonoran Desert plots were grasses or DV/MA. The mallows eaten by tortoises were desert rose mallow *(Hibiscus coulteri)*, globe mallow, *Herissantia crispa*, Indian mallow, and *Sida* sp. For the Bonanza Wash, Sand

Tank Mountain, and Tortilla Mountain samples, grasses were dominant (60.8–84.2 percent; especially threeawn, 46.0–58.3 percent) food plants, with DV/MA secondary (15.7–38.3 percent). In the Granite Hills, samples were dominated by desert vine (61.5–80.9 percent) with lesser amounts of grasses (5.3–21.3 percent), although an April sample lacking desert vine had 94.2 percent threeawn. At Four Peaks and Little Shipp Wash, the two most mesic sites, dominance fluctuated between DV/MA and grasses. The importance of the herbaceous perennials *Evolvulus* and *Sida* is notable in these samples. In the xeric Eagletail Mountains, tortoises were eating mostly desert vine (27.8–68.2 percent) as well as Indian mallow (2.8–29.3 percent), wild buckwheat (11.3 percent), and rose mallow (11.5 percent). Grasses were less important (13.2–19.5 percent). Grasses and DV/MA together make up the bulk of the tortoise diets at all of the sites: Bonanza Wash (94.3 percent), Eagletail Mountains (59.8–93.2 percent), Four Peaks (50.6–99.9 percent), Granite Hills (95.8–97.6 percent), Little Shipp Wash (55.1–93.2 percent), Sand Tank Mountains (97.8 percent), and Tortilla Mountains (98.6 percent). In the Picacho Mountains, however, DV/MA and grasses combined accounted for only 20.4 percent of the dry weights of plant foods, 36.0 percent with the addition of lanceleaf ditaxis (Vaughan 1984). In desert grassland in the Tortolita Mountains, grasses and DV/MA together comprised 24.9 percent in spring and 96.6–99.0 percent later in summer and fall. These food plants were mostly grasses, because desert vine was absent and mallows *(Abutilon* and *Hibiscus)* were only important in fall (17.3 percent).

The significance of many of the plants eaten by tortoises is not clear because the bulk of the desert tortoise diet is comprised by a relatively few species. The many species of annual plants consumed may reflect some kind of dietary supplements—extra protein, trace minerals, and vitamins deficient in the grass and DV/MA staples (Oftedal, ch. 9 of this volume)—or just tasty treats.

Another enigma is that it is unlikely that some of the plant foods were eaten deliberately, raising the possibility that tortoises eat miscellaneous plant debris accumulated under plants or in other protected spots, as has been observed in the Mohave Desert. Examples include the rare seeds or fruit of such plants as brittlebush, saguaro, and triangleleaf bursage.

There are obvious relationships between diet and the seasonality of rainfall and tortoise activity (Averill-Murray et al., ch. 6 of this volume).

In the biseasonal Sonoran Desert and more tropical areas to the south, tortoises are primarily active during the summer rainy season. The majority of the annual dietary intake occurs from July to September. In contrast, in the winter-rainfall climate of the Mohave Desert, tortoises are primarily active during the spring, passing most of the remainder of the year in burrows. Tortoises do most of their feeding in the spring on living annuals (Jennings 1993; Esque 1994). During drought, tortoises lose body mass, body water contents decline, and solute concentrations of blood plasma increase to very high levels (Peterson 1996a, 1996b). Nagy and Medica (1986) found that the ability to drink rainwater in the occasional warm-season storms helped to relieve physiological stress by flushing excess potassium and salts from the bladder and allowing tortoises to utilize dry grasses and forbs to accumulate energy. Well-hydrated tortoises achieve positive energy balances and are able to store lipids as potential energy reserves.

Our results suggest that even in the biseasonal Sonoran Desert, where tortoises achieve positive water balance more often and most of the foraging behavior is from late July to October, the consumption of dried plants is important. For several weeks after the beginning of the summer monsoon, the only plants available to well-hydrated tortoises are those from previous seasons. In drier years, sporadic rainfall is enough to hydrate tortoises but not to allow substantial plant growth. Except in the driest years, tortoises are less likely to be critically stressed by drought because they rehydrate in spring and/or summer, and dried or fresh grasses, DV/MA, and dead annuals are generally available. Critical physiological stress is much more likely in the uniseasonal Mohave Desert, where the winter-only rainfall presents greater difficulty in eating dried plants in the summer and fall due to longer periods of dehydration.

Biologists reported limited activity and feeding displayed by tortoises, especially in males, in the spring (Martin 1995), in part because relatively few studies have been conducted in the spring. Vaughan (1984) observed a tortoise in the Picacho Mountains engulfing lupine flowers on 16 March 1983. Recently, Averill-Murray and Klug (2000) found that spring foraging in females at Sugarloaf Mountain (northeast of Phoenix, Maricopa County; 549–853 m; Arizona Upland vegetation) was correlated to rainfall. In years with adequate winter-spring rainfall and abundant spring annuals, these female tortoises can be as active in spring as in summer. In the

Suizo and Tortolita Mountains (north of Tucson, Pima County; 760–855 m; Arizona Upland vegetation), the summer monsoon is the primary feeding time for tortoises. Spring foraging is more opportunistic, with both male and female tortoises feeding vigorously in wet winters and springs and showing little activity in dry years (Roger Repp, pers. comm. 2001).

Peterson (1996a, 1996b) concluded from his studies in the western and eastern Mohave Desert in California that the desert tortoise is not physiologically adapted to live in the desert, but is a tenuous relic of a less rigorous climate. This conclusion is consistent with the idea that the Mohave tortoise evolved from Sonoran or Sinaloan tortoises in summer-rainfall environments, eating grasses, mallows, and various other herbs in the diverse floras of tropical and subtropical vegetation (Van Devender, ch. 2 of this volume).

Aseasonal Foods

Plant taxa were found in the diets of tortoises outside the normal growth season for that plant, and thus are considered aseasonal. It is assumed under those circumstances that the tortoise found desiccated portions of the plant from a previous season and consumed them. In the AGFD scat study, the abundance in fresh scat (either fresh samples or fresh pellets in mixed samples) of remains of plant species from another season was notable. A total of 224 records of fifty-three taxa were food plants eaten out of season. In appendix 8.1, sixteen of twenty-nine annuals (55.2 percent) and three of eight annual grasses (37.5 percent) observed eaten by tortoises were spring species eaten in summer or fall.

The plants eaten out of season were mostly spring taxa (94.3 percent: thirty-nine annual dicots, five annual grasses, four subshrubs, one shrub, one tree). The only summer taxa found in spring scat were bristlelobe sandmat *(Euphorbia setiloba)*, saguaro, and spiderling *(Boerhavia intermedia)*. Fully 98.2 percent (forty-one taxa) of the aseasonal feeding records were spring taxa. The most common taxa eaten out of season in descending order were: combbur (fifteen), red brome (fourteen), nievitas (twelve), desert catchfly (twelve), Indian wheat (eleven), filaree (ten), deer vetch (ten), and six-weeks fescue (ten).

The abundance of dead spring annuals eaten in the summer and fall

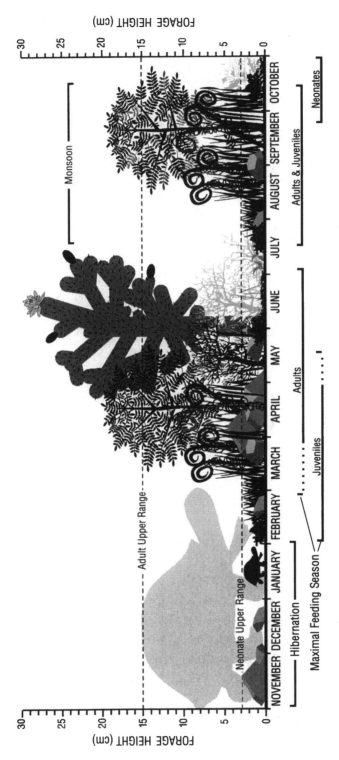

Figure 8.2. Forage availability for the desert tortoise in the Sonoran Desert. Silhouettes contrast size differences between juveniles less than one year old and adults, reflecting differences in foraging seasons and food availability and accessibility. Winter and monsoon rainy season plants are dark, whereas dry season forage is light. Short vertical black blades represent small annual forage, including nonnative grasses and Indian wheat (*Plantago ovata*). Large hooked shapes represent fiddleneck (*Amsinckia* spp.) annuals. Compound-leaf plants are shrub legumes, such as catclaw acacia (*Acacia greggii*) and honey mesquite (*Prosopis glandulosa*), and annual lupine (*Lupinus sparsiflorus*). The thick, stippled branches represent cacti and their fruit and flowers. Illustration courtesy of David J. Morafka.

that result from scat analysis is even more impressive than feeding observations. For example, red brome florets were present in 82–100 percent of the twenty-five pellets in four samples of fresh pellets collected at Four Peaks in September and October of 1995. There were 1–164 florets per pellet; three to fourteen pellets per sample had 20 or more florets.

Forage Availability

Ideally, diet studies could be calibrated by studying the availability of plant foods. In the wet-winter/dry-summer Mohave Desert, where tortoises primarily live in relatively homogeneous flat terrain of valley bottoms and lower bajadas, the frequency, density, or biomass of fresh spring annuals can readily be measured. In the topographically complexity and biseasonal precipitation regimes of Sonoran desert tortoise habitats, assessing availability is more challenging (fig. 8.2).

The Tortolita Mountains diet study in desert grassland provides an excellent example (Brent E. Martin and Thomas R. Van Devender, unpubl. data). Only 32.9 percent of the 152 species of plants recorded in the local flora were also identified in fragment analyses of fecal pellets. This is not surprising considering that most of the long-lived trees, shrubs, and succulents are too tall, woody, fibrous, or spiny for tortoises to eat. The only shrubs eaten were catclaw acacia and fairy duster, and 80 percent of the food plants were herbs or grasses (i.e., tortoise-sized plants). In contrast, 66.7 percent of annual grasses, 35.3 percent of the perennial grasses, and 47.9 percent of dicot annuals (38.2 percent spring, 72.7 percent summer) were eaten. The last percentages are misleading because there are many more spring ($n = 34$) than summer ($n = 11$) annuals.

Percentages of the flora eaten by tortoises are only the first level of forage availability. Abundance and biomass of food plants are profoundly influenced by heterogeneity in the habitat, annual variations in seasonal precipitation, and where individual tortoises spend most of their time. In the Arizona Upland, the vegetation is usually a complex mosaic with nonrandom distributions of plants and tortoise sheltersites. Relatively mesic northerly slopes often have higher densities of grasses, and southerly slopes more succulents and heat-tolerant shrubs. The bare soils and shady crevices that favor annuals are not uniformly distributed but are well-known to the tortoises.

The availability of short-lived plants, especially annuals, is highly variable, tracking rainfall that varies locally and regionally on seasonal, annual, and longer time scales. Considering the variability in the sizes of home ranges (Averill-Murray et al., ch. 7 of this volume), the food plants available to individual tortoises must be variable as well.

Considering the generally inverse relationships between importance in the diet and the size of the plant, traditional measures of vegetation such as coverage and biomass may have little to do with the forage available to tortoises. For example, the total biomass of shrubby catclaw acacias and succulent prickly pears is much greater than the biomass of the respective leaflets and fruits that tortoises consume. Estimates of the forage availability of catclaw acacia and prickly pear would have to be restricted to those leaflets and fruits within reach of tortoises.

Introduced Species in Tortoise Diets

There has been concern that the tortoise's preferences for introduced exotic annuals, such as Mediterranean grass in the Mohave Desert, might lead to nutrient deficiencies and health problems (Avery 1995). Introduced species were found in all of the diet studies in the Sonoran Desert in Arizona and in tropical deciduous forest in Sonora. In the Picacho Mountains, the only introduced species were filaree, Mediterranean grass, and red brome, accounting for 6.5 percent of the taxa and 10.0 percent dry weight of the material identified in the microhistological analyses. In the Harcuvar Mountains and Little Shipp Wash, introduced annuals including red brome and filaree accounted for less than 5 percent of the total observed bites. In the AGFD scat study, only five introduced species were identified in fragment analyses of 603 fecal pellets from seven Sonoran Desert sites: filaree, Malta starthistle *(Centaurea melitensis)*, Mediterranean grass, red brome, and Sahara mustard *(Brassica tournefortii)*. Low levels of brome (0.3–5.5 percent), filaree (0.3–0.8 percent), and Russian thistle *(Salsola tragus*, 1.3 percent) were identified in the microhistological analyses of the combined pellets. In the Maricopa Mountains, the only introduced species identified were filaree (present in 12.7 percent of seventy-five pellets) and red brome (3.8 percent). In the Tortolita Mountains, tortoises ate filaree, Mediterranean grass, red brome, and yellow sweet clover *(Melilotus indicus)*. In tropical deciduous forest in the Sierra de Alamos, the African Natal grass was the only introduced species identified in forty-four fecal pellets.

With only eight exotic species of 222 food plants (3.6 percent) identified at Sonoran Desert and Sonoran sites, introduced species are not important in tortoise diets from a floristic point of view. Only red brome and filaree are frequently eaten and are at times important. Nagy et al. (1998), Van Devender and Schwalbe (2000), and Oftedal (ch. 9 of this volume) reported that the nutrients of exotic species are similar to those of native species in the same taxonomic groups and life forms: red brome and Mediterranean grass versus six-weeks fescue and six-weeks threeawn, and filaree versus Texas filaree *(Erodium texanum)*. Filaree was prized as high-quality cattle forage in Arizona ranges. The only potential impact on tortoise diets in Arizona Upland might be a general reduction of species abundance and diversity in other annual food plants because red brome germinates readily and grows in dense clumps in cool winter soils. Red brome maturing in late January or early February may have absorbed the water and nutrients needed by native annuals to grow into their mid-March peak flowering. A far more serious threat is habitat conversion resulting from the clearing of native vegetation and the deliberate introduction of exotic grasses; Sonoran desertscrub, thornscrub, and tropical deciduous forest plants and animals are severely impacted by the resultant grass fires in communities without fire adaptations (Esque et al., ch. 13 of this volume).

ACKNOWLEDGMENTS

Funds were provided to Thomas R. Van Devender, Brent E. Martin, and Howard E. Lawler by the Roy Chapman Andrews Fund of the Arizona-Sonora Desert Museum to study desert tortoise diet and nutrition. The Arizona Game and Fish Department Heritage Fund supported Van Devender and Cecil R. Schwalbe on projects on the diet of free-ranging tortoises and analyses of the nutrient and mineral contents of tortoise plant foods. Ana Lilia Reina G. helped in the collection of tortoise scat and plant samples to be analyzed for the Heritage Fund diet and nutrition studies. Jesse Piper helped in the AGFD and Maricopa Mountains scat studies. The comments and careful, constructive editing of James L. Jarchow were very helpful. I thank Tom Wiewandt for the use of his photo of a feeding tortoise.

APPENDIX 8.1. Plants observed eaten by Sonoran Desert tortoises from Vaughan (1984; Picacho Peak); Johnson et al. (1990); Schneider (1981), Snider (1993), and Dickinson et al. (unpubl. data; bite-count studies, Harcuvar Mountains, Little Shipp Wash); Martin (1995; Tortolita Mountains); Holm and Wirt (unpubl. data; Rincon and Tucson Mountains); and Averill-Murray and Klug (2000; various permanent plots, Sugarloaf Mountain, Maricopa County). Most observations were made in summer and fall. Site codes: A, Arrastra Mountains; B, Bonanza Wash; E, Eagletail Mountains; G, Granite Hills; HA, Harcuvar Mountains; HQ, Harquahala Mountains; HU, Hualapai Mountains foothills; L, Little Shipp Wash; MA, Maricopa Mountains; N, New Water Mountains; RM, Rincon Mountains; S, San Pedro River Valley; SL, Sugarloaf Mountain; SP, San Pedro River Valley (Ogden 1993); TL, Tortilla Mountains; TM, Tucson Mountains; TO, Tortolita Mountains; WI, Wickenberg Mountains; and WS, West Silverbell Mountains. Numbers are abbreviated years of observations; sequences of years are separated with a dash and individual years with a slash. An asterisk indicates an introduced species.

Taxa (Common Name)

TREES AND SHRUBS
Calliandra eriophylla (fairy duster): L91–94, SL96–99, SP80–90
Eriogonum fasciculatum (California buckwheat): SL96–99, SP80–90
Krameria erecta (range ratany): L91–93
Krameria sp. (ratany): SL96–99
Lycium berlandieri (wolfberry): G93
Lycium sp. (wolfberry): TM96
Olneya tesota (ironwood): TM97
Parkinsonia microphylla (foothills paloverde): G93
Simmondsia chinensis (jojoba): SP80–90
Viguiera parishii (goldeneye): SL96–99
  $N = 10$

SUBSHRUBS AND WOODY VINES
Adenophyllum porophylloides (San Felipe dyssodia): WS00
Ambrosia deltoidea (triangleleaf bursage): G92
Ambrosia dumosa (white bursage): E93
Ditaxis lanceolata (lanceleaf ditaxis): E91–93, G92/94, HA91–93, HQ94, L94, WS95/00
Fagonia californica: A97
Galium stellatum (desert bedstraw): MA90
Hibiscus coulteri (desert rose mallow): HA93/97
Janusia gracilis (desert vine): A97, B97, HA88/91–93/97, L91–93/98, M88/90/00, RM96, S95, SL96–99, SP80–90, TM96–97, WS91/00
Lotus rigidus (desert rock pea): L91–93, SL96–99
Porophyllum gracile (odora): SP80–90

APPENDIX 8.1. Continued

Taxa (Common Name)

Sphaeralcea ambigua (globe mallow): G94, L91, SL96–99
Sphaeralcea sp. (globe mallow): G91, S89, TO90–91
  $N = 12$

SUCCULENT

Opuntia engelmannii (prickly pear): G90/93/98, L90–94/98, SL96–99, SP80–90,
  TM96–97, TO90–91, WI91/00
  $N = 1$

PERENNIAL GRASSES

Aristida arizonica (Arizona threeawn): SP80–90
Aristida purpurea (purple threeawn): HA91–93, HQ94, L91–93/98, SP80–90, TO90–91
Aristida ternipes (spider grass): WS00
Bouteloua curtipendula (sideoats grama): B97, L94, SP80–90, TO90–91, WI00
Bouteloua hirsuta (hairy grama): TO90–91
Bouteloua repens (slender grama): SP80–90, TO90–91
Bouteloua rothrockii (Rothrock grama): SP80–90
Cynodon dactylon (Bermuda grass)*: SP80–90
Digitaria californica (Arizona cottontop): A97
Erioneuron pulchellum (fluffgrass): L91–93, S89/91/95, SP80–90, TO90–91, WS00
Hilaria belangeri (curly mesquite grass): SP80–90, TM96–97
Muhlenbergia porteri (bush muhly): HA97, MA88, WS95/00
Pleuraphis mutica (tobosa): L91–93
Pleuraphis rigida (big galleta grass): B92, E91, HA88/91–93/97, L91–93
Setaria macrostachya (plains bristlegrass): SP80–90
Tridens muticus (slim tridens): SP80–90
  $N = 16$

ANNUAL GRASSES

Aristida adscensionis (six-weeks threeawn): E94, G92, SP80–90, TL96, TO90–91, WS95
Bouteloua aristidoides (six-weeks needle grama): E91/94/98, L90–91, SP80–90, To–91,
  WS00
Bouteloua barbata (six-weeks grama): E90–92/94/98, HA88/97, HU91, L91–93/98, S89,
  SP80–90, TO90–91
Brachiaria arizonica (Arizona panicgrass): S95, SP80–90
Bromus rubens (red brome; all dried)*: B92, E91–94, G92/94, HA88/91–93, HU96,
  L81/91–93, SP80–90
Muhlenbergia microsperma (littleseeds muhly): SP80–90
Schismus barbatus (Mediterranean grass; all dried)*: G93–94, HA91–93, L81/91–93,
  SL96–99
Vulpia octoflora (six-weeks fescue; all dried): G92, HA93, L91–93, SP80–90
  $N = 8$

APPENDIX 8.1. Continued

Taxa (Common Name)

PERENNIAL/ANNUAL GRASSES

Aristida sp. (threeawn): G90/94, L90–91, L93, TM96

Bouteloua sp. (grama): L92, SL96–99

  $N = 2$

HERBACEOUS PERENNIALS

Abutilon parishii (Parish Indian mallow): TO90–91

Acourtia nana (desert holly): SP80–90

Acourtia wrightii (brownfoot): SP80–90

Ayenia filiformis: L91–94, SP80–90

Cirsium neomexicanum (thistle): TO90–91

Euphorbia capitellata (spurge): WS95

Evolvulus alsinoides: L92/94

Selaginella arizonica (spikemoss): WS91

Senna bauhinioides (twoleaf desert senna): SP80–90

Sida neomexicana: SP80–90

  $N = 10$

PERENNIAL/ANNUAL HERBS

Allionia incarnata (windmills): E94, S95, TO90–91

Astragalus sp. (milkvetch): HA91–93

Ditaxis sp.: TM96–97

Euphorbia polycarpa (smallseed sandmat): G94, WS00

Euphorbia sp. (spurge): G92, L91, MA90, SL96–99, SP80–90, TO90–91

Marina parryi: E91

Oenothera sp.: TO90–91

  $N = 7$

ANNUALS

Amaranthus palmeri (pigweed): SP80–90

Amsinckia sp. (fiddleneck): G92 (dried)

Boerhavia coulteri (spiderling): SP80–90

Boerhavia wrightii (spiderling): E94, WS00

Boerhavia sp. (spiderling): E90–91, L90, MA90

Brassica tournefortii (Sahara mustard)*: E93 (dried)

Castilleja exserta (owl clover; all dried): L81, HA91–93

Cryptantha barbigera (nievitas): TO90–91

Cryptantha nevadensis (nievitas): G92 (dried)

Cryptantha sp. (nievitas; all dried): L91–93, HA91–93, SL96–99

Descurainia pinnata (tansy mustard): L91 (dried)

Eriogonum deflexum (skeleton weed): SP80–90

Erodium cicutarium (filaree; all dried)*: L81/91–93, SL96–99, SP80–90

Euphorbia florida (spurge): WS00

APPENDIX 8.1. Continued

Taxa (Common Name)

Euphorbia hyssopifolia (spurge): TO90–91
Euphorbia setiloba (bristlelobe sandmat): WS95
Lepidium lasiocarpum (peppergrass; all dried): G92, SL96–99, TO90–91
Lotus humistratus (deer vetch): TO90–91
Lotus strigosus (deer vetch): L81 (dried)
Lotus sp. (deer vetch): G92, TO90–91
Lupinus sparsiflorus (lupine; all dried): G92/98, HA91–93, HU93, L91–93, P84, SL96–99
Lupinus sp. (lupine): S91 (dried)
Phacelia sp. (caterpillar weed): G92 (dried)
Plagiobothrys arizonicus (popcorn flower): L91 (dried)
Plantago ovata (Indian wheat; all dried): E92–94, HA91–93, L81/91–93, N99, SL96–99
Plantago patagonica (Indian wheat): TO90–91
Portulaca oleracea (purslane): TO90–91
Thysanocarpus curvipes (lacepod): L91–93 (dried)
Tidestromia lanuginosa: E94
   $N = 29$

TOTAL $= 95$

APPENDIX 8.2. Plants identified in fragment analyses that have not been observed eaten by Sonoran tortoises in Arizona (Appendix 8.1). Records are from Cordery et al. (1993; Tucson Mountains), Van Devender and Schwalbe (1999), Martin and Van Devender (unpubl. data; Tortolita Mountains), and Van Devender (unpubl. data; Tucson and Waterman Mountains, Ragged Top Peak). Site codes as in Appendix 8.1 plus: F, Four Peaks; RP, Ragged Top Peak; ST, Sand Tank Mountains; WM, Waterman Mountains. An asterisk indicates an introduced species.

Taxa (Common Name)

TREES AND SHRUBS

Acacia greggii (catclaw acacia): F

Crossosoma bigelovii (ragged rock flower): TM

Encelia farinosa (brittlebush): E, F, G, L, ST, TL

Ephedra nevadensis (Mormon tea): TM

Ephedra trifurca (Mormon tea): TM

Fouquieria splendens (ocotillo): E, F

Hyptis albida (desert lavender): E, F, G, L, ST, TL

Krameria grayi (range ratany): G, TM

Larrea divaricata (creosotebush): E, G, L, TM, WM

Prosopis glandulosa (honey mesquite): L

Prosopis velutina (velvet mesquite): F, TM

Quercus turbinella (shrub live oak): L

 $N = 12$

SUBSHRUBS

Abutilon incanum/malacum (Indian mallow): TM

Ambrosia eriocentra (woollyfruit bursage): L

Ayenia microphylla: L

Brickellia coulteri (brickell bush): RP, TM, TO

Carlowrightia arizonica: L

Eriogonum wrightii (wild buckwheat): TO

Justicia longii: TM

Psilostrophe cooperi (paper flower): L

Sphaeralcea laxa (globe mallow): TM

Trixis californica: TM

 $N = 10$

PERENNIAL GRASSES

Enneapogon desvauxii (spike pappusgrass): TM, WM

Heteropogon contortus (tanglehead): TM

Pappohorum vaginatum (pappusgrass): TM

Setaria cf. *leucopila* (bristlegrass): TM

 $N = 4$

APPENDIX 8.2. Continued

Taxa (Common Name)

ANNUAL GRASSES

Leptochloa panicea (red sprangletop): TM
Panicum hirticaule: TO
Poa bigelovii (Bigelow bluegrass): G
Setaria grisebachii (bristlegrass): TO
Vulpia microstachys (small fescue): B, F, L
  $N = 5$

SUCCULENTS

Carnegiea gigantea (saguaro): F, TM
Dasylirion wheeleri (sotol): F
Echinocereus fasciculatus (hegdhog cactus): TM
Ferocactus cylindraceus (California barrel cactus): F, G
Ferocactus wislizeni (fishhook barrel cactus): TO
Mammillaria grahamii (pincushion fishhook cactus): L, TM
Opuntia acanthocarpa (buckhorn cactus): TL, TM
Opuntia spinosior (cane cholla): TO
Opuntia versicolor (staghorn cholla): TM
  $N = 9$

HERBACEOUS PERENNIALS

Astrolepis cochisensis (Cochise cloak fern): TM, WM
Boerhavia diffusa (red spiderling): G
Cheilanthes parryi (cloak fern): G
Eriogonum inflatum (desert trumpet): WM
Euphorbia arizonica (spurge): F, ST?
Euphorbia eriantha (desert poinsettia): E
Euphorbia melanadenia (spurge): F, L, ST, TM
Euphorbia pediculifera (spurge): TM
Herissantia crispa: F, MM, TL, TM
Pellaea truncata (cliff brake): TM
Physalis crassifolia (ground cherry): TM
Portulaca suffrutescens: G
Sida abutifolia: TO
Tiquilia canescens: WM
  $N = 14$

ANNUALS

Amaranthus fimbriatus: T
Amsinckia tessellata (fiddleneck): E, F, G
Astragalus nuttalianus (locoweed): E, F, G, L
Boerhavia erecta (spiderling): WM
Boerhavia intermedia (spiderling): B, E, F, L, ST

APPENDIX 8.2. Continued

Taxa (Common Name)

Boerhavia megaptera (spiderling): TM

Bowlesia incana (hairy bowlesia): F, TM

Calycoseris wrightii (white tackstem): E, ST

Camissonia sp. (primrose): F

Centaurea melitensis (Malta starthistle)*: F

Chenopodium sp. (goosefoot): L

Chorizanthe brevicornu (brittle spineflower): E, F, G, L, ST, TM

Cryptantha maritima (nievitas): E

Cryptantha pterocarya (wingnut nievitas): TM, WM

Daucus pusillus (wild carrot): E, F, G, L, TL, TM

Ditaxis neomexicana: TM

Draba cuneifolia: TM, TO, WM

Eucrypta chrysanthemifolia: E, F, G, L, ST, TL, TM

Eucrypta micrantha: TM

Euphorbia abramsiana (spurge): E, G, TM

Filago arizonica: RP

Filago / Stylocline: F, G, L, TM

Galium proliferum: TM

Gilia sp.: TM

Harpagonella palmeri: E, L, TL

Kallstroemia sp. (summer poppy): TM, TO

Lepidium cf. *virginicum* (peppergrass): F

Linanthus bigelovii: RP, ST, TM

Lotus salsuginosus (deer vetch): E, F, G, L, TL

Lupinus bicolor (lupine): TO

Lupinus concinnus (lupine): E, F, G, L

Melilotus indicus (yellow sweet clover)*: TO

Microseris linearifolia (silver puffs): L

Mollugo cerviana (carpet weed): F

Mollugo verticillata (Indian chickweed): TO

Monoptilon bellioides (Mohave desert star): RP

Parietaria hespera (pellitory): E, F, G, L, TM

Pectocarya platycarpa (combbur): G

Pectocarya recurvata (combbur): B, E, F, G, L, ST, TL, TM, WM

Pectis cylindrica: TM

Pectis papposa (chinchweed): E

Perityle emoryi (rock daisy): E, L,

Phacelia crenulata (caterpillar weed): E, L, TM

Phacelia distans (caterpliiar weed): F, G, TM

Portlaca halimoides: G

Rafinesquia neomexicana (desert dandelion): F, TM?

APPENDIX 8.2. Continued

Taxa (Common Name)

Silene antirrhina (sleepy catchfly): B, E, F, G, L, ST, TL, TM, WM
Spermoepis echinata: RP
Sphaeralcea cf. *coulteri* (globe mallow): E
Streptanthus carinatus (silverbell): TM
   $N = 50$

TOTAL = 104

LITERATURE CITED

Averill-Murray, R. C., and C. M. Klug. 2000. *Monitoring and ecology of Sonoran desert tortoises in Arizona.* Nongame and Endangered Wildlife Program Technical Report no. 161, Arizona Game and Fish Dept., Phoenix.

Avery, H. W. 1995. *Digestive performance of the desert tortoise* (Gopherus agassizii) *fed native and non-native desert vegetation.* Report to California Dept. of Parks and Recreation, Off-highway Motor Vehicle Recreation Division, Sacramento.

Barrett, S. L. 1990. Home range and habitat use of the desert tortoise *(Xerobates agassizii)* in the Picacho Mountains, Pinal County, Arizona. *Herpetologica* 46:202–6.

Cordery, T. E., Jr., T. A. Duck, T. C. Esque, and J. J. Slack. 1993. Vegetation needs of the desert tortoise *(Gopherus agassizii)* in the Sonoran and Mojave Deserts. Pp. 61–80 *in* D. D. Young, ed. *Proceeding of the symposium on vegetation management of hot desert rangeland ecosystems.* Arizona Section of the Society for Range Management, Bureau of Land Management, and the University of Arizona, Phoenix.

Esque, T. C. 1994. The diet and diet selection of the desert tortoise *(Gopherus agassizii)* in the northeast Mojave Desert. M.S. thesis, Colorado State University, Fort Collins.

Free, J. C., R. M. Hansen, and P. L. Sims. 1970. Estimating the dry weights of food plants in feces of herbivores. *Journal of Range Management* 23:300–302.

Germano, D. J., R. B. Bury, T. C. Esque, T. H. Fritts, and P. A. Medica. 1994. Range and habitats of the desert tortoise. Pp. 73–84 *in* R. B. Bury and D. J. Germano, eds. *Biology of North American tortoises.* Fish and Wildlife Research no. 13, U.S. Dept. of the Interior National Biological Survey, Washington, D.C.

Hansen, R. M., D. G. Peden, and R. W. Rice. 1973. Discerned fragments in feces indicates diet overlap. *Journal of Range Management* 26:103–5.

Hansen, R. M., M. K. Johnson, and T. R. Van Devender. 1976. Foods of the desert tortoise, *Gopherus agassizii*, in Arizona and Utah. *Herpetologica* 32:247–51.

Jennings, W. B. 1993. Foraging of the desert tortoise *(Gopherus agassizii)* in the western Mojave Desert. M.S. thesis, University of Texas, Austin.

Johnson, T. B., N. M. Ladehoff, C. R. Schwalbe, and B. K. Palmer. 1990. *Summary of literature on the Sonoran Desert population of the desert tortoise.* Report to U.S. Fish and Wildlife Service, Phoenix.

Lowe, C. H. 1964. *The vertebrates of Arizona*. University of Arizona Press, Tucson.

Martin, B. E. 1995. Ecology of the desert tortoise *(Gopherus agassizii)* in a desert-grassland community in southern Arizona. M.S. thesis, University of Arizona, Tucson.

McClaran, M. P., and T. R. Van Devender. 1995. *The desert grassland*. University of Arizona, Tucson.

Murray, R. C. 1993. Mark-recapture methods for monitoring Sonoran populations of the desert tortoise *(Gopherus agassizii)*. M.S. thesis, University of Arizona, Tucson.

Nagy, K. A., and P. A. Medica. 1986. Physiological ecology of desert tortoises in southern Nevada. *Herpetologica* 42:73–92.

Nagy, K. A., B. T. Henen, and D. B. Vyas. 1998. Nutritional quality of native and introduced food plants of wild desert tortoises. *Journal of Herpetology* 32:260–67.

Ogden, P. R. 1993. Diets of desert tortoise *(Xerobates agassizii)* along the San Pedro River watershed of Arizona. Pp. 107–13 *in* D. D. Young, ed. *Proceeding of the symposium on vegetation management of hot desert rangeland ecosystems*. Arizona Section of the Society for Range Management, Bureau of Land Management, and the University of Arizona, Phoenix.

Ortenburger, A. I., and R. D. Ortenburger. 1927. Field observations on some amphibians and reptiles of Pima County, Arizona. *Proceedings of the Oklahoma Academy of Science* 6:101–21.

Peterson, C. C. 1996a. Ecological energetics of the desert tortoise *(Gopherus agassizii)*: Effects of rainfall and drought. *Ecology* 77:1831–44.

———. 1996b. Anhomeostasis: Seasonal water and solute relations in two populations of the desert tortoise *(Gopherus agassizii)* during chronic drought. *Physiological Zoology* 69:1324–58.

Rondeau, R., T. R. Van Devender, C. D. Bertelsen, P. Jenkins, R. K. Wilson, and M. A. Dimmitt. 1996. Annotated flora of the Tucson Mountains, Pima County, Arizona. *Desert Plants* 12:3–46.

Schneider, P. B. 1981. *A population analysis of the desert tortoise* (Gopherus agassizii) *in Arizona*. Bureau of Land Management, Phoenix.

Sellers, W. D., and R. H. Hill. 1974. *Arizona Climate, 1931–1972*. University of Arizona Press, Tucson.

Shields, T., S. Hart, J. Howland, N. Ladehoff, T. Johnson, K. Kime, D. Noel, B. Palmer, D. Roddy, and C. Staab. 1990. *Desert tortoise population studies at four plots in the Sonoran Desert, Arizona*. Report to Arizona Game and Fish Dept. and U.S. Fish and Wildlife Service, Albuquerque.

Snider, J. R. 1993. Foraging ecology and sheltersite characteristics of Sonoran desert tortoises. *Proceedings of the Desert Tortoise Council Symposium* 1992:82–84.

Sparks, D. R., and J. C. Malechek. 1968. Estimating percent dry weight in diets using a microscope technique. *Journal of Range Management* 21:264–65.

Turner, R. M., and D. E. Brown. 1982. Sonoran desertscrub. *Desert Plants* 4:121–81.

Van Devender, T. R., and C. R. Schwalbe. 1999. *Diet of free-ranging desert tortoises* (Gopherus agassizii) *in the northeastern Sonoran Desert, Arizona*. Report on IIPAM Project no. 195044 to Heritage Program, Arizona Game and Fish Dept., Phoenix.

———. 2000. *Nutritional analyses of desert tortoise* (Gopherus agassizii) *plant foods in the northeastern Sonoran Desert, Arizona.* Report on IIPAM Project no. 195043 to Heritage Program, Arizona Game and Fish Dept., Phoenix.

Van Devender, T. R., K. B. Moodie, and A. H. Harris. 1976. The desert tortoise *(Gopherus agassizii)* in the Pleistocene of the northern Chihuahuan Desert. *Herpetologica* 32:298–304.

Van Devender, T. R., C. H. Lowe, and H. E. Lawler. 1994. Factors influencing the distribution of the Neotropical vine snake *(Oxybelis aeneus)* in Arizona and Sonora, Mexico. *Herpetological Natural History* 2:25–42.

Van Devender, T. R., A. C. Sanders, R. K. Wilson, and S. A. Meyer. 2000. Flora and vegetation of the Río Cuchujaqui, a tropical deciduous forest near Alamos, Sonora. Pp. 36–101 *in* R. H. Robichaux and D. Yetman, ed. *The tropical deciduous forest of the Alamos: Biodiversity of a threatened ecosystem.* University of Arizona Press, Tucson.

Vaughan, S. L. 1984. Home range and habitat use of the desert tortoise *(Gopherus agassizii)* in the Picacho Mountains, Pinal County, Arizona. M.S. thesis, Arizona State University, Tempe.

Ward, A. L. 1970. Stomach content and fecal analysis: Methods of forage identification. *U.S. Forest Service Rocky Mountain Forest and Experimental Station Miscellaneous Publication* 1147:146–58.

Wirt, E. B., and P. A. Holm. 1997. *Climatic effects on survival and reproduction of the desert tortoise* (Gopherus agassizii) *in the Maricopa Mountains, Arizona.* Report to Arizona Game and Fish Dept., Phoenix.

Woodbury, A. M., and R. Hardy. 1948. Studies of the desert tortoise, *Gopherus agassizii. Ecological Monographs* 18:145–200.

Woodman, P., and T. Shields. 1988. *Some aspects of the ecology of the desert tortoise in the Harcuvar study plot, Arizona.* Report to U.S. Bureau of Land Management, Phoenix.

CHAPTER 9

Nutritional Ecology of the Desert Tortoise in the Mohave and Sonoran Deserts

OLAV T. OFTEDAL

Desert tortoises are obligate herbivores, only rarely consuming such foods as insect larvae or carrion (Jennings 1993). In any year or season, the amount of food available depends on plant germination and growth. Both the quantity and diversity of desert plants varies greatly in response to rainfall, soil characteristics, elevation, and geography (Beatley 1969, 1976). Thus, to understand the nutritional ecology of desert tortoises, it is necessary to examine both the ecology of desert plants and the physiological adaptations of the desert tortoise.

Although most studies of the nutritional ecology of tortoises have been in the Mohave Desert, there are important differences between the Mohave and Sonoran Deserts and within each of these deserts. Unfortunately, little is known about the nutritional ecology of the tortoise in Mexico, where it extends into thornscrub and tropical deciduous forest habitats (Bury et al., ch. 5 of this volume). By necessity, this chapter will be restricted to tortoises in true desertscrub habitats, with an emphasis on tortoises in the Mohave Desert.

Biomass and Diversity in Desert Plant Communities

Deserts generally have low rates of plant biomass production, primarily due to low rainfall (Beatley 1969; Patten 1978; Bowers 1987; Turner and Randall 1989). High temperatures and frequent winds exacerbate evaporative losses of water from soil, contributing to very low soil moisture, and thus highly negative soil water potential, during much of the year (Rundel and Gibson 1996; Smith et al. 1997). Desert plants must either have physiologic and anatomic mechanisms to cope with water shortage, or they must compress their physiologic activity into those relatively brief periods when conditions are more favorable, after periods of rainfall.

Relatively few desert plants, such as agaves, cacti, yuccas, and other water-storing succulents, can maintain photosynthetic activity during the very dry periods that characterize much of the year and that may continue for years during drought (von Willert et al. 1992). Among plant species, there is a wide range of tolerance to water stress, usually measured by soil water potential (Smith et al. 1997). Shrubs that can cope with very low soil water potential can sustain aboveground photosynthetic parts, such as leaves, for prolonged periods. Their leaves are often small and narrow, may be covered with resin, or may be densely hairy—all morphological adaptations to minimize water loss (Rundel and Gibson 1996; Smith et al. 1997). Such shrubs also tolerate very low tissue water potential, as they must to extract water from relatively dry soils.

In contrast, the majority of desert plants abandon active photosynthesis when soil water potential becomes very low. During dry periods most desert shrubs and trees drop their leaves, subshrubs die back to the woody bases, herbaceous perennials die back to root crowns at ground level, and annual plants senesce entirely, surviving as seed in the soil. Desert herbs are ephemeral, appearing aboveground only in response to wetting of the soil by rain by regrowing from roots or germinating from seeds.

In the Sonoran and Mohave Deserts, there has been a general evolutionary trend from herbaceous perennials to annuals, obscuring the difference between annual and perennial. There are both annual and perennial gramas (*Bouteloua* spp.), deer vetches (*Lotus* spp.), *Ditaxis* spp., globe mallows (*Sphaeralcea* spp.), milkvetches (*Astragalus* spp.), muhlys (*Muhlenbergia* spp.), spiderlings (*Boerhavia* spp.), spurges (*Euphorbia* spp.), and threeawns (*Aristida* spp.). There are even species that can be perennial or annual in different portions of their range or by tracking fluctuations in rainfall. For example, fluffgrass *(Erioneuron pulchellum)* is a tufted perennial in biseasonal rainfall areas from Texas to Arizona that persists for many years in safe sites. Plants grow, flower, and germinate in response to both spring and summer rainfall. Seedlings in open areas do not live long, essentially functioning as annuals, but contribute to the seed bank. Summer rainfall declines to the northwest in the transition into the winter-rainfall-dominated Mohave Desert (Mitchell 1976). In southern Nevada, where there is little summer rainfall, fluffgrass is strictly a spring annual (Beatley 1976). Thus, biologists would consider this species and other perennial/annuals including windmills

(Allionia incarnata) and many spurges as perennial in the Sonoran Desert and annual in the Mohave Desert.

Because lignified plant tissues are resistant to microbial fermentation in vertebrate guts, tortoises are presumably unable to digest woody structures (Stevens 1988; Van Soest 1994). Thus, the only potential foods that are available during dry periods are succulent plants (such as cacti) and senescent, nonwoody plant material (such as senescent grasses). However, tortoises that eat dry plant materials lose more water in their feces than they obtain from their food (Nagy et al. 1998), a consequence that is very costly to an animal facing dehydration. The nutritional fate of desert tortoises depends on the rather unpredictable rainfall and the ensuing growth of new tissues by both perennial plants and newly germinated annuals.

Which particular plant species germinate and grow following rain depends on the timing of the rain: Annuals that germinate in response to rains during winter or early spring are termed *winter annuals* (although they flower in spring), whereas those that germinate in response to summer or early fall rains are termed *summer annuals*. The active periods of these two plant groups usually do not overlap, leading to completely different plant assemblages depending on when the rain comes, even in the same locale.

Winter and summer annuals differ not only in their season but also in typical physiological and morphological features. Winter annuals, many of which are derived from temperate floras, germinate following heavy rains under cold conditions and typically emerge as small, low-lying rosettes or tufts that can take advantage of solar heating of the substrate (Mulroy and Rundel 1977). Their maturation accelerates as temperatures rise. A few flowers and fruits may be produced even when the plants are small; this provides reproductive insurance against the mortality that may result from early onset of high temperatures and soil desiccation. However, if moist soil conditions persist, winter annuals may continue to increase in size and reproductive output, even though basal rosettes and lower leaves wither.

Summer annuals germinate in response to rains when ambient temperatures are typically high; their rates of growth and development are high from the outset. They typically lack basal rosettes and are often large in stature (Mulroy and Rundel 1977). Because high temperature and high photosynthetic rate imply high transpirational water loss, efficiency of water use is at a premium. One primary adaptation to such conditions is employment of the C_4 photosynthetic pathway. C_4 refers to the initial carbon fixation product that contains four carbon atoms rather that the initial three-carbon structure in most plants in temperate climates (C_3 plants). Tissue morphology and biochemistry of C_4 plants allow carbon dioxide (CO_2) to be captured into organic compounds with high net efficiency by minimizing unneeded respiratory CO_2 loss (termed *photorespiration*). C_4 plants incorporate more CO_2 per gram of water used and thus have a competitive advantage in hot, dry climates. Not surprisingly, in the Mohave and Sonoran Deserts most summer annuals are C_4 plants derived from subtropical floras, whereas most winter annuals are C_3 plants (Mulroy and Rundel 1977; Raven and Axelrod 1978).

In a particular location, the relative abundance and diversity of annual species are a function of the prevailing patterns of rainfall. Winter rains derive from large storm systems that develop over the North Pacific and traverse the deserts from west to east, covering wide areas but dropping more rain to the west and at higher elevations (Thorne et al. 1981). These frontal systems may extend eastward to Nevada, Utah, and Arizona and southward to Baja California and Sonora, Mexico, but the average amounts of rain decrease to the east and south. The predominance of winter rains in the west, particularly in the Mohave Desert, leads to great species diversity of winter annuals in this area. To the east and south the diversity of winter annuals falls. Shreve and Wiggins (1964) estimate that only one-seventh of the winter ephemerals of the Mohave Desert extend eastward into the Arizona Upland of the Sonoran Desert. Only a few winter annuals extend as far south as Alamos in southern Sonora (Van Devender et al. 2000).

Summer rains have a nearly inverse distribution, because they originate in the tropical oceans. In summer a low-pressure cell typically develops over Arizona, resulting in a monsoonal indraft of moisture in the form of southeasterly winds that cross Mexico from the Gulf of Mexico (Thorne et al. 1981) or southerly winds that bring moisture from the Gulf of California. Summer precipitation is highest in the eastern Sonoran Desert and increases southward into the New World tropics. The northern limits of effective summer rainfall are in the eastern Mohave Desert, where rain falls mostly as localized thunderstorms (Mitchell 1976). This precipitation may be supplemented in late summer or early fall by the relics of powerful tropical storms *(chubascos)* driving north from the Gulf of California. The preponderance and reliability of summer rains to the south and east result in diverse and abundant populations of summer annuals in the eastern and central Sonoran Desert. Shreve and Wiggins (1964) noted that more than eighty species of summer ephemerals grow in the Arizona Upland and Plains of Sonora subdivisions of the Sonoran Desert, but only a handful grow in the western part of the arid lower Colorado River Valley subdivision (often called the Colorado Desert), which receives little rain in either season.

Although summer annuals are generally less abundant and less diverse in the Mohave Desert, they exhibit a similar pattern of east-to-west decline. In southern Nevada we have observed twenty-three species of summer annuals (Olav T. Oftedal and T. Christopher, pers. obs.), many of which appear to be restricted to the extreme southern tip of the state, where summer storms are more frequent.

These are mostly C_4 plants (some of which are introduced) near the edge of their range. They are widely distributed to the south (in the Sonoran Desert) but do not extend into the western Mohave Desert. The seeds of summer annuals must retain viability for prolonged periods in the Mohave Desert because many years may pass between successive summer rains at any particular locale. This may limit which species can adapt to this environment (Mulroy and Rundel 1977). The infrequent germination of these plants makes them unpredictable as a food source and complicates efforts to delimit their range. For example, some of the species we observed in southern Nevada in 1999, such as wingnut spiderling *(Boerhavia pterocarpa)*, smallflower summer poppy *(Kallstroemia parviflora)*, and sticky sprangletop *(Leptochloa viscida)* have not previously been reported in Nevada; undoubtedly others remain to be discovered.

The gradients in summer rainfall also influence perennial plants. Warm-season perennial grasses, such as big galleta *(Pleuraphis rigida)*, bush muhly *(Muhlenbergia porteri)*, and threeawns, respond to summer rains by rapid growth, using the C_4 photosynthetic pathway to capture carbon. These grasses flourish in areas of regular summer rains, but become sparse or absent to the west, where summer rains are rare. Warm-season grasses do not grow leaves at low temperatures and thus cannot take advantage of cold winter rains (Cable 1975). Some herbaceous or suffrutescent perennials, such as desert marigold *(Baileya multiradiata)* and perennial globe mallows, produce new growth in response to both winter and summer rains and thus prosper in locations with both types of rainfall, such as in the eastern Mohave Desert and in the northeastern Sonoran Desert, but are not restricted to these areas. Cacti also increase in stature and abundance from west to east and from north to south, reflecting both increased summer rainfall and reduced risk of freezing.

Thus, the potential foods that tortoises may encounter vary greatly from east to west and from north to south in the Mohave and Sonoran Deserts. Rainfall is not only sparse in desert habitats, it is also extremely variable from year to year, producing great annual variation in food availability. For example, in the Mohave Desert the biomass production of winter annual plants can range from zero in a drought year to more than 500 kg/ha in a wet year (Esque 1994; Rundel and Gibson 1996). Tortoises face very different foraging opportunities and choices in such conditions. In areas of both winter and summer rainfall, it is possible to distinguish at least four differing scenarios based on plant responses to rainfall patterns.

YEARS OF DROUGHT. Deserts are characterized by prolonged periods of barely measurable rainfall (Smith et al. 1997). In much of the winter-rainfall Mohave Desert droughts of eighteen months or more occur regularly. Absence of rain precludes germination of annuals or regrowth of perennials. At such times the desert is virtually devoid of food for tortoises, with the exception of some of the smaller, less armored cacti that tortoises can eat, such as beavertail cactus *(Opuntia basilaris)* and diamond cholla *(Opuntia ramosissima)*, and whatever nonwoody senescent material that has not disintegrated or blown away, including dried grasses such as the annual Mediterranean or split grass *(Schismus barbatus)* and the perennial big galleta grass.

YEARS OF LOW WINTER RAINFALL. In the absence of large (about 2.5 cm) rainfall events at the appropriate time, most species of winter annuals do not germinate (Beatley 1976; Bowers 1987; Rundel and Gibson 1996). However, a small number of weedy annual species do germinate following modest winter or spring rains in the Mohave Desert. These include several nonnative species, such as filaree *(Erodium cicutarium)*, red brome *(Bromus rubens)*, and Mediterranean grass, as well as such native species as fiddleneck *(Amsinckia tessellata)* and woolly plantain *(Plantago ovata)*. In sandy areas, redroot nievitas *(Cryptantha micrantha)* and woolly marigold *(Baileya pleniradiata)* may germinate (Olav T. Oftedal and T. Christopher, pers. obs.).

Some herbaceous perennials also respond to limited rains, including desert globe mallow *(Sphaeralcea ambigua)*, desert marigold, desert trumpet *(Eriogonum inflatum)*, and Layne milkvetch *(Astragalus layneae)*. These plants do not attain the size, lushness, or quantity of seed as in wet years, but their presence offers some foraging opportunity for tortoises.

YEARS OF HIGH WINTER RAINFALL. Large winter rains trigger massive germination of annual plants. Although the exact species composition (including representation of specific families) depends on the temperature and time of year when the rains fall (Beatley 1976; Rundel and Gibson 1996), the large number of species provide tortoises with a wide array of choices. For example, following heavy winter and spring rains due to the 1997–1998 El Niño, we observed sixty-eight species of annuals blooming in northern Piute Valley, sixty-seven species in Eldorado Valley, and fifty-five species in eastern Mormon Mesa, all in Clark County, Nevada. There were four or more species of composites (Asteraceae), borages (Boraginaceae), mustards (Brassicaceae), gilias and allies (Polemoniaceae), and buckwheats (Polygonaceae) in each area, but the legumes (Fabaceae) were more variable, with nine or ten species in the southern sites (Piute and Eldorado Valleys) but only two in eastern Mormon Mesa.

YEARS OF HIGH SUMMER RAINFALL. In the northeastern Sonoran Desert and to a lesser degree in the eastern Mohave Desert, summer storms may bring substantial downpours that trigger germination of summer annuals. These summer annuals constitute a large and diverse group of primarily Sonoran plants (Shreve and Wiggins 1964; Mulroy and Rundel 1977), a subset of which extends northward into the Mohave Desert. At the same time herbaceous perennials (including C_4 grasses) respond by growth of new vegetative and reproductive parts. The summer thunderstorms can be very intense but are often localized. In 1999 massive germination of summer annuals and robust growth of perennials occurred in an area measuring 10 by 13 km in Piute Valley, Nevada, but immediately outside this local area very few plants germinated or grew (Olav T. Oftedal and T. Christopher, pers. obs.).

Although these scenarios represent types of years, a single chronological year may have different rainfall patterns in different regions. Winter storms are typically large eastward-trending fronts, but the average amounts of rain decrease to the east and south. At the extremes, tortoise populations in parts of the southeastern Sonoran Desert receive minimal winter rain (Reyes O. and Bury 1982), whereas tortoise populations in the westernmost Mohave Desert rarely receive summer rain (Peterson 1996).

Foraging Choices of Desert Tortoises

The great geographic and annual variation in plant germination and growth complicates any description or analysis of the nutritional ecology of desert tortoises. Although several comprehensive studies have examined the foods eaten by tortoises, either by direct observation of bites taken or by analysis of plant parts or tissue fragments in feces (see Van Devender et al., ch. 8 of this volume), most are restricted to one or a few years at one or a few sites in one region of the Mohave or Sonoran Deserts. It is not possible to determine how much of the substantial variation among studies is due to, for example, variation in regional floras, differences in the amount and timing of rainfall in the years of study, or local variation in plant abundance and diversity associated with soil texture and parent material, surface run-off, and rain-shadow effects. The extent to which tortoise populations differ in their foraging behavior independent of differences in plant availability is not known.

Behavioral observations permit a great degree of specificity in describing diet because it is possible to ascertain not only the species eaten but also the phenological stage of these plants and the relative proportions of different parts (such as leaves, stems, flowers and fruit) that are consumed. Such details are of great nutritional importance because of the wide variation in nutrient composition among phenological stages (see below) and among plant parts (Olav T. Oftedal and Duane E. Ullrey, unpubl. data; Oftedal et al., in press). However, behavioral studies are very time-consuming and require field observers with detailed botanical knowledge if they are to distinguish among similar plant species or among succeeding phenological stages. Visual barriers may obscure feeding activity in rocky or densely vegetated terrain, leading to biases in what is observed. Failure to distinguish between bite attempts and successful bites also can generate error. Another criticism is that counts of bites will not accurately reflect the amounts of specific foods eaten if the amount ingested per bite varies greatly from one food type to another, as it may in primates and ungulates (Oftedal 1991; Jhala 1997).

Fecal samples reflect the consequences of actual food ingestion over an extended period of time, are relatively easy to collect, and thus provide a valuable opportunity to compare diets among sites or between years. Unfortunately, it may not be possible to discriminate among related species if those parts or tissue fragments that survive digestion are insufficiently different. Identifications based on macroscopic parts inevitably emphasize distinctive structures such as seeds and structural components of fruits. Plants eaten in early phenological stages may be overlooked because these structures have yet to develop, and senescent plants that have dropped fruit may be missed. If tortoises eat the fruits of one species but the leaves of another, fecal fragment analysis may misrepresent their relative contribution. Microhistological analysis of tissue fragments overcomes much of this bias, as long as the plants eaten have distinctive siliceous or lignified tissues that survive digestion and can be identified. The fact that microhistological and fragment analysis of the same feces can produce strikingly different results (Van Devender et al., ch. 8 of this volume) suggests that further study of potential errors is needed. It would be especially useful to study the feces of captive tortoises fed known mixtures of various plant species at different phenological stages.

A list of some of the major plant species eaten by tortoises can be extracted from various studies, with the caveat that such a list incorporates any biases in these studies as well as the limited range of coverage (table 9.1). Only studies in which it was possible to estimate the relative importance of different foods were included (Woodbury and Hardy 1948; Burge and Bradley 1976; Hansen et al. 1976; Coombs 1979; Jennings 1993; Esque 1994; Henen 1994; Avery 1998; Van Devender and Schwalbe 2000; Oftedal et al., in press). Most studies are from the Mohave Desert during periods of low to moderate rainfall, although Jennings's (1993) work encompassed a year (1992) in which a very wide array of annuals germinated and Van Devender et al. (ch. 8 of this volume) compared seven areas of the northeastern Sonoran Desert. Three of the studies refer to the northeastern Mohave, defined herein as the part in Utah and Arizona north of the Virgin Mountains (Coombs 1979; Esque 1994); four studies encompass sites within the eastern Mohave in California, Nevada, and Arizona (Burge and Bradley 1976; Hansen et al.

TABLE 9.1. Major plant foods of the desert tortoise in the Mohave and Sonoran Deserts.[*]

Family Species	Common Name	Seasonality and Growth Form	Mohave W	C	E	NE	Sonoran NW	NE
ASTERACEAE								
Malacothrix glabrata	desert dandelion	wHA	O	X	X	O	O	O
Prenanthella exigua	bright white	wHA	X	O	O	O	O	O
Stephanomeria exigua	small wire lettuce	wHA	O	O	O	X	O	O
BORAGINACEAE								
Cryptantha angustifolia	narrowleaf nievitas	wHA	O	X	X	O	O	O
Cryptantha circumscissa	capped nievitas	wHA	X	O	X	X	O	—
Cryptantha micrantha	redroot nievitas	wHA	O	O	O	X	O	O
Cryptantha nevadensis	Nevada nievitas	wHA	O	O	O	X	O	O
Pectocarya recurvata	combbur	wHA	O	O	X	O	O	X
BRASSICACEAE								
Descurainia pinnata	tansy mustard	wHA	O	O	O	X	O	O
Lepidium lasiocarpum	peppergrass	wHA	O	O	O	O	O	X
CACTACEAE								
Opuntia basilaris	beavertail cactus	SS	O	O	X	X	O	O
Opuntia ramosissima	diamond cholla	SS	O	O	X	?	O	O
EUPHORBIACEAE								
Euphorbia albomarginata	rattlesnake weed	HP	X	O	X	X	O	O
Euphorbia micromera	sandmat	sHA	—	?	X	O	O	X
FABACEAE								
Astragalus didymocarpus	twoseed milkvetch	wHA	X	O	O	—	O	O
Astragalus layneae	Layne milkvetch	HP	X	O	O	—	—	—
Lotus humistratus	deer vetch	wHA	X	O	O	O	O	O
Lotus oroboides	longbract trefoil	HP	—	—	—	X	—	O
Lotus strigosus	deer vetch	wHA	O	O	X	—	O	O
GERANIACEAE								
Erodium cicutarium[*]	filaree	wHA	X	X	O	X	O	O
LOASACEAE								
Mentzelia albicaulis	whitestem stickleaf	wHA	O	O	X	O	O	O
MALPIGHIACEAE								
Janusia gracilis	desert vine	PV	—	—	—	—	—	X

TABLE 9.1. Continued

Family Species	Common Name	Seasonality and Growth Form	Mohave W	C	E	NE	Sonoran NW	NE
MALVACEAE								
Abutilon spp.	Indian mallow	HS	—	—	—	—	—	X
Sphaeralcea ambigua	desert globe mallow	HS	O	O	X	X	O	X
NYCTAGINACEAE								
Allionia incarnata	windmills	P/A	O	O	X	O	O	O
Mirabilis laevis	wishbone bush	HS	X	O	O	O	O	O
ONAGRACEAE								
Camissonia boothii	woody bottlebrush	wHA	X	O	X	O	O	.O
Camissonia claviformis	browneye primrose	wHA	O	X	O	O	O	O
Oenothera deltoides	dune primrose	wHA	—	O	O	X	O	O
PLANTAGINACEAE								
Plantago ovata	Indian wheat	wHA	O	O	X	X	O	X
Plantago patagonica	Indian wheat	wHA	—	?	O	X	O	O
POACEAE								
Aristida adscensionis	six-weeks threeawn	bGA	O	O	X	X	O	X
Bouteloua barbata	six-weeks grama	sGA	—	—	O	O	O	X
Bromus rubens•	red brome	wGA	X	O	X	X	O	X
Bromus tectorum•	cheatgrass	wGA	O	O	O	X	—	—
Erioneuron pulchellum	fluffgrass	GX	O	O	O	X	O	X
Muhlenbergia porteri	bush muhly	GP	—	O	O	X	O	X
Oryzopsis hymenoides	Indian ricegrass	GP	O	O	O	X	O	O
Pleuraphis rigida	big galleta	GP	—	O	X	X	O	O
Schismus barbatus•	Mediterranean grass	wGA	O	O	X	X	O	O
Sporobolus flexuosus	mesa dropseed	GP	—	O	O	O	O	X
Vulpia microstachys	small fescue	wGA	—	—	O	O	O	X
Vulpia octoflora	six-weeks fescue	wGA	O	O	X	X	O	X

Growth forms: GA, annual grass; GP, perennial grass; GX, perennial/annual grass; HA, dicot annual; HP, herbaceous perennial; HS, subshrub; P/A, perennial/annual herb; PV, perennial vine; SS, stem succulent (cactus). Seasonality: b, summer or winter; s, summer; w, winter. Regions: Mohave Desert: W, west of Victorville; C, between Victorville and Baker; E, southern Nevada, southeastern California, and northwestern Arizona; NE, extreme northwestern Arizona and southwestern Utah. Sonoran Desert: NW, southeastern California; NE, northwestern Arizona. X, major plant food; O, occurs in area; —, does not grow in area. An asterisk indicates an introduced species.

1976; Henen 1994; Avery 1998); and one study was restricted to the western Mohave (Jennings 1993). Reference is also made to a recent study of the foraging behavior of juvenile tortoises in natural enclosures at Fort Irwin, California, in April and May 1998, a year of high rainfall (Oftedal et al., in press). Plant distributions were based on personal observation and regional floras (Kearney and Peebles 1960; Shreve and Wiggins 1964; McDougall 1973; Munz 1974; Hickman 1993; Welsh et al. 1993).

Of the forty-three major food species in table 9.1, more than half (twenty-six) grow throughout the Mohave and Sonoran Deserts and thus could be of wider importance as tortoise food. Only eight species were reported as major foods in both deserts, however, and only five in both the western and eastern (or northeastern) Mohave Desert. The differences in species eaten among areas are striking and cannot be solely attributed to regional patterns of plant distribution (table 9.1). Part of the difference between Mohave and Sonoran tortoises may result from different patterns of habitat use: Sonoran tortoises typically inhabit rocky and hilly terrain quite different from the broad basins and *bajadas* where Mohave tortoises have been studied. The local floras of such sites may differ. Much of the variation reported is probably due to the limited duration and habitat coverage of most studies. The abundance and diversity of annual plants varies so greatly from year to year and even from one local site to another that it is difficult—if not misleading—to characterize an average tortoise diet.

An alternative approach is to examine differences among years of varying rainfall. In drought years tortoises typically remain in their burrows, where they sustain a much reduced metabolic rate, and rarely emerge to feed (Nagy and Medica 1986; Peterson 1996; Henen et al. 1998). When rains finally arrive and if temperatures permit, tortoises that emerge from their burrows to drink may also begin to feed on the only foods available, cacti and senescent material such as dry grasses (e.g., big galleta and Mediterranean grass) and dry nievitas (*Cryptantha* spp.; Medica et al. 1982; Nagy and Medica 1986; Henen 1994).

In years of low winter rainfall, foraging choices are limited by the small number of plant species that germinate and grow and the fact that many of these are introduced weedy species. At two sites in the northeastern Mohave Desert, introduced annual grasses (cheatgrass *[Bromus tectorum]*, Mediterranean grass, and red brome) and the introduced filaree accounted for 78–80 percent of all tortoise feeding bites in a year (1990) with very low biomass production of annuals, but as biomass production increased in subsequent years these species accounted for only 34–40 percent (1991) and 34–64 percent (1992; Esque 1994). Tortoises can sample a greater variety of foods as biomass and diversity of plants increase. For example, at City Creek, Utah, only eleven species (including the four introduced species) individually represented 1 percent or more of the diet in 1990, but this number doubled in 1991 (twenty species) and 1992 (twenty-one species; Esque 1994).

Years of exceptional winter rainfall, such as when the El Niño pattern develops, may be especially important because tortoises can afford to be selective, bypassing abundant species and focussing on uncommon species that are particularly palatable. In the 1992 El Niño year, when rainfall at the Desert Tortoise Research Natural Area (DTRNA; Kern County, California), was double the annual average (Henen et al. 1998), tortoises were observed to feed extensively on a variety of species that were of relatively low abundance, including evening primroses *(Camissonia boothii, C. palmeri)* and legumes (deer vetch *[Lotus humistratus]*, Layne milkvetch, and twoseed milkvetch *[Astragalus didymocarpus]*; Jennings 1993). This was especially remarkable in May, when one legume (deer vetch) accounted for 64 percent of tortoise feeding bites even though this species was so patchily distributed that it did not even appear in plant abundance surveys (Jennings 1993). In this season DTRNA tortoises were very active and ate a lot of high-moisture plants, as indicated by high field metabolic rates and high water influx rates, respectively (Henen et al. 1998). They tended to move up and down washes and washlets where preferred food plants such as deer vetch could be found (Jennings 1993).

Less is known about tortoise foraging choices in response to the mass germination of summer annuals and growth of perennials that occur following summer rains. Woodbury and Hardy (1948) were the first to point out that perennial grasses, and especially bush muhly, may be important summer and fall foods for tortoises in the northeastern Mohave Desert. A subset of grasses in which initial capture of CO_2 is into four-carbon compounds (the C_4 grasses) are especially efficient at maintaining high net rates of photosynthesis under ambient conditions of high temperature and limited water (Hattersley and Watson 1992). C_4 grasses that may be important summer foods for tortoises in the eastern Mohave and Sonoran Deserts include big galleta, bush muhly, fluffgrass, six-weeks grama *(Bouteloua barbata)*, and six-weeks threeawn *(Aristida adscensionis*; table 9.1; Waller and Lewis 1979; Olav T. Oftedal, unpubl. data). The summer annuals also employ the C_4 pathway (Mulroy and Rundel 1977), as do such herbaceous perennials as evolvulus (*Evolvulus* sp.), various spurges *(Euphorbia albomarginata, E. poly-carpa)*, and windmills (Downton 1975; Olav T. Oftedal, unpubl. data). Although tortoises forage on these C_4 perennials (table 9.1), the extent to which they ingest C_4 summer annuals other than grasses is unclear. Tortoises have been observed feeding on annual purslane (*Portulaca* spp.), spiderling, and spurge (*Euphorbia* spp.; Esque 1994; Henen 1994; Van Devender et al., ch. 8 of this volume), and fecal analysis indicates that tortoise eat chickweed (*Mollugo* spp.), chinchweed *(Pectis papposa)*, *Ditaxis*, fringed amaranth (*Amaranthus* spp.), purslane, spiderling, spurge and summer poppy (Kallstroemia) at least on occasion (Van Devender et al., ch. 8 of this volume). However, fresh and senescent grasses tend to dominate summer and fall diets in both the eastern Mohave and northeastern Sonoran Deserts (Woodbury and Hardy 1948; Nagy and Medica 1986; Henen 1994; Van Devender et al., ch. 8 of this volume). The relationships among summer rainfall patterns, biomass and diversity of plants, and tortoise diet habits warrant further investigation.

Van Devender et al. (ch. 8 of this volume) report that fresh tortoise scat collected in summer and fall contain a high frequency of parts and fragments from spring plants (including winter annuals) and conclude that tortoises feed extensively on senescent plants left over from the earlier season. Some of this may reflect carryover of digesta derived from plants ingested earlier in the year (Brent E. Martin and Thomas R. Van Devender, unpubl. data). Marker studies indicate mean digesta retention times of fourteen days for large grass particles, but particles are still excreted in feces after more than a month (Barboza 1995a). Captive tortoises that are fasting and inactive may retain fecal residues for five to six months, and in a few instances grass fragments have been found in feces of active, nonfasting animals several months after the grass was last fed (Olav T. Oftedal and T. Christopher, pers. obs.). Thus, it is not clear how much of the out-of-season plant material in the feces studied by Van Devender et al. (ch. 8 of this volume) was carryover across the late spring–early summer inactive period and how much was more recently ingested senescent material. The finding of out-of-season plants even in feces produced late in the summer-fall season (Thomas R.

Van Devender, pers. comm. 1999), the fact that spring feeding appears to be minimal in these areas (Thomas R. Van Devender, pers. comm. 2000), and observations of hydrated tortoises feeding on senescent material in other areas (e.g., Nagy and Medica 1986; Henen 1994) are all consistent with the latter interpretation. Clearly, further research is needed on digesta retention and the importance of ingestion of senescent plants in tortoises in the wild, especially in relation to inactive periods and opportunities to drink.

Nutritional Constraints: The Problem of Potassium

A wide range of herbivorous vertebrates live in North American deserts, including iguanine lizards, rabbits, rodents, ruminants, and tortoises. All face the challenge of acquiring sufficient water, energy, and nutrients while minimizing intake of potentially toxic plant constituents. The latter include such compounds as tannins, which interfere with protein digestion, phytates and oxalates, which interfere with mineral absorption, and alkaloids and terpenoids, which may cause neurologic and other dysfunction in herbivores (Kingsbury 1964; Rosenthal and Janzen 1979). However, tortoises are uniquely vulnerable to one constituent that is ubiquitous and abundant in plant materials: potassium.

Potassium is an electrolyte that readily dissociates into an ionic state in an aqueous medium. As such, it plays an important role in inter- and intracellular physiology and must be maintained within a relatively narrow concentration range (usually expressed as mequiv or mmol/L) in animal tissues or metabolic derangements occur (Bentley 1976; Minnich 1979). Unlike calcium and phosphorus, which precipitate in crystalline structure in bone, potassium cannot be stored to a significant extent in the body. Any excess ingested must be rapidly excreted if toxic effects are to be avoided.

The elongation of the nephron as a loop of Henle in the mammalian kidney enables countercurrent exchange of ions and thus the production of urine with high concentrations of electrolytes, including potassium (Maloiy 1979). However, reptilian nephrons lack loops of Henle and cannot produce urine that is more concentrated in total osmotic constituents than is blood plasma. Dehydrated tortoises and those eating plants high in potassium (K), are unable to produce urine averaging more than about 160 mmol K/L water (Minnich 1979; Nagy and Medica 1986; Peterson 1996). The average potassium in fluid urine of captive tortoises fed high-potassium diets (3.7 percent potassium on a dry matter basis) was 163 mmol/L (Olav T. Oftedal, unpubl. data). Yet, most desert plants, including many major tortoise foods, contain much more than this (table 9.2). A herbivorous reptile that must rely on urine to excrete potassium will lose more water in urine than it obtains in its food. This situation is not viable in an arid environment where drinking water is only rarely available.

Many reptiles have evolved salt glands that they can use to excrete electrolyte loads, including chuckwalla *(Sauromalus obesus)*, desert iguana *(Dipsosaurus dorsalis)*, land iguana *(Conolophus subcristata)*, and other arid-adapted iguanine lizards that excrete excess potassium via nasal salt glands (Dunson 1976). However, the only turtles known to have evolved salt glands are those that must cope with excess sodium in a marine environment, including sea turtles *(Caretta, Chelonia,* and *Lepidochelys)* and diamondback terrapins *(Malaclemys terrapin)*. Thus, turtles (including tortoises) lack salt glands that can excrete potassium.

Tortoises do have a means of coping with potassium loads, but it has a profound effect on their nutrient needs. They produce uric acid that precipitates with cations such as ammonium, potassium, and sodium (Minnich 1972, 1977). In tortoises uric acid is a normal end-product of protein metabolism, and thus its production would be expected to track protein intake. However, when tortoises ingest high levels of potassium, without an increase in protein intake, both the amounts of urate precipitated in the bladder and the concentration of potassium in these precipitates increase (Oftedal et al. 1994). The mechanism by which this occurs is not known. By storing these precipitates in the bladder and voiding bladder contents infrequently, tortoises are able to excrete potassium with minimal water losses (Minnich 1977; Peterson 1996).

Because urates contain about 30 percent nitrogen, a critical side effect of urate production is the removal of nitrogen from the body. Urate production is usually considered part of the nitrogenous waste disposal system, and in reptiles with high protein intakes, such as carnivorous species, this is certainly a primary function (Coulson and Hernandez 1970; Allen and Oftedal 1994). Urates have the advantage over ammonia (the primary product in marine turtles) and urea (the primary product in many semi-aquatic turtles) that they can be expelled as semisolid precipitates with little

TABLE 9.2. Nutrient composition and PEP index of tortoise plant foods.

Family Species	Stage	Site	Date	N	DM (%)	Water (g/g DM)	Crude Protein (%DM)	K (%DM)	K (mmol/L)	PEP (g/kg DM)
ASTERACEAE										
Malacothrix glabrata	fl	DTNA	02 Jun 98	2	43.6	1.29	8.6	1.17	231	5.3
Prenanthella exigua	fl	DTNA	21 Apr 98	2	19.2	4.25	8.7	2.67	162	9.4
Stephanomeria exigua	fl	MM	23 Apr 97	2	38.4	1.6	9.2	2.34	373	-4.0
BORAGINACEAE										
Cryptantha angustifolia	im fr	FISS	20 Apr 98	2	28.2	2.57	11.2	2.75	275	0.1
Cryptantha circumscissa	fl	DTNA	12 Mar 98	2	30.5	2.28	7.1	1.91	215	2.6
Cryptantha micrantha	fl	PV	11 Mar 97	2	30.2	2.32	14.5	2.36	263	5.5
Cryptantha nevadensis	fl	DTNA	12 Mar 98	2	24.2	3.14	9.7	2.32	189	6.6
Pectocarya recurvata	im fr	DTNA	26 Mar 98	2	34.7	1.88	7.1	1.72	233	2.0
BRASSICACEAE										
Descurainia pinnata	im fr	LVV	08 Mar 95	1	27.0	2.70	25.6	2.58	244	16.7
Lepidium lasiocarpum	im fr	PV	21 Mar 98	2	24.2	3.14	20.2	2.11	172	19.1
CACTACEAE										
Opuntia basilaris	im fr	MM	17 Jun 92	4	15.0	5.74	3.7	1.85	84	22.4
Opuntia ramosissima	fl	MM	03 Jun 92	1	28.4	2.52	6.3	1.03	105	12.2
EUPHORBIACEAE										
Euphorbia albomarginata	fl	DTNA	21 Apr 98	2	30.4	2.29	13.2	1.40	151	14.2
Euphorbia micromera	fl	EV	02 Sep 98	2	31.2	2.20	12.8	1.09	127	15.9

				n						
FABACEAE										
Astragalus didymocarpus	fl	DTNA	01 Apr 98	2	22.0	3.55	19.3	1.73	125	24.6
Astragalus layneae	fl	DTNA	21 Mar 98	2	17.6	4.67	23.2	3.08	169	22.2
Lotus humistratus	fl	DTNA	26 Mar 98	2	17.3	4.80	15.4	1.98	106	26.4
Lotus oroboides	fl	CC	18 Apr 99	2	22.1	3.53	17.2	2.32	168	16.6
Lotus strigosus	fl	PV	12 Apr 98	2	17.3	4.79	14.8	2.50	134	20.6
GERANIACEAE										
Erodium cicutarium•	fl	MM	28 Mar 98	2	19.1	4.26	16.6	2.40	144	19.9
LOASACEAE										
Mentzelia albicaulis	im fr	EV	18 May 98	2	24.5	3.08	12.0	1.67	139	15.0
MALPIGHIACEAE										
Janusia gracilis	fl	GH	20 Aug 96	1	38.2	1.62	18.2	1.86	294	9.7
MALVACEAE										
Abutilon parishii	lf	LSW	25 Aug 96	1	48.2	1.07	18.8	2.11	502	4.2
Sphaeralcea ambigua	fl bd	MM	16 Apr 98	2	27.5	2.64	15.4	2.26	219	9.6
NYCTAGINACEAE										
Allionia incarnata	im fr	MM	26 May 98	2	32.4	2.09	12.3	1.81	222	7.5
Mirabilis laevis	fl bd/fl	MM	16 Apr 98	4	16.2	5.28	17.0	3.87	190	12.2
ONAGRACEAE										
Camissonia boothii	fl	DTNA	20 May 98	2	27.9	2.58	8.4	1.27	126	12.3
Camissonia claviformis	im fr	DTNA	26 Mar 98	2	17.1	4.84	10.0	2.28	121	18.4
Oenothera deltoides	fl	EV	25 Mar 98	2	17.3	4.77	7.1	1.88	101	19.1
PLANTAGINACEAE										
Plantago ovata	fl	DTNA	11 Mar 98	2	26.9	2.72	11.2	1.47	138	13.9
Plantago patagonica	fl	MM	15 Mar 97	2	25.7	2.89	14.7	1.77	157	15.3

TABLE 9.2. Continued

Family Species	Stage	Site	Date	N	DM (%)	Water (g/g D)	Protein (%DM)	K (%DM)	K (mmol/L)	PEP (g/kg DM)
POACEAE										
Aristida adscensionis	im fr	EV	21 Oct 97	2	33.1	2.02	13.3	1.25	158	13.6
Bouteloua barbata	fl	PV	01 Sep 98	2	25.6	2.90	16.3	2.12	187	13.6
Bromus rubens•	fl/im fr	MM	12 Apr 95	4	40.2	1.49	6.8	1.21	208	4.2
Bromus tectorum•	fl	DLV	29 Mar 95	4	28.8	2.52	9.2	2.35	239	1.9
Erioneuron pulchellum	fl	EV	02 Sep 98	2	41.3	1.42	12.8	1.27	228	9.1
Muhlenbergia porteri	fl	EV	02 Sep 98	2	38.0	1.64	14.3	1.92	300	5.4
Oryzopsis hymenoides	fl	MM	12 Apr 95	4	35.5	1.82	9.8	1.36	191	7.8
Pleuraphis rigida	fl	FISS	20 Apr 98	2	34.7	1.89	11.7	1.57	213	8.0
Schismus barbatus•	fl	DTNA	21 Mar 98	2	43.8	1.29	7.9	0.92	184	6.9
Sporobolus flexuosus	fl	GB	24 Sep 98	2	38.0	1.63	10.9	1.56	244	5.7
Vulpia microstachys	im fr	TM	*07 Apr 93*	*1*	*43.4*	*1.30*	*6.4*	*0.34*	*67*	*11.3*
Vulpia octoflora	im fr	MM	06 Apr 97	2	42.9	1.33	6.8	0.85	163	6.7

NOTES: 1. Compositional data are means. *N*, number of samples. Data in italics from Van Devender and Schwalbe (2000). Duplicate assays of other samples were: DM (dry matter) and water measured by oven-drying at 55°C; CP (crude protein) was 6.25 times nitrogen measured by the Kjeldahl method or CHN elemental analyses calibrated to Kjeldahl results; potassium (K) measured by atomic absorbtion spectroscopy after hot-plate digestion in perchloric and nitric acids or after microwave digestion at elevated pressure in nitric acid. Potassium excretion potential (PEP), expressed as g/kg DM, is calculated as follows: PEP = [6.5 × water (g/g DM)] + [0.976 × CP (%DM)] − [10 × K (%DM)].

2. Entire aboveground plants or representative branches were collected except *Abutilon parishii* (leaves only). Phenological stages reflect most mature reproductive part on plants sampled: fl, flowers without fruit; fl bd, well-developed flower buds; im fr, immature fruit without mature fruit; lf, leaves only. An asterisk indicates an introduced species.

3. Collection sites: CC, City Creek, Washington Co., UT; DLV, Dry Lake Valley, Clark Co., NV; DTRNA, Desert Tortoise Research Natural Area, Kern Co., CA; EV, Eldorado Valley, Clark Co., NV; FISS, Fort Irwin, San Bernardino Co., CA; GB, Gold Butte area, Clark Co., NV; GH, Granite Hills, Pinal Co., AZ; LSW, Little Shipp Wash, Yavapai Co., AZ; LVV, Las Vegas Valley, Clark Co., NV; MM, Mormon Mesa, Clark Co., NV; PV, Piute Valley, Clark Co., NV; TM, Tortolita Mountains, Pinal Co., AZ.

water loss (Khalil and Haggag 1955; Coulson and Hernandez 1970). In tortoises small amounts of urea are produced, especially during fasting and dehydration when rising blood urea contributes to an increase in blood osmolarity (Minnich 1977; Christopher et al. 1999). This rise may help offset the increased osmolarity of bladder urine, preventing net flux of water from the body into the bladder. However, urates are the primary nitrogenous excreta of tortoises, and they increase in the blood when animals feed on high-potassium plants in spring (Minnich 1977; Nagy and Medica 1986; Peterson 1996). Because plant materials are much lower in protein than animal tissues, herbivorous reptiles must maintain nitrogen retention and recycling as much as nitrogen excretion (Guard 1980). The need to excrete urates to dispose of excess potassium places a tremendous burden on the nitrogen economy of a tortoise that does not have access to drinking water.

This can be seen in data from a study in which juvenile tortoises were fed diets varying in potassium concentration from 0.5 to 3.7 percent (on a dry matter basis), all of which contained about 20 percent protein (or 3.6 percent nitrogen; Oftedal et al. 1994). The amounts of food ingested were similar in all treatments because the amounts of food offered were restricted; tortoises given an ad libitum choice ingest much more of the lower-potassium diets. Tortoises ingesting high-potassium diets increased water intake (measured by hydrogen isotope turnover) and fluid urine output, but the increased urinary potassium output could not, by itself, compensate for the higher potassium intakes. The amounts of nitrogen excreted in urates increased dramatically as dietary potassium levels increased, with the net effect that animals on the highest potassium level could not retain nitrogen for growth even though the protein level of the diet was high (fig. 9.1).

Foraging Choice: The Potassium Excretion Potential Index

At any point in time, a foraging tortoise selects particular plants among the species available (Jennings 1993; Esque 1994). Given the adverse effects of dietary potassium on nitrogen retention, one would expect tortoises to avoid high-potassium foods, if possible. Studies with captive animals offered a choice among formulated feeds varying only in the proportions

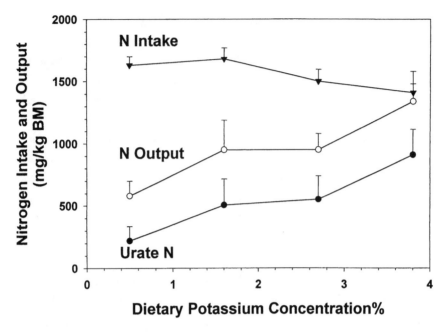

Figure 9.1. Intake and output of nitrogen (N; expressed per kilogram body mass) by juvenile tortoises fed four levels of dietary potassium. The diets were identical except for substitutions of potassium salts for sucrose; the amounts offered were restricted so that intakes would be equivalent. Tortoises were offered water ad libitum twice per week. Each point and error bar represents the mean and standard error of six animals. Note the increase in urate N output and decline in N retention (defined as $N_{intake} - N_{output}$) as dietary potassium increased. Details of the methodology are described by Oftedal et al. (1994).

of potassium salts demonstrate that tortoises can detect, and do avoid, feeds of higher potassium content (Oftedal et al. 1995). Determining the factors underlying food choice in the wild is more difficult, however.

Food choices in the wild are complicated by the fact that plants vary in a wide array of nutrients, not just potassium. For example, tortoises might benefit from eating plants that are high in water and protein, even though they are high in potassium, as long as the amounts of water and protein are sufficient for production of the urine and urates needed to excrete the potassium. Tortoises face a difficult balancing act between one potentially toxic resource (potassium) and two scarce resources (water and protein). Logically, one would expect the relative proportions of potassium, water, and protein to be important, not just the absolute amounts of each.

One way to evaluate the relative amounts of these three constituents is to consider how much of each is lost during excretion. Tortoises confronted with a surfeit of potassium and scarcities of water and protein should minimize the amounts of water and nitrogen used to excrete excess potassium. For any food, it is possible to estimate how much potassium could potentially be excreted based on the amounts of water and nitrogen in the food and to compare this with the amount of potassium contained in the food. In other words, it is possible to calculate a potassium excretion potential (PEP) index integrating all three constituents. A positive PEP index indicates there is more water and nitrogen in the food than is needed to excrete the potassium, whereas a negative PEP index indicates there is insufficient water and nitrogen in the food to excrete the potassium. Tortoises are predicted to avoid foods with a negative PEP index unless they can find additional sources of water and nitrogen. Calculated PEP indices for major tortoise foods are presented in table 9.2. To avoid the confounding effect of phenological stage, these samples are all at or near the transition from flowering to first fruit (see below). Among the major tortoise foods with high PEP indices (greater than 15) are evening primroses (Onagraceae), filaree (Geraniaceae), legumes (Fabaceae), mustards (Brassicaceae), and spurges (Euphorbiaceae) at least at this phenological stage.

The equation for calculation of the PEP index (see footnote to table 9.2) is based on the observation that the maximal urinary potassium concentration achieved by tortoises is about 6.5 g K/kg water (165 mmol/L) and the maximal urate potassium is 0.61 g K/g urate nitrogen (Olav T. Oftedal, unpubl. data). These values, obtained for captive tortoises fed high–potassium diets, are rarely attained by wild tortoises (Minnich 1977; Nagy and Medica 1986; Peterson 1996) and may overestimate the ability of tortoises to excrete potassium under natural conditions.

The PEP index also makes no allowance for losses of water, nitrogen, or potassium in feces; water generated by metabolic processes; or evaporative water losses from respiratory and other surfaces (such as the eyes). The biggest error in applying the PEP index to any particular plant species is probably due to variation in protein digestibility. The apparent digestibility of protein has been shown to vary from 54 to 81 percent among a few herbaceous plants (desert dandelion *[Malacothrix glabrata]*, filaree, Mediterranean grass, and desert globe mallow) but to be close to zero for senescent grasses (Indian rice-grass *[Oryzopsis hymenoides]* and Mediterranean grass; Meienberger et al. 1993; Barboza 1995b; Nagy et al. 1998). The PEP index overestimates potassium excretion potential for senescent grasses both because so little dietary nitrogen can be digested and because dietary water intake is inadequate to cover fecal water losses, let alone urine production (Meienberger et al. 1993; Nagy et al. 1998).

The PEP index varies greatly both among plant species (table 9.2) and within a plant species according to phenological stage. This can be demonstrated by data obtained from sequential collections of three annual plant species in the western Mohave (DTRNA) in 1998. This was an El Niño year with repeated rain events throughout the spring, so that these species persisted until early summer before senescing (fig. 9.2A). The proportions of water (fig. 9.2A) and protein (fig. 9.3A) generally dropped as plants matured and senesced, even in a good year such as the 1998 El Niño. Potassium was more variable and did not show significant seasonal trends for all species (fig. 9.3B). The combined effect on the PEP index varied among the three species. The nitrogen-fixing legume, deer vetch, maintained a high PEP index from March through May but thereafter the index declined rapidly (fig. 9.3B). Capped nievitas *(Cryptantha circumscissa)*, a relatively low-protein species (fig. 9.3A), had a low PEP index even in flower bud, and this gradually declined over the spring. Filaree had a high PEP index when in flower bud and early flower, but with the onset of fruiting the PEP values fell

Figure 9.2. Concentrations of water (A) and potassium (B) in three plant species eaten by desert tortoises in the western Mohave Desert. Plants were collected at the Desert Tortoise Research Natural Area, Kern County, California, during the 1998 El Niño. Deer vetch *(Lotus humistratus)* is a native annual legume, filaree *(Erodium cicutarium)* is an introduced annual in Geraniaceae, and capped nievitas *(Cryptantha circumscissa)* is a native annual borage. Concentrations are expressed per unit dry matter (DM). Calendar year day is the number of days since 31 December 1997. Number labels on the curves refer to first observations of phenological stages: 1, prebloom; 2, flowering; 3, immature fruit; 4, mature fruit; 5, senescent.

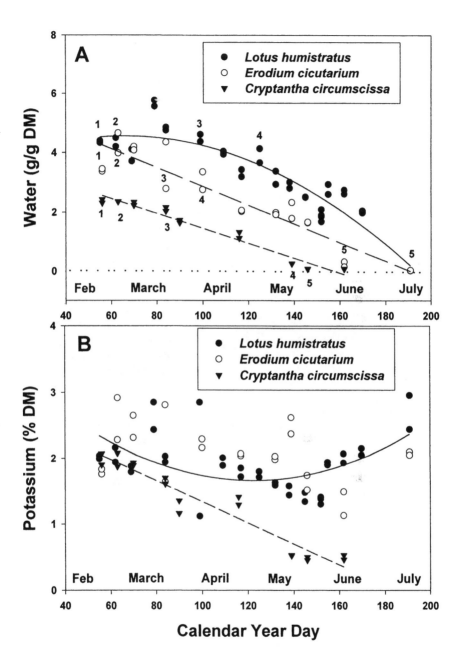

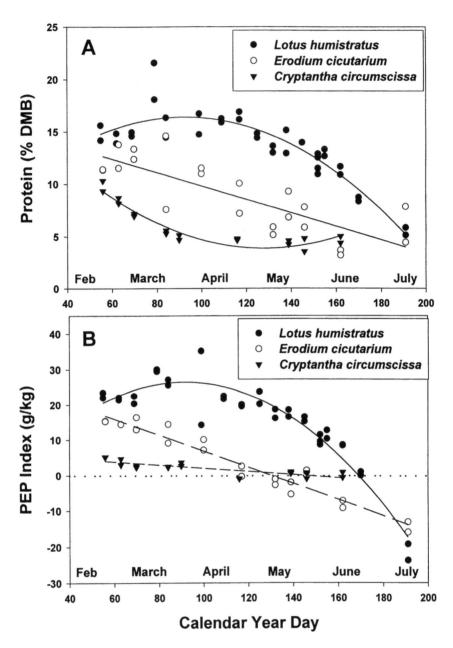

Figure 9.3. Protein concentration (A) and the potassium excretion potential (or PEP) index (B) for three plant species eaten by desert tortoises in the western Mohave Desert. Concentrations are expressed per unit dry matter (DM). Calendar year day is the number of days since 31 December 1997. The equation for the PEP index is provided in a footnote to table 9.2.

markedly (for phenological stages, see fig. 9.2A). Filaree had PEP values close to zero by the end of April and highly negative values thereafter (fig. 9.3B).

These data illustrate the danger of drawing conclusions about the PEP status of plant species from only one or a few samples and the need to compare plants at similar phenological stages. The complexity facing a foraging tortoise should be readily apparent, because the PEP index of each species is changing in different ways (fig. 9.3B) and the number of available species may be great, especially in a year of high rainfall. The array of PEP values in plants confronted by tortoises undoubtedly varies greatly among years, among seasons, and even by week within season as plants mature and senesce.

There is some evidence that tortoises seek out high-PEP plants, at least in wet years. The top four plants eaten by desert tortoises in the DTRNA in April–June 1992, accounting for 60 percent of all bites, were deer vetch, Layne milkvetch, rattlesnake weed *(Euphorbia albomarginata)*, and wishbone bush *(Mirabilis laevis [M. bigelovii]*; Jennings 1993). These species have PEP indices ranging from 12 to 26 at flowering (table 9.2). In contrast, the average PEP index for the ten most abundant species in April was 4.6 and in May was 4.2 (Olav T. Oftedal and W. J. Jennings, unpubl. data for samples collected at DTRNA in 1998). Thus, the PEP index provides a possible explanation for the highly selective foraging behavior reported by Jennings (1993). Certainly, the fact that deer vetch can maintain a high PEP index for such a prolonged time when soil moisture is favorable makes this a potentially valuable forage plant, although it has a patchy distribution in washes.

Much more research is needed to determine the extent to which tortoises make choices that match the predictions of the PEP index. Specifically, do tortoises usually choose those plants with the highest PEP indices available at a particular time of year? Do they eat plant species at the physiological stage when PEP is maximal? Do they eat the plant parts that have the highest PEP index (e.g., Oftedal et al., in press)? Are the differences in foraging choices made by tortoises in dry versus wet years due to differences in the availability of high-PEP plants?

The PEP index does not take into account other factors, such as alkaloids and other toxic compounds that could deter feeding on high-PEP plants. It is also possible that tortoises will choose plant material with a low PEP index if it contains scarce nutrients of importance, such as phosphorus (see below).

Physiological Strategies and Nutrient Composition of Plants

Are there characteristics of plants that make it possible to predict that they will be low or high in potassium, nitrogen, or PEP? A preliminary evaluation suggests there may be important physiological traits of plants that influence their nutrient composition.

Desert plants adopt a variety of physiological strategies that enable them to cope with a desert environment (Smith et al. 1997). For example, some shrubs are able to tolerate very low soil moisture levels, whereas many annuals cannot. Drought-tolerant shrubs and rapidly growing annuals represent opposite ends of a spectrum in terms of water use strategies.

Plants require relatively large amounts of water (compared to animals) because of the water losses associated with photosynthesis (Kramer and Boyer 1995). Plants must capture carbon from the air, but can only do so if they open their surface pores (stomata), allowing CO_2 to diffuse inward. However, whenever the stomata are open, water vapor diffuses outward, and the water loss is exacerbated if the air is dry or the leaf surface is hot (Smith et al. 1997). This water must be replaced by root uptake of water from the soil or the plant will lose turgidity and wither. Water can only move into roots if the water in the roots is at a lower energy state (measured as water potential) than is the water in the soil (Kramer and Boyer 1995). As the soil dries, the water potential of the soil drops steeply. Plants that are unable to compensate (such as many annuals) lose their ability to extract water from the soil and die.

Desert shrubs have a number of adaptations to cope with dropping water potential in the soil (Rundel and Gibson 1996; Smith et al. 1997). They may conserve water by keeping the stomata closed much of the time, but this necessitates a low photosynthetic rate. Equally important, they accumulate osmotic constituents, such as potassium, in tissues that cause the water potential of their fluids to drop. This osmotic adjustment allows plants to create a water potential difference between the roots and the soil, which in turn permits net movement of water into the roots (Turner and Jones 1980; Kramer and Boyer 1995).

There are several ways plants can increase the fluid concentrations of osmotic constituents, including synthesis of small organic compounds, increased uptake of potassium from soil, and allowing tissue water to decline by failing to replace water lost through stomata. Because the PEP index decreases as water decreases or potassium increases in plant tissues, one would expect plants that make osmotic adjustments to have low PEP indices. Plants making osmotic adjustments may also reduce photosynthetic capacity (Bennert and Mooney 1979), including a reduction in the amounts of the photosynthetic enzyme rubisco (ribulose bisphosphate carboxylase-oxygenase). Rubisco accounts for so much of the protein in plants that a low level of this enzyme is often accompanied by low leaf protein content. To the extent that drought-adapted plants are also low in protein, they will be even lower in PEP index. The suite of physiological responses to low soil moisture appears to result in plants of low PEP that are probably poor food for tortoises. This may be one reason tortoises rarely eat desert shrubs in the Mohave Desert.

Many annuals take an opposite approach—one of high water use. Their strategy is to complete their life cycle before soils dry out by maximizing photosynthesis and thus growth rates when conditions remain suitable. Some of the highest photosynthetic rates ever measured in vascular plants have been for desert annuals (Mooney et al. 1976; Smith et al. 1997). A high rate of photosynthesis requires that the stomatal openings in leaves remain open for long periods to allow CO_2 to enter, although great amounts of water are lost. Such plants cannot sustain high photosynthetic rates if soil water potential drops. In the absence of osmotic adjustments, they are predicted to contain only modest concentrations of potassium (relative to tissue water) and relatively high tissue water content. A high rate of photosynthesis also requires an abundance of rubisco. Thus, annual plants that grow quickly are predicted to have high leaf rubisco concentrations, and consequently high leaf protein concentrations (Mooney et al. 1976; Ehleringer 1983). This combination of low potassium, high water, and high protein implies that these plants should have high PEP indices.

These two examples represent extremes, but many desert shrubs undoubtedly adhere to the low photosynthesis–low PEP approach, whereas many annuals adhere to the high photosynthesis–high PEP approach. However, there are many species-specific variants both among shrubs and among annuals, and many herbaceous perennials fall in the middle (Forsyth et al. 1984; Smith et al. 1997). The hypothesis that tortoises choose plant species with high photosynthetic rates but low drought tolerance, and that these are scarce or absent in years of low rainfall, warrants further investigation. If, in fact, many high-PEP plants only germinate and grow in years of unusually high rainfall, such as during El Niño events, selective foraging may have its greatest reward in such times.

The Nutritional Consequences of Drinking

The greater regularity and amount of summer rain in the eastern and southern Sonoran Desert permit an increased diversity of summer annuals and greater abundance and growth of summer-active perennials, including grasses and cacti. As compared to Mohave tortoises, tortoises in these areas encounter a different array of plants and, of equal importance, have more frequent opportunities to drink following rains.

In the Mohave Desert, tortoises must obtain most water from winter annuals and spring-active herbaceous perennials, but summer drinks still can have a major impact on annual water and potassium budgets. These episodes allow tortoises to dump potassium-rich urates and urine from their bladders, reduce circulating potassium levels, and refill their bladders with copious amounts of dilute, hypo-osmotic urine (Nagy and Medica 1986; Peterson 1996; Henen et al. 1998). Such rehydrated tortoises are able to eat senescent grasses, deposit body fat, and are better prepared to withstand subsequent droughts (Henen et al. 1998).

In the northeastern Sonoran Desert, the primary activity period is during the summer rainy season from July to September (Averill-Murray et al., ch. 7 of this volume). Significant numbers of female tortoises emerge and feed in spring, whereas most males remain in their shelters until late May or even July, when annuals and many herbaceous perennials are in late phenological stages. Most foraging, especially for males, appears to occur after summer rains, when tortoises eat senescent plants, grasses, and a variety of fresh annual and perennial plants (Averill-Murray et al., ch. 7 of this volume; Van Devender et al., ch. 8 of this volume). Thus, in areas of regular summer rains foraging may be closely linked to the intake of free water by drinking.

Drinking has a great impact on electrolyte excretion. The amount of water ingested during rains can be substantial, equivalent to 11–28 percent of body mass (Minnich 1977; Nagy and Medica 1986; Peterson 1996). A copious amount of urine containing as little as 6 mmol potassium (Minnich 1977) is stored in the large bladder. This urine serves as a sink into which potassium can be concentrated, both by sodium-potassium exchange at the bladder wall and by water resorption following further renal excretion (Bentley 1976). A 3-kg tortoise in which 20 percent of body mass is bladder urine can store nearly 4 g (100 mmoles) potassium in fluid urine, assuming maximal urinary concentration of 165 mmol/L K (Nagy and Medica 1986; Olav T. Oftedal, unpubl. data).

The potassium sink provided by dilute bladder urine enables a hydrated tortoise to ingest more potassium than could be excreted if it had to rely solely on the water and nitrogen in its food. In other words, the constraints of the PEP index are temporarily relaxed in a hydrated tortoise. The potassium concentration in potential food is still of great importance, because the sink is of limited capacity. The amount of food that can be eaten without exceeding this capacity will be a function of food potassium concentration as well as the quantities of urates formed.

This can be illustrated by considering the situation in which a hydrated tortoise eats dry, senescent material that is so low in water and nitrogen (Nagy et al. 1998; Olav T. Oftedal, unpubl. data) that the potassium excretion potential associated with these constituents can be ignored. For example, if bladder water stores allow a tortoise to dispose of 4 g potassium, the tortoise could eat 200 g of material containing 2 percent potassium, 400 g of material containing 1 percent potassium, and 1,000 g of material containing 0.4 percent potassium. This calculation is only approximate, because it does not take into account fecal potassium losses, but it may explain why hydrated tortoises readily ingest senescent grasses that have low or negative PEP indices and are poorly digestible, but contain little potassium (fig. 9.4).

Tortoises in the eastern Sonoran Desert appear to eat substantial quantities of senescent plants following summer rains (Van Devender et al., ch. 8 of this volume), but the extent to which foraging choice is correlated to nutrient composition is not known. In captivity, hydrated tortoises offered a choice among dry feeds containing different levels of potassium choose the lower-potassium feeds (Oftedal et al. 1995; Olav T. Oftedal, unpubl. data) and may do so in the wild. If this is the case, a hydrated tortoise that had to

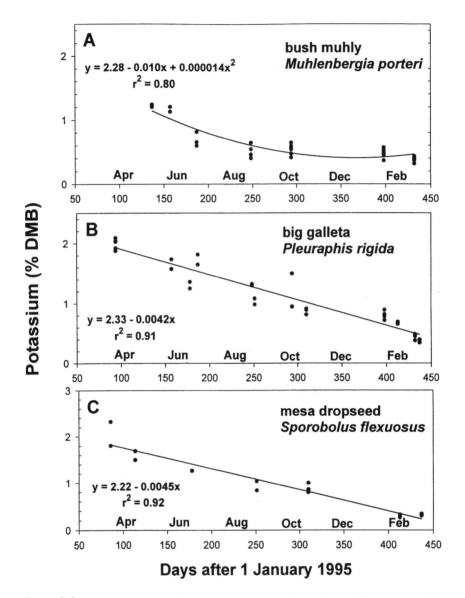

Figure 9.4. Decline over time in the concentration of potassium of three perennial grasses collected in Eldorado Valley, Mormon Mesa, and Piute Valley, Clark County, Nevada: (A) bush muhly, (B) big galleta, and (C) mesa dropseed. Only material that began growth in 1995 (including leaves, stems, and inflorescences) was sampled. Big galleta and mesa dropseed had cast seed and were senescent by September and bush muhly by October 1995; subsequent compositional changes were apparently due to leaching.

choose among senescent deer vetch, filaree, and nievitas would be predicted to select the latter due to its lower potassium content (fig. 9.2B).

Little is known about the factors influencing the potassium content of senescent plants. Rainfall might cause potassium to leach out of senescent desert plants, as it does in senescent forage plants of wetter areas (Karn and Clanton 1977). Leaching may underlie the progressive winter decline in the potassium concentration of senescent perennial grasses in the Mohave Desert (fig. 9.4). (Note that figure 9.4 depicts a winter [1995–1996] of minimal rainfall; greater declines might occur in wetter years.) Can tortoises discriminate between recently senescent material and more leached material from prior seasons? This rarely has been considered, but in the western Mohave Desert Jennings (1993) observed that in June tortoises took more than five hundred bites of dried annual plants from the previous year. This was during an El Niño year when the senescent plants had been heavily rained on. In the eastern Sonoran Desert potassium could be leached due to both winter and summer rains.

Nutrient Composition of Summer Diets

Once summer annuals germinate and perennial food plants regrow, tortoises undoubtedly switch from feeding on senescent material. As in spring, in the summer tortoises can choose among plants that employ a wide range of physiological adaptations, including rapidly growing annuals with high photosynthetic rates; perennial C_4 grasses that are adapted to hot, arid environments; and highly drought-tolerant shrubs (Smith et al. 1997). The variation in water, crude protein, and potassium concentrations among summer-active annuals result in PEP indices ranging from at least 6 to 25 (fig. 9.5), but whether any of these species is an important food plant is not known. Relatively few data are available on such perennial food plants as desert globe mallow, desert vine *(Janusia gracilis)*, and Parish Indian mallow *(Abutilon parishii*; table 9.2). In desert globe mallow, flowers are higher in water but lower in protein than are leaves, but concentrations of both change over time, presumably in response to such factors as phenology and water stress (Olav T. Oftedal and Duane E. Ullrey, unpubl. data).

In summer, tortoises feed extensively on warm season grasses. Compositional data for big galleta, bush muhly, fluffgrass, mesa dropseed *(Sporo-*

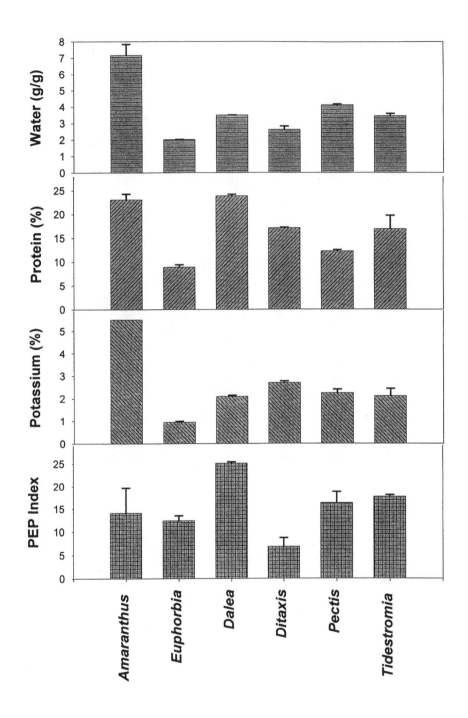

bolus flexuosus), six-weeks grama, and six-weeks threeawn are presented in table 9.2 and figure 9.4. In comparison to cool-season C_3 grasses, these C_4 grasses all have relatively high protein concentrations when in flower and immature fruit. Based on these data and unpublished results for six-weeks needle grama *(Bouteloua aristidoides)* and littleseed muhly *(Muhlenbergia microsperma)*, C_4 desert grasses from southern Nevada ($n = 8$) are significantly higher in protein (14.9 ± 3.51 SEM vs. 8.1 ± 1.37; $t = -4.07$, $P = 0.002$) and PEP (11.7 ± 5.46 vs. 5.5 ± 1.08 SEM; $t = 2.35$, $P = 0.038$) than C_3 grasses from the same area ($n = 5$). This finding stands in contrast to the usual statements that C_4 grasses are lower in protein than C_3 grasses (Minson 1990; Van Soest 1994), but this comparison is based on tropical versus temperate species and may be confounded by climate and soils. Desert C_4 grasses may require substantial investment in photosynthetic enzymes to achieve the very high photosynthetic rates observed during brief periods following rains (Nobel 1980).

There were no significant differences between C_3 and C_4 grasses in water or potassium concentrations. It is interesting that bush muhly, an important summer and fall food for tortoises in the northeastern Mohave and Sonoran Deserts (Woodbury and Hardy 1948; Hansen et al. 1976), appears to contain lower concentrations of potassium than do other perennial C_4 grasses at comparable time periods (fig. 9.4). The predominance of C_4 grasses in tortoise diets in areas of summer rainfall is probably related to the plants' superior protein and PEP values, rather than digestible energy content. The anatomic structure of C_4 grass leaves, involving lignified bundle

Figure 9.5. Nutrient composition of six species of summer annuals in southern Nevada. Two samples of entire plants (except roots) were collected for each species and assayed in duplicate. The species, phenological stage, date, and site of collection are as follows: fringed amaranthus *(Amaranthus fimbriatus)* in flower on 26 September 1997, Eldorado Valley; spurge *(Euphorbia setiloba)* with immature fruit on 27 September 1997, Eldorado Valley; *Dalea mollissima* in flower on 27 September 1997, Eldorado Valley; *Ditaxis neomexicana* with immature fruit on 30 September 1997, Eldorado Valley; chinchweed with flowers and first immature fruit on 8 October 1997, Eldorado Valley; hierba lanuda *(Tidestromia lanuginosa)* in flower on 28 September 1997, Mormon Mesa. Methods of analysis as in table 9.2. Error bars represent standard error of the mean.

sheaths about vascular bundles, renders them less digestible than C_3 grass leaves, at least to mammalian herbivores (Minson 1990; Van Soest 1994).

The large changes in nutrient composition that occur as grasses flower, set seed, and senesce are well known to hay farmers and animal nutritionists (Minson 1990; Van Soest 1994). The rapidity with which annual warm-season grasses progress through phenological stages is indicated by such names as six-weeks grama and six-weeks threeawn; changes in nutrient composition occur with corresponding speed. In 1997, we sampled six-weeks grama at intervals following rain that fell on 1–3 September on Mormon Mesa and in southern Eldorado Valley, Nevada. The grama was in flower in about three weeks, formed mature fruit by five weeks, and was senescent by eight weeks (fig. 9.6A). Although potassium levels dropped quickly over the period from three to five weeks, so did water and protein, with the result that the PEP index declined dramatically (figs. 9.6, 9.7). The accelerated speed of these changes is especially evident if compared to winter annuals (figs. 9.2, 9.3). Hot surface temperatures and rapid drying of the soil force an accelerated pace on summer annuals, with the consequence that earlier, more nutritious phenological stages are available to tortoises for only a brief period of time.

The transitory availability of summer annuals may be one reason that tortoises in the Sonoran Desert appear to rely heavily on perennial plant species, including C_4 grasses, mallows (*Abutilon* spp., *Sphaeralcea* spp.), and desert vine (Van Devender et al., ch. 8 of this volume). Relatively little is known about the relation of tortoise foraging behavior to phenological stage or nutrient composition in these perennial species, but it is unlikely that the composition changes as rapidly as in summer annuals. Perennial plants have the additional advantage that they are more predictable from year to year in terms of location and abundance. The fact that summer-active tortoises are often hydrated due to drinking may also allow them to feed on plants of lower PEP indices.

Other Important Nutrients

A fundamental concept of animal nutrition is that animals, at the tissue level, have similar nutrient needs (Allen and Oftedal 1994, 1996). As determined for commercially reared fish, poultry, livestock, companion

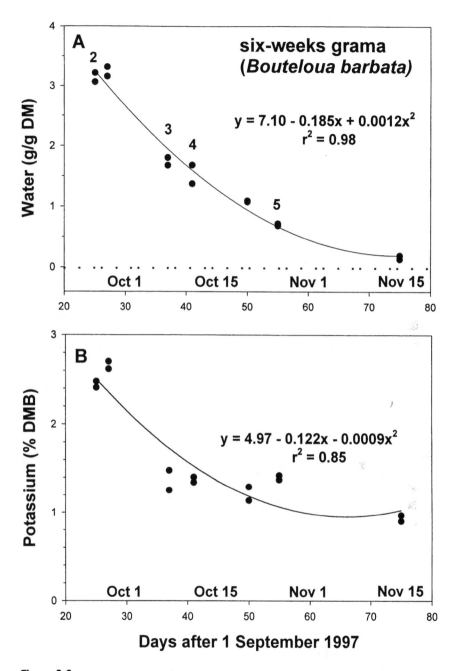

Figure 9.6. Water (A) and potassium (B) concentrations of six-weeks grama, a summer annual grass. Sequential samples of entire plants (except roots) were collected at Eldorado Valley and Mormon Mesa following rains that fell 1–3 September 1997. Number labels on the curves refer to first observations of phenological stages: 2, flowering; 3, immature fruit; 4, mature fruit; 5, senescent. Note that plants were in mature fruit by six weeks and senescent by eight weeks.

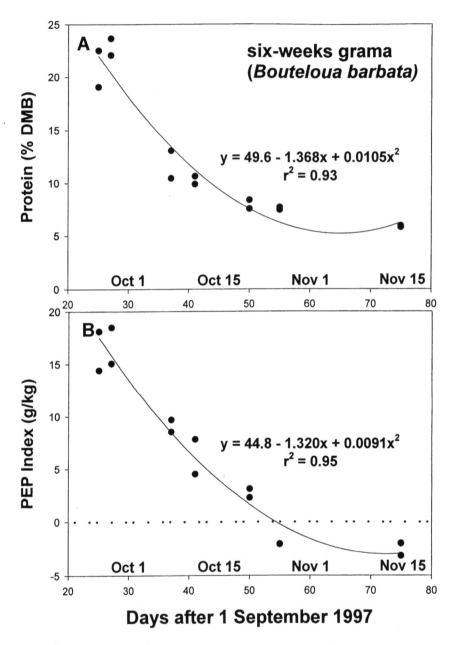

Figure 9.7. Protein concentration (A) and PEP index (B) of six-weeks grama, a summer annual grass. See figure 9.6 legend for details on samples. Equation for the PEP index is provided in a footnote to table 9.2.

animals, and humans, nutrient needs are relatively similar, allowing a fair degree of certainty in predicting the nutritional requirements of other vertebrate taxa. Differences that do exist are primarily a consequence of variation in digestive tract anatomy and function, growth and reproductive rates relative to metabolic rate, and intermediary metabolic pathways.

Although it cannot be proven at present, tortoises are presumed to have nutrient requirements that resemble those of other obligate herbivores, such as ruminants, horses, rabbits, and iguanas (Allen and Oftedal 2002). As a working hypothesis, any nutrient that exists in food plants at such low or high levels as to be outside the normal range tolerated by other herbivores may be considered potentially problematic. Based on an evaluation of about two hundred desert plant samples representing more than fifty species, Olav T. Oftedal and Duane E. Ullrey (unpubl. data) considered that protein, phosphorus, sodium, and possibly copper and zinc were potentially deficient in most tortoise food plants, whereas potassium and calcium were potentially excessive.

The interactions of potassium, protein (nitrogen), and water have already been discussed. The other nutrient pair of particular interest is calcium and phosphorus. In many plant species, the entire plant and/or its vegetative parts are high (greater than 1 percent) or very high (greater than 2 percent) in calcium but low (less than 0.5 percent) or very low (less than 0.2 percent) in phosphorus (Olav T. Oftedal and Duane E. Ullrey, unpubl. data). In vertebrates, high levels of calcium usually impede absorption of phosphorus, thereby aggravating phosphorus deficiency. Thus, this combination of nutrient levels poses the possibility that phosphorus may be a limiting nutrient in diets of tortoises in nature. If so, phosphorus deficiency may contribute to some of the bone pathology seen in some tortoise populations (Wronski et al. 1992; Dickinson et al., this vol., chap. 10). Research is needed on the requirements for phosphorus, the signs of phosphorus deficiency in tortoises, and on how these are influenced by calcium intake.

One way in which tortoises might avoid deficiency is to choose species, stages, or parts of plants that are high in phosphorus. An appetite for phosphorus has been demonstrated in phosphorus-deficient ruminants, but has not been looked for in tortoises. Seed plants invest a considerable amount of phosphorus in their seeds to provide the developing embryo and seedling a source of available phosphorus (Marschner 1995); thus, tortoises may be able to enhance phosphorus intakes by ingesting such plant parts as flowers and fruits. For example, the phosphorus content of reproductive parts (flower buds, flowers, and leaves) of desert globe mallow was 65–120 percent higher than that of leaves collected from the same plants (Olav T. Oftedal and Duane E. Ullrey, unpubl. data). Much of the variation in phosphorus content of plants over their life cycle reflects the early movement of phosphorus from seed and soil into vegetative parts (causing higher initial levels), followed by translocation of much of the phosphorus into new reproductive structures, and finally by casting of the seeds, leaving behind senescent material that is very low in phosphorus.

Tortoises' ability to digest and absorb phosphorus and many other mineral elements may vary considerably from one plant species to another and even among plant parts. In seeds, phosphorus is predominantly stored in a bound form as phytate that must be hydrolyzed by enzymes before the phosphorus can be absorbed (Marschner 1995). In ruminants, microbes in the foregut produce phytases that break down phytates, but it is not known if a similar process occurs in the tortoise hindgut or if phosphorus so released can be absorbed.

Trace-element concentrations in plants tend to vary from one site to another, largely due to differences in soil type and composition (Mertz 1987; Marschner 1995). If tortoises suffer from trace-element deficiencies due to low dietary concentrations, they are apt to be local in distribution. For example, in much of the arid southwestern United States, plants tend to be low in copper but high in molybdenum, with the result that copper deficiency may occur in grazing or browsing ruminants if copper-containing supplements are not provided (McDowell 1985). However, there appear to be species-specific differences in susceptibility to copper deficiency and toxicity. Studies are needed of tortoise requirements for copper and of signs of copper deficiency. The high calcium content in many desert plants is also of potential importance to trace-element metabolism because high calcium may interfere with absorption of some trace elements, including zinc.

It is not known whether the so-called mining behavior of desert tortoises, whereby they repeatedly ingest soil at specific sites with apparent deliberation (Marlow and Tollestrup 1982), is related to mineral appetites or deficiencies. Marlow and Tollestrup (1982) noted that tortoises would target exposed strata containing calcium carbonate accretions and excavate soil high in calcium content. However, other minerals were not tested, and tortoises in the wild are not apt to be calcium-depleted (see above). Soils often contain substantial amounts of iron, but iron is so ubiquitous in the environment and predicted requirements are so low that iron deficiency is unlikely. Among other herbivores, sodium appetite is most commonly implicated in soil ingestion, or geophagia (Denton 1982; Kreulen and Jager 1984). We have been able to demonstrate sodium avoidance but not sodium appetite in captive desert tortoises (K. Chu and Olav T. Oftedal, unpubl. data). Desert tortoises also avidly ingest weathered bone, which led Esque and Peters (1994) to suggest that the bone may be valuable as a source of minerals. As with geophagia, however, there are no controlled studies to support this hypothesis.

Nutritional Ecology and Tortoise Conservation

Desert herbivores must contend with scarcities of water and nitrogen, but an excess of dietary potassium. Tortoises have the unusual distinction of requiring nitrogen from dietary protein to eliminate potassium, because they are unable to excrete potassium via salt glands or to produce hyperosmotic urine. Desert iguanids also use nitrogen to synthesize urates as a means of potassium excretion, but considerable amounts of potassium are eliminated in their salt gland secretions (Minnich 1970; Nagy 1972, 1975).

In the western Mohave and northwestern Sonoran (or Colorado) Deserts, tortoises rarely have opportunity to drink because winter rains fall when tortoises are inactive. In the absence of drinking, tortoises must integrate water, protein, and potassium budgets via foraging in spring. Tortoises feeding on high-moisture foods in spring are often potassium-loaded and show an increase in circulating potassium levels (Nagy and Medica 1986; Peterson 1996). In years of drought or low winter rainfall, the PEP of available food material may be too low to allow tortoise to excrete ingested potassium and still remain in positive water and nitrogen balance. In such years tortoises gradually lose body water and dry fat-free lean mass (which is predominantly protein), especially if they produce eggs (Henen 1994, 1997). However, in good years tortoises can seek out high-PEP plants that require substantial rains to germinate, such as annual legumes. Body water and protein content may increase while feeding on winter annuals, despite the potassium burden (e.g., Nagy and Medica 1986). This balancing of nutrient, electrolyte, and water budgets over a period of years depends on the frequency of years of high rainfall. Juvenile tortoises may be especially sensitive to the PEP index of food plants, given their small size (and thus small nutrient reserves) and the importance of investing water and protein in growth.

At the other extreme, such as the southern Sonoran Desert, virtually all rain falls in summer. Even modest rains may permit tortoises to drink, even though plant productivity is low. Drinking allows tortoises to escape the tight links between food water, nitrogen, and potassium, both because bladder contents can be voided and because the store of dilute bladder urine provides a sink for excess potassium that is subsequently ingested. This allows tortoises to expand their dietary repertoire to include senescent plants and probably other forage with low or even negative PEP indices. It is possible that tortoises can attain positive energy, electrolyte, and water balance even with modest rainfall. Plane of nutrition is no doubt improved in years of high rainfall, which induces good growth of C_4 grasses and favored perennials.

In the northeastern Sonoran Desert and, to a lesser degree, in the eastern Mohave desert tortoises have the benefit of dual rainy seasons. Even if winter rains fail, summer showers may arrive or vice versa. The extent to which tortoises rely on winter annuals and spring-active perennials, as opposed to summer annuals and perennials, probably varies from site to site and year to year. Tortoises feeding on senescent material following summer rains obtain little if any nitrogen, but because they are in positive energy balance, the animals can deposit body fat as an energy reserve (Henen 1997; Nagy et al. 1998).

It is intriguing that large die-offs of desert tortoises have occurred in the western deserts, resulting in the designation of tortoise populations north and west of the Colorado River as threatened with extinction. Even if disease, such as upper respiratory tract disease or cutaneous dyskeratosis (Homer et al. 1998), is the proximal cause of mortality, are these tortoise populations more prone to such morbidity and mortality due to their nutritional ecology? Are tortoises in areas of low summer rainfall in a particularly precarious position, being dependent on high-PEP winter annuals and perennials? A reduction in the frequency of high rainfall years might tax the ability of tortoise populations to recover from the loss of condition that arises when rainfall and high-PEP winter annuals are sparse.

Rainfall patterns in the southwestern United States have changed over time and may continue to do so, especially if global warming becomes pronounced. Paleobotanical studies, including analyses of packrat middens, indicate that rainfall patterns have changed in the desert Southwest since the end of the Pleistocene, about 10,000–12,000 years ago (Spaulding 1990; Van Devender 1990). At the height of the Wisconsin glaciation, desertscrub in the present United States was apparently restricted to low-lying areas including the Amargosa Valley, Death Valley, and the lower Colorado River Valley (Spaulding 1990; Van Devender 1990). With warming and changes in rainfall, pinyon-juniper-oak woodlands gave way to expansion of desertscrub, producing the current Mohave Desert and the Arizona Upland subdivision of Sonoran Desert. In general, winter rainfall has decreased since the Pleistocene as summer rainfall increased (Van Devender 1990). Summer rainfall, at least in the eastern Sonoran Desert, peaked in the middle Holocene (about 4,000–9,000 years ago) and has decreased since. A trough of particularly low rainfall now extends from Death Valley southeast to Yuma, Arizona, and southward into Baja California and Sonora (Smith et al. 1997). Here aridity reflects the combined effects of the Sierra Nevada rain shadow, limited eastward penetration of winter frontal systems, and minimal incursion of summer storms from the southeast. There may be a limit to the ability of tortoises to cope with low rainfall, especially if they are not able to drink.

Habitat management decisions should take both the quantity and quality of nutritional resources into account. Particular attention should be paid to factors affecting the distribution and abundance of high-PEP plants, especially in western areas of limited summer rains. Avery (1998) observed that cattle grazing in the eastern Mohave Desert led to reduced densities of the winter annual desert dandelion, and thus reduced tortoise foraging on an important food plant. Because of their high PEP value (23 g/kg; Oftedal et al., in press), desert dandelion leaves may have a disproportionate influence on the nutritional quality of tortoise diets. Although counter-intuitive, it may be particularly important to protect tortoise food resources from livestock grazing in years of high winter rainfall, because high-PEP plants may only be abundant under such conditions. This is contrary to recommendations based on plant biomass production, in which the assumption is made that there is excess biomass in years of high rainfall.

To the east and especially in the Arizona Upland of the Sonoran Desert, the dietary importance of high-PEP winter annuals may be usurped by summer plants, including C_4 grasses. The overgrazing-induced decline in C_4 grasses, including bush muhly, in parts of the eastern Mohave and Sonoran Deserts may have adverse nutritional consequences for tortoises. The replacement of C_4 perennial grasses by invading annual C_3 grasses, such as Mediterranean and brome grasses, may also impact nutritional status of tortoises given the lower protein and PEP content of C_3 desert grasses and their rapid phenological maturation with an associated decline in nutrient concentrations.

Much more research is needed on the nutritional ecology of the desert tortoise, especially in areas of summer rainfall and in areas undergoing habitat change, whether due to livestock grazing, invading plants, or changing patterns of rainfall. Research to date has highlighted the nutritional susceptibilities of this tortoise species, but the impact of these susceptibilities upon tortoise populations may vary greatly from place to place and year to year.

ACKNOWLEDGMENTS

The research reported here would not have been possible without the consistent support of Sid Slone of the Las Vegas District of the Bureau of Land Management. Sid championed the importance of the research, was instrumental in obtaining funding, and ensured that the facilities at the Desert Tortoise Conservation Center met the needs of the research. I am indebted to the many individuals who contributed time, energy and ideas to this effort. Ideas presented here have been developed in discussions with a variety of collaborators and colleagues, including Brian Henen, Mary Allen, Duane Ullrey, Terry Christopher, and David Hellinga. Tom Van Devender challenged me to consider the scenarios presented in the Sonoran Desert, and both he and Jim Jarchow made valuable comments on a draft of this chapter. Terry Christopher participated in most plant collections and tortoise trials; David Hellinga supervised much of the laboratory work; and Michael Jakubasz provided essential logistic and administrative support. Thanks to the research assistants, including Alice Chung, Katie Flickinger, Denise Freitas, Amy Hunt, Julie Keene, Patrina Merlino, Laurie Nelson, and Randall Reed. Denise LaBerteaux and Mark Bagley oversaw plant collections at the Desert Tortoise Research Natural Area, and Terry Knight and Wes Niles assisted in plant identification. Perry Barboza collaborated in the first two years of this research.

I gratefully acknowledge the multiple-year financial backing provided by the Bureau of Land Management, the Clark County Desert Conservation Plan, and The Nature Conservancy. Additional funds were provided by the Smithsonian Office of Fellowships and Grants and by the Friends of the National Zoo.

LITERATURE CITED

Allen, M. E., and O. T. Oftedal. 1994. The nutrition of carnivorous reptiles. Pp. 71–82 *in* B. Murphy, K. Adler, and J. T. Collins, eds. *Captive management and conservation of amphibians and reptiles*. Society for the Study of Amphibians and Reptiles, Ithaca, N.Y.

———. 1996. Essential nutrients in mammalian diets. Pp. 117–28 *in* D. G. Kleiman, M. E. Allen, K. V. Thompson, S. Lumpkin, and H. Harris, eds. *Wild mammals in captivity: Principles and techniques*. University of Chicago Press, Chicago.

———. 2002. The nutritional management of an herbivorous reptile, the green iguana *(Iguana iguana)*. *In* E. R. Jacobson, ed. *Biology, husbandry and medicine of the green iguana*. Krieger, Huntington, N.Y.

Avery, H. W. 1998. Nutritional ecology of the desert tortoise *(Gopherus agassizii)* in relation to cattle grazing in the Mojave desert. Ph.D. diss., University of California, Los Angeles.

Barboza, P. S. 1995a. Digesta passage and functional anatomy of the digestive tract in the desert tortoise *(Xerobates agassizii)*. *Journal of Comparative Physiology B* 165:193–202.

———. 1995b. Nutrient balances and maintenance requirements for nitrogen and energy in desert tortoises *(Xerobates agassizii)* consuming forages. *Comparative Biochemistry and Physiology A* 112:537–45.

Beatley, J. C. 1969. Biomass of desert winter annual plant populations in southern Nevada. *Oikos* 20:261–73.

———. 1976. *Vascular plants of the Nevada Test Site and central-southern Nevada: Ecologic and geographic distributions*. Report no. TID-26881, Energy Research and Development Administration, U.S. Dept. of Commerce, Springfield, Va.

Bennert, W. H., and H. A. Mooney. 1979. The water relations of some desert plants in Death Valley, California. *Flora* 168:405–27.

Bentley, P. J. 1976. Osmoregulation. Pp. 365–412 *in* C. Gans, ed. *Biology of the Reptilia*, vol. 5, *Physiology A*. Academic Press, New York.

Bowers, M. A. 1987. Precipitation and the relative abundance of desert winter annals: A 6-year study in the northern Mohave Desert. *Journal of Arid Environments* 12:141–49.

Burge, B. L., and W. G. Bradley. 1976. Population density, structure and feeding habits of the desert tortoise, *Gopherus agassizii*, in a low desert study area in southern Nevada. *Proceedings of the Desert Tortoise Council Symposium* 1976:51–74.

Cable, D. R. 1975. Influence of precipitation on perennial grass production in the semidesert Southwest. *Ecology* 56:981–86.

Christopher, M. M., K. H. Berry, I. R. Wallis, K. A. Nagy, B. T. Henen, and C. C. Peterson. 1999. Reference intervals and physiologic alterations in hematologic and biochemical values of free-ranging desert tortoises in the Mojave Desert. *Journal of Wildlife Diseases* 35:212–38.

Coombs, E. M. 1979. Food habits and livestock competition with the desert tortoise on the Beaver Dam Slope, Utah. *Proceedings of the Desert Tortoise Council Symposium* 1979:132–47.

Coulson, R. A., and T. Hernandez. 1970. Nitrogen metabolism and excretion in the living reptile. Pp. 639–710 *in* J. W. Campbell, ed. *Comparative biochemistry of nitrogen metabolism.* Vol. 2, *The vertebrates.* Academic Press, New York.

Denton, D. A. 1982. *The hunger for salt.* Springer-Verlag, New York.

Downton, W.J.S. 1975. The occurrence of C_4 photosynthesis among plants. *Photosynthetica* 9:96–105.

Dunson, W. A. 1976. Salt glands in reptiles. Pp. 413–45 *in* C. Gans, ed. *Biology of the Reptilia.* Vol. 5, *Physiology A.* Academic Press, New York.

Ehleringer, J. 1983. Ecophysiology of *Amaranthus palmeri*, a Sonoran Desert summer annual. *Oecologia* 57:107–12.

Esque, T. C. 1994. Diet and diet selection of the desert tortoise *(Gopherus agassizii)* in the northeast Mojave Desert. M.S. thesis, Colorado State University, Fort Collins.

Esque, T. C., and E. L. Peters. 1994. Ingestion of bones, stones and soil by desert tortoises. Pp. 105–11 *in* R. B. Bury and D. J. Germano, eds. *Biology of North American tortoises.* Fish and Wildlife Research no. 13, U.S. Dept. of the Interior National Biological Survey, Washington, D.C.

Forsyth, I. N., J. R. Ehleringer, K. S. Werk, and C. S. Cook. 1984. Field water relations of Sonoran desert annuals. *Ecology* 65:1436–44.

Guard, C. L. 1980. The reptilian digestive system: General characteristics. Pp. 43–51 *in* K. Schmidt-Nielsen, L. Bolis, and C. R. Taylor, eds. *Comparative physiology: Primitive mammals.* Cambridge University Press, New York.

Hansen, R. M., M. K. Johnson, and T. R. Van Devender. 1976. Foods of the desert tortoise, *Gopherus agassizii*, in Arizona and Utah. *Herpetologica* 32:247–51.

Hattersley, P. W., and L. Watson. 1992. Diversification of photosynthesis. Pp. 38–116 *in* G. P. Chapman, ed. *Grass evolution and domestication.* Cambridge University Press, Cambridge, U.K.

Henen, B. T. 1994. Seasonal and annual energy and water budgets of female desert tortoises *(Xerobates agassizii)* at Goffs, California. Ph.D. diss., University of California, Los Angeles.

———. 1997. Seasonal and annual energy budgets of female desert tortoises *(Gopherus agassizii)*. *Ecology* 78:283–96.

Henen, B. T., C. C. Peterson, I. R. Wallis, K. H. Berry, and K. A. Nagy. 1998. Effects of climatic variation on field metabolism and water relations of desert tortoises. *Oecologia* 117:365–73.

Hickman, J. C., ed. 1993. *The Jepson manual: Higher plants of California.* University of California Press, Berkeley.

Homer, B. L., K. H. Berry, M. B. Brown, G. Ellis, and E. R. Jacobson. 1998. Pathology of diseases in wild desert tortoises from California. *Journal of Wildlife Diseases* 34:508–23.

Jennings, W. J. 1993. Foraging ecology of the desert tortoise *(Gopherus agassizii)* in the western Mojave Desert. M.S. thesis, University of Texas, Arlington.

Jhala, Y. J. 1997. Seasonal effects on the nutritional ecology of blackbuck, *Antelope cervicapra. Journal of Applied Ecology* 34:1348–58.

Karn, J. F., and D. C. Clanton. 1977. Potassium in range supplements. *Journal of Animal Science* 45:1426–34.

Kearney, T. H., and R. H. Peebles. 1960. *Arizona flora*. University of California Press, Berkeley.

Khalil, F., and G. Haggag. 1955. Ureotelism and uricotelism in tortoises. *Journal of Experimental Zoology* 130:423–32.

Kingsbury, J. M. 1964. *Poisonous plants of the United States and Canada*. Prentice-Hall, Englewoood Cliffs, N.J.

Kramer, P. J., and J. S. Boyer. 1995. *Water relations of plants and soils*. Academic Press, San Diego, Calif.

Kreulen, D. A., and T. Jager. 1984. The significance of soil ingestion in the utilization of arid rangelands by large herbivores, with special reference to natural licks on the Kalahari pans. Pp. 204–21 *in* F.M.C. Gilchrist and R. I. Mackie, eds. *Herbivore nutrition in the subtropics and tropics*. Science Press, Craighall, South Africa.

Maloiy, G. M. O., ed. 1979. *Comparative physiology of osmoregulation in animals*. Academic Press, New York.

Marlow, R. W., and K. Tollestrup. 1982. Mining and exploitation of natural mineral deposits by the desert tortoise, *Gopherus agassizii. Animal Behaviour* 30:475–78.

Marschner, H. 1995. *Mineral nutrition of higher plants*. Academic Press, San Diego.

McDougall, W. B. 1973. *Seed plants of northern Arizona*. Museum of Northern Arizona, Flagstaff.

McDowell, L. R., ed. 1985. *Nutrition of grazing ruminants in warm climates*. Academic Press, New York.

Medica, P. A., C. L. Lyons, and F. B. Turner. 1982. A comparison of 1981 populations of the desert tortoise *(Gopherus agassizii)* in grazed and ungrazed areas in Ivanpah Valley, California. *Proceedings of the Desert Tortoise Council Symposium* 1982:99–124.

Meienberger, C. M., I. R. Wallis, and K. A. Nagy. 1993. Food intake rate and body mass influence transit time and digestibility in the desert tortoise *(Xerobates agassizii). Physiological Zoology* 66:847–62.

Mertz, W., ed. 1987. *Trace elements in human and animal nutrition*. 5th ed. Academic Press, New York.

Minnich, J. E. 1970. Water and electrolyte balance of the desert iguana, *Dipsosaurus dorsalis*, in its natural habitat. *Comparative Biochemistry and Physiology* 35:921–33.

———. 1972. Excretion of urate salts by reptiles. *Comparative Biochemistry and Physiology A* 41:535–49.

———. 1977. Adaptive responses in the water and electrolyte budgets of native and captive desert tortoises, *Gopherus agassizii*, to chronic drought. *Proceedings of the Desert Tortoise Council Symposium* 1977:102–29.

———. 1979. Comparison of maintenance electrolyte budgets of free-living desert and gopher tortoises *(Gopherus agassizii and G. polyphemus). Proceedings of the Desert Tortoise Council Symposium* 1979:166–74.

Minson, D. J. 1990. *Forage in ruminant nutrition.* Academic Press, San Diego, Calif.

Mitchell, V. L. 1976. The regionalization of climate in the western United States. *Journal of Applied Meteorology* 15:920–27.

Mooney, H. A., J. Ehleringer, and J. A. Berry. 1976. High photosynthetic capacity of a winter annual in Death Valley. *Science* 194:322–24.

Mulroy, T. W., and P. W. Rundel. 1977. Annual plants: Adaptations to desert environments. *Bioscience* 27:109–14.

Munz, P. A. 1974. *A flora of southern California.* University of California Press, Berkeley.

Nagy, K. A. 1972. Water and electrolyte budgets of a free-living desert lizard, *Sauromalus obesus. Journal of Comparative Physiology* 79:39–62.

———. 1975. Nitrogen requirement and its relation to dietary water and potassium content of the lizard *Sauromalus obesus. Journal of Comparative Physiology* 104:49–58.

Nagy, K. A., and P. A. Medica. 1986. Physiological ecology of desert tortoises in southern Nevada. *Herpetologica* 42:73–92.

Nagy, K. A., B. T. Henen, and D. B. Vyas. 1998. Nutritional quality of native and introduced food plants of wild desert tortoises. *Journal of Herpetology* 32:260–67.

Nobel, P. S. 1980. Water vapor conductance and CO_2 uptake for leaves of a C_4 desert grass, *Hilaria rigida. Ecology* 6:252–58.

Oftedal, O. T. 1991. The nutritional consequences of foraging in primates: The relationship of nutrient intakes to nutrient requirements. *Philosophical Transactions of the Royal Society London B* 334:161–70.

Oftedal, O. T., M. E. Allen, A. L. Chung, R. C. Reed, and D. E. Ullrey. 1994. Nutrition, urates and desert survival: Potassium and the desert tortoise *(Gopherus agassizii). Proceedings American Association of Zoo Veterinarians Annual Meeting* 1994:308–13.

Oftedal, O. T., M. E. Allen, and T. E. Christopher. 1995. Dietary potassium affects food choice, nitrogen retention, and growth of desert tortoises. *Proceedings of the Desert Tortoise Council Symposium* 1995:58–61.

Oftedal, O. T., S. Hilliard, and D. J. Morafka. In press. Selective spring foraging by juvenile desert tortoises *(Gopherus agassizii)* in the Mojave Desert: Evidence of an adaptive nutritional strategy. *Chelonian Conservation and Biology.*

Patten, D. T. 1978. Productivity and production efficiency of an Upper Sonoran Desert ephemeral community. *American Journal of Botany* 65:891–95.

Peterson, C. C. 1996. Anhomeostasis: Seasonal water and solute relations in two populations of the desert tortoise *(Gopherus agassizii)* during chronic drought. *Physiological Zoology* 69:1324–58.

Raven, P. H., and D. I. Axelrod. 1978. *Origin and relationships of the California flora.* University of California Publications in Botany no. 72, University of California Press, Berkeley.

Reyes O., S., and R. B. Bury. 1982. Ecology and status of the desert tortoise *(Gopherus agassizii)* on Tiburón Island, Sonora. Pp. 39–49 *in* R. B. Bury, ed. *North American tortoises: Conservation and ecology.* Wildlife Research Report no. 12, U.S. Fish and Wildlife Service, Washington, D.C.

Rosenthal, G. A., and D. H. Janzen, eds. 1979. *Herbivores: Their interaction with secondary plant metabolites.* Academic Press, New York.

Rundel, P. W., and A. C. Gibson. 1996. *Ecological communities and processes in a Mojave Desert ecosystem: Rock Valley, Nevada.* Cambridge University Press, Cambridge, U.K.

Shreve, F., and I. L. Wiggins. 1964. *Vegetation and flora of the Sonoran Desert*. Stanford University Press, Stanford, Calif.

Smith, S. D., R. K. Monson, and J. E. Anderson. 1997. *Physiological ecology of North American desert plants*. Springer-Verlag, Berlin.

Spaulding, W. G. 1990. Vegetational and climatic development of the Mojave Desert: The last glacial maximum to the present. Pp. 166–69 *in* J. L. Betancourt, T. R. Van Devender, and P. S. Martin, eds. *Packrat middens: The last 40,000 years of biotic change*. University of Arizona Press, Tucson.

Stevens, C. E. 1988. *Comparative physiology of the vertebrate digestive system*. Cambridge University Press, New York.

Thorne, R. F., B. A. Prigge, and J. Henrickson. 1981. A flora of the higher ranges and the Kelso Dunes of the eastern Mojave Desert in California. *Aliso* 10:71–186.

Turner, F. B., and D. C. Randall. 1989. Net production by shrubs and winter annuals in southern Nevada. *Journal of Arid Environments* 17:23–36.

Turner, N. C., and M. M. Jones. 1980. Turgor maintenance by osmotic adjustment: A review and evaluation. Pp. 87–103 *in* N. C. Turner and P. J. Kramer, eds. *Adaptations of plants to water and high temperature stress*. John Wiley & Sons, New York.

Van Devender, T. R. 1990. Late Quaternary vegetation and climate of the Sonoran Desert, United States and Mexico. Pp. 134–65 *in* J. L. Betancourt, T. R. Van Devender, and P. S. Martin, *Packrat middens: The last 40,000 years of biotic change*. University of Arizona Press, Tucson.

Van Devender, T. R., and C. R. Schwalbe. 2000. *Nutritional analyses of desert tortoise* (Gopherus agassizii) *plant foods in the northeastern Sonoran Desert, Arizona*. Report on IIPAM Project no. 195043 to Heritage Program, Arizona Game and Fish Dept., Phoenix.

Van Devender, T. R., A. C. Sanders, R. K. Wilson, and S. A. Meyer. 2000. Flora and vegetation of the Río Cuchujaqui, a tropical deciduous forest near Alamos, Sonora. Pp. 36–101 *in* R. H. Robichaux and D. Yetman, ed. *The tropical deciduous forest of the Alamos: Biodiversity of a threatened ecosystem*. University of Arizona Press, Tucson.

Van Soest, P. J. 1994. *Nutritional ecology of the ruminant*. 2d ed. Cornell University Press, Ithaca, N.Y.

von Willert, D. J., B. M. Eller, M.J.A. Werger, E. Brinckmann, and H.-D. Ihlenfeldt. 1992. *Life strategies of succulents in deserts with special reference to the Namib Desert*. Cambridge University Press, Cambridge, U.K.

Waller, S. S., and J. K. Lewis. 1979. Occurrence of C_3 and C_4 photosynthetic pathways in North American grasses. *Journal of Range Management* 32:12–28.

Welsh, S. L., N. D. Atwood, S. Goodrich, and L. C. Higgins, eds. 1993. *A Utah flora*. Brigham Young University, Provo, Utah.

Woodbury, A. M., and R. Hardy. 1948. Studies of the desert tortoise, *Gopherus agassizii*. *Ecological Monographs* 18:146–200.

Wronski, T. J., C.-F. Yen, and E. R. Jacobsen. 1992. Histomorphometric studies of dermal bone in the desert tortoise, *Gopherus agassizii*. *Journal of Wildlife Diseases* 28:603–9.

CHAPTER 10

Are Free-ranging Sonoran Desert Tortoises Healthy?

VANESSA M. DICKINSON, JAMES L. JARCHOW,

MARK H. TRUEBLOOD, AND JAMES C. deVOS

Desert tortoises *(Gopherus agassizii)* are long-lived herbivorous reptiles found in the deserts of the southwestern United States and northern Mexico (Germano et al. 1994). In southeastern California and western Arizona, the Colorado River separates the major genetic populations in the Mohave and Sonoran Deserts (Lamb and McLuckie, ch. 4 of this volume). The Sonoran tortoise ranges from central Arizona south to central Sonora (Bury et al., ch. 5 of this volume). In Arizona, the prime tortoise habitats are found in the scenic saguaro *(Carnegiea gigantea)*–foothills paloverde *(Parkinsonia microphylla)* desertscrub on the rocky boulder slopes of the Arizona Upland subdivision of the Sonoran Desert (Van Devender, ch. 1 of this volume).

The Mohave tortoise population, which was listed as threatened in 1990 (U.S. Fish and Wildlife Service 1990), has been intensely studied, stimulating interest in the desert tortoise throughout its range. Most of our knowledge about the health of free-ranging tortoises came from the Mohave Desert in eastern California (Rosskopf 1982; Nagy and Medica 1986; Jacobson et al. 1991; Christopher et al. 1994, 1997; O'Connor et al. 1994; Rostal et al. 1994) and southern Nevada (O'Connor et al. 1994). Knowledge of the health of tortoises is important for understanding changes in populations in response to habitat loss, forage competition with domestic livestock, off-road-vehicle use, drought, and other environmental stresses (Nagy and Medica 1986).

In humans, health is generally thought of in terms of physical and mental well-being. Good health generally refers to the lack of disease and the harmony of good physical and mental condition. Considering the difficulty in understanding the mental state of desert tortoises, assessment of their health is, of necessity, related to physical condition. Here, we use the term *health* to reflect the ability of a tortoise to meet its physiological needs for metabolism, growth, behavior, and reproduction. In deserts, climate is variable on daily, seasonal, annual, and longer time scales. Healthy tortoises display a complex of physiological responses to environmental conditions and events, especially extremes in heat and aridity (Oftedal, ch. 9 of this volume). In this chapter, we present information on the health status of free-ranging tortoises in the Sonoran Desert.

In the Wild Tortoise Clinic

Imagine yourself in a doctor's office discussing the results of your physical examination. The doctor asks about your health and examines you physically, noting your height, weight, body temperature, and skin condition. He looks into your eyes, ears, nose, mouth, and other body orifices. A sample of your blood is analyzed for indicators of abnormal health conditions. A fecal sample is examined for parasites and cultured for bacteria. The doctor mentions that your cholesterol level is higher than normal and that you are overweight. He recommends that you reduce your intake of animal fats and prescribes some medicine and vitamin dietary supplements.

Veterinarians use the same diagnostic methods in their clinics, except that most pets cannot describe how they feel. Evaluating the health of free-ranging desert tortoises is more difficult, however. The biologist is more like a traveling physician ministering to a mute aboriginal tribe than a clinical veterinarian. Unlike pets, wild tortoises are not accustomed to handling, view the biologist as a predator, and must be treated carefully. Like a doctor, the biologist routinely records the physical parameters of each tortoise including shell size, weight, and any distinctive characteristics. Blood and fecal samples can be examined for additional indicators of health. Following individual tortoises over seasons and years presents a picture of how tortoises are affected by reproduction, food and water availability, and disease. A detailed picture of a tortoise's normal blood parameters and bacterial tests assists researchers in the diagnosing disease and evaluating the general health of the population (Dickinson et al. 1996).

The Arizona Game and Fish Department Health Study

In 1990 the research branch of the Arizona Game and Fish Department (AGFD) began a study to better understand the health status of free-ranging tortoises in Arizona Upland habitats on established study plots at Little Shipp Wash near Bagdad and in the Harcuvar Mountains near Aguila, Arizona. The objectives of this five-year study were to collect baseline data on hematology (the study of blood cells) and blood chemistry; determine if there were physiological differences related to site, sex, season, or year; infer seasonal tortoise activities from physiological parameter values, Sonoran tortoise nutrition studies, and local weather data; and establish normal reference ranges for Sonoran tortoise hematologic and biochemical parameters.

Most study animals were adult tortoises (greater than 208 mm median carapace length) captured in 1990 in their sheltersites or out in the open. We used gel epoxy to affix radio-transmitters to their anterior marginal scutes. The transmitters weighed 47–53 g, measured 4.1 × 2.4 × 2.0 cm, and had an active life of nine to eighteen months. For additional identification, marginal scutes were notched following the system used on Bureau of Land Management tortoise monitoring plots in Arizona, California, and Nevada (Berry 1988). We determined the sex (male, female, unknown) of tortoises based on plastron indentation, tail morphology, and gular size (Woodbury and Hardy 1948).

We recaptured from five to twenty adult tortoises during each sampling trip; these trips were made to Little Shipp Wash and the Harcuvar Mountains twice in 1990 (September, November) and then three times a year (May, July, September) from 1991 through 1994. Upon recapture, each tortoise was weighed with a 5-kg Pesola scale and measured with a caliper and 24-cm ruler. We handled all tortoises with latex gloves, changed gloves between tortoises, and kept tortoises in clean individual cardboard boxes to minimize the probability of disease transfer among animals.

Tortoises were immobilized four to six hours after capture with 15 mg/kg of ketamine hydrochloride injected intramuscularly into a rear leg using a 25-gauge needle. Twenty minutes after immobilization, we collected 6.0 ml of whole blood by jugular venipuncture using a 22-gauge needle. We placed 0.6 ml of whole blood in a lithium heparin microtainer; the sample was mixed for five minutes, kept on ice, and sent to the laboratory within twenty-four hours. To determine packed cell volume, fibrinogen, and hemo-globin, we used several drops of whole blood to fill two heparinized micro-hematocrit capillary tubes. We also used whole blood to make two air-dried blood smears, which were sent within two days to APL Veterinary Laboratory in Las Vegas, Nevada. In the laboratory, smears were stained with modified Wright's stain and examined for white blood cell (WBC) estimate; differen-tial WBCs (heterophils, lymphocytes, monocytes, azurophils, eosinophils, and basophils); platelet estimate; red blood cell morphology; blood parasites; and evidence of anisocytosis (red cell size variation), polychromasia (variation in red cell hemoglobin content), and anemia. The WBC estimate was calculated by counting the number of WBCs in ten fields under a 500 × magnification and then multiplying the count by 2,000. We calculated the number of each WBC type (e.g., heterophil) by multiplying the percentage of each type by the WBC estimate.

We placed the remaining whole blood in a lithium heparin vacu-tainer to obtain plasma, mixed the sample for five minutes, and then cen-trifuged for five minutes. The plasma was pipetted off and divided into three aliquots. In 1990 we placed the plasma in red top vacutainers and placed them on dry ice, whereas from 1991 to 1994 we placed the plasma in cryogenic vials and immediately froze them in liquid nitrogen. We sent plasma samples on dry ice within two days to Animal Diagnostic Lab, Inc., in Tucson, Arizona.

The first plasma aliquot (1.0 ml) was analyzed for twenty-four blood variables: glucose, blood urea nitrogen, creatinine, uric acid, total protein, albumin, total globulins, bile acids, aspartate aminotransferase, alanine aminotransferase, alkaline phosphatase, calcium, cholesterol, triglycerides, total bilirubin, direct bilirubin, indirect bilirubin, phosphorus, sodium, po-tassium, chloride, total carbon dioxide, anion gap, and osmolality. Total glob-ulin was calculated using the following formula:

$$\text{total globulin} = \text{total protein} - \text{albumin}$$

The second aliquot (1.5 ml) was analyzed for vitamin A, vitamin E, cop-per, selenium, and zinc by the Arizona Veterinary Diagnostic Laboratory in Tucson, Arizona. Vitamin levels were measured by high-pressure chroma-tography, selenium levels by gas chromatography, and copper and zinc by atomic absorption. The third aliquot (1.0 ml) was analyzed for testosterone, estradiol, and corticosterone by radio-immunoassay at the San Diego Zoo, San Diego, California (Lance et al. 1985).

After sampling, we rehydrated tortoises to replace any fluids voided during handling by injecting a fluid volume equivalent to 1–2 percent body mass of equal parts Normosol and 2.5 percent dextrose in 0.45 percent sodium chloride into the body cavity of each tortoise with a 20-gauge needle. Tortoises were released at the point of capture during early morning of the day following health assessment, more than ten hours after injection of ketamine hydrochloride.

Diagnosing Disease

Knowing the normal ranges of human blood cholesterol enables your doctor to accurately diagnose your high cholesterol level — your reading was above the normal range. An important result of the AGFD health study was to establish normal reference ranges for red and WBC numbers, various blood chemistry parameters, and types of bacteria colonizing nasal and cloacal mucous membranes for free-ranging Sonoran tortoises (table 10.1). Reference ranges are values within two standard deviations of the mean and are considered normal for a healthy animal (Hoffman 1971). These values can assist researchers in pinpointing dehydration (elevated osmolality, reduced body mass), vitellogenesis (yolk formation; elevated cholesterol, triglycerides, vitamins A and E), and inflammation (elevated WBC counts), to name a few.

Blood Parameters

Blood parameters used to diagnose tortoise conditions and diseases are useful in assessing the physiological status of a population. In terms of physiological condition, increased levels of vitamins A and E and body mass suggests the tortoises are foraging. Elevated uric acid (a by-product of protein metabolism) together with increased body mass also indicates the tortoises are eating. During below-average rainfall, when tortoises become water deprived, there is a different blood chemistry scenario. Dehydrated tortoises have lower body mass, uric acid is usually decreased, and electrolytes (sodium and potassium) and osmolality levels are increased. In female

TABLE 10.1. Means, standard deviations, and reference ranges for body mass, median carapace length, and hematological and plasma biochemical parameters in Sonoran Desert tortoises from Little Shipp Wash ($n = 3$) and Harcuvar Mountains ($n = 23$), Arizona, in 1990–1994.

Parameter	Reference Range $\bar{x} \pm 2$ SD (n)
BODY MASS (kg)	
male	3.1 ± 0.68 (105)
female	3.1 ± 0.63 (98)
MEDIAN CARAPACE LENGTH (mm)	
male	258.8 ± 22.1 (105)
female	256.6 ± 18.1 (98)
PACKED CELL VOLUME FIELD (%)	
male	25.0 ± 3.6 (98)
female	24.3 ± 3.2 (85)
HEMOGLOBIN (g/dl)	
male	10.2 ± 1.5 (93)
female	10.3 ± 1.3 (85)
FIBRINOGEN (mg/dl)	
male	131.6 ± 32.2 (35)
female	130.2 ± 35.1 (43)
WHITE BLOOD CELL ESTIMATE (k/μl)	
male	8.5 ± 3.7 (80)
female	7.8 ± 3.7 (75)
HETEROPHILS (k/μl)	
male	582.3 ± 309.2 (79)
female	483.1 ± 307.3 (74)
LYMPHOCYTES (k/μl)	
male	100.7 ± 62.6 (81)
female	93.9 ± 64.8 (75)
MONOCYTES (k/μl)	
male	13.0 ± 16.6 (77)
female	9.9 ± 11.7 (72)
AZUROPHILS (k/μl)	
male	6.3 ± 12.3 (79)
female	4.2 ± 37.6 (75)
EOSINOPHILS (k/μl)	
male	26.5 ± 33.0 (75)
female	16.8 ± 23.4 (68)

TABLE 10.1. Continued

Parameter	Reference Range $\bar{x} \pm 2$ SD (n)
BASOPHILS (k/μl)	
male	83.2 ± 46.6 (82)
female	116.1 ± 79.4 (75)
GLUCOSE (mg/dl)	
male	132.6 ± 32.2 (49)
female	127.1 ± 34.9 (59)
BLOOD UREA NITROGEN (mg/dl)	
male	1.6 ± 2.4 (102)
female	1.1 ± 2.0 (92)
CREATININE (mg/dl)	
male	0.24 ± 0.13 (81)
female	0.25 ± 0.10 (80)
URIC ACID (mg/dl)	
male	4.8 ± 2.1 (46)
female	5.4 ± 1.8 (57)
TOTAL PROTEIN (g/dl)	
male	3.4 ± 0.43 (104)
female	3.9 ± 0.69 (96)
ALBUMIN (g/dl)	
male	1.7 ± 0.5 (109)
female	1.7 ± 0.5 (91)
TOTAL GLOBULINS (g/dl)	
male	1.7 ± 0.5 (107)
female	2.1 ± 0.5 (90)
BILE ACIDS (μmol/l)	
male	2.2 ± 3.1 (25)
female	2.1 ± 3.3 (36)
ASPARTATE AMINOTRANSFERASE (iu/l)	
male	73.2 ± 32.5 (67)
female	63.5 ± 21.0 (69)
ALANINE AMINOTRANSFERASE (iu/l)	
male	2.9 ± 2.0 (46)
female	2.7 ± 2.4 (59)

TABLE 10.1. Continued

Parameter	Reference Range $\bar{x} \pm 2$ SD (n)
ALKALINE PHOSPHATASE (iu/l)	
male	72.5 ± 29.4 (34)
female	107.3 ± 68.8 (45)
CALCIUM (mg/dl)	
male	10.3 ± 1.0 (103)
female	13.2 ± 1.5 (92)
CHOLESTEROL (mg/dl)	
male	77.1 ± 20.8 (106)
female	175.8 ± 56.9 (94)
TRIGLYCERIDES (mg/dl)	
male	18.7 ± 25.1 (108)
female	237.4 ± 187.7 (92)
TOTAL BILIRUBIN (mg/dl)	
male	0.1 ± 0.1 (34)
female	0.5 ± 0.4 (44)
DIRECT BILIRUBIN (mg/dl)	
male	0.02 ± 0.01 (33)
female	0.02 ± 0.01 (39)
INDIRECT BILIRUBIN (mg/dl)	
male	0.1 ± 0.1 (34)
female	0.5 ± 0.4 (44)
COPPER (ppm)	
male	0.6 ± 0.1 (53)
female	0.5 ± 0.2 (44)
SELENIUM (ppm)	
male	0.03 ± 0.01 (67)
female	0.04 ± 0.01 (49)
ZINC (ppm)	
male	1.9 ± 1.5 (11)
female	2.2 ± 1.5 (11)
VITAMIN A (μg/ml)	
male	0.4 ± 0.2 (91)
female	0.4 ± 0.2 (86)

TABLE 10.1. Continued

Parameter	Reference Range $\bar{x} \pm 2$ SD *(n)*
VITAMIN E (μg/ml)	
male	4.3 ± 3.6 (90)
female	7.9 ± 5.1 (82)
TESTOSTERONE (ng/ml)	
male	127.0 ± 106.8 (27)
female	1.5 ± 1.6 (33)
ESTRADIOL (ng/ml)	
male	
female	113.6 ± 102.7 (46)
CORTICOSTERONE (ng/ml)	
male	6.9 ± 5.4 (41)
female	5.1 ± 3.4 (48)
PHOSPHORUS (meq/l)	
male	1.6 ± 0.6 (86)
female	4.4 ± 1.9 (82)
SODIUM (meq/l)	
male	129.2 ± 6.8 (102)
female	130.4 ± 5.1 (92)
POTASSIUM (meq/l)	
male	4.1 ± 0.6 (99)
female	4.2 ± 0.5 (87)
CHLORIDE (meq/l)	
male	104.4 ± 7.8 (36)
female	101.9 ± 8.0 (45)
TOTAL CARBON DIOXIDE (meq/l)	
male	34.6 ± 6.0 (34)
female	33.0 ± 6.8 (44)
ANION GAP (meq/l)	
male	−2.6 ± 10.8 (35)
female	3.5 ± 10.0 (42)
OSMOLALITY (mos/kg)	
male	268.1 ± 22.9 (82)
female	275.8 ± 24.5 (82)

tortoises, higher levels of albumin, calcium, phosphorus, and cholesterol indicate ongoing egg formation. In male tortoises, high levels of aspartate aminotransferase, an enzyme released with cell damage or necrosis, in the spring and fall are possibly due to increased activity and male aggression associated with mating.

Blood parameters may also reveal disease. Increased numbers of WBCs (part of an animal's immune response) indicate an inflammation. Different types of WBCs tell different stories: Increased heterophils may indicate a bacterial infection or tissue injury, whereas increased lymphocytes may indicate wound healing, parasitic infections, or a viral disease. Liver disease in tortoises may be indicated by prolonged increases in the enzyme alanine aminotransferase and direct bilirubin (the orange-yellow bile pigment produced from the breakdown of hemoglobin in old red blood cells). High levels of phosphorus probably indicate decreased kidney function; elevated aspartate aminotransferase with decreased body mass may indicate a muscular catabolic process.

Peterson (1996a) found most of the variation in field metabolic rates in free-ranging Mohave desert tortoises was due to a single climatic factor — rainfall. We found a similar response: Seasonal and yearly changes in rainfall had a major effect on Sonoran desert tortoise blood chemistry values. The amount of rainfall determined the amount of forage available to the Sonoran tortoises and drove the water metabolism strategy.

The Sonoran Desert has rainfall in the winter and summer (Turner and Brown 1982). In our study, there were above- and below-average years in terms of rainfall. Above-average rainfall in 1992 and 1993 resulted in more available forage, whereas below-average rainfall in 1994 resulted in less available forage. Cable (1975) found high precipitation led to high perennial grass production. Increased body mass and elevated levels of vitamins A and E in 1992 suggests better-quality forage or more foraging by tortoises. All tortoises with elevated body mass and elevated levels of blood proteins and vitamins were observed foraging at least one week before blood collection (Vanessa M. Dickinson, unpubl. data). From 1991 to 1993 increased body mass and elevated uric acid in Little Shipp Wash tortoises compared to Harcuvar Mountain tortoises reflected greater foraging activity. Uric acid elevations have been associated with potassium and protein intake in Mohave tortoises (Christopher et al. 1997, 1999) and protein intake in carnivorous reptiles (Maixner et al. 1987). From 1991 to 1993 Sonoran tortoises did not exhibit seasonal reciprocal fluctuations in blood urea nitrogen and uric acid, as was reported for Mohave tortoises (Christopher et al. 1997), and blood levels of sodium and potassium (important ions for muscle contractions and nerve impulses) and blood osmolality were not affected.

A different blood chemistry scenario was found in Sonoran tortoises in 1994, the year of below-average rainfall. As tortoises became water deprived (and foraged less), they lost body mass and uric acid levels decreased, whereas blood urea nitrogen, electrolytes (sodium and potassium), and osmolality increased. Increased levels of cholesterol during 1994 were probably a result of fat catabolism. In addition, some Harcuvar Mountain female tortoises had higher levels of total bilirubin and indirect bilirubin in 1993, a year of below-average rainfall. This may reflect hepatic change resulting in reduced rates of free bilirubin conjugation in years of below-average rainfall in females whose physiology is affected by vitellogenesis. O'Conner et al. (1994) found higher levels of sodium, potassium, and osmolality in water-stressed captive Mohave tortoises. Similarly, Peterson (1996b) reported that free-ranging Mohave tortoises in drought conditions had increased urea, osmolality, and chloride.

Tortoise sex influenced seasonal changes in clinical chemistry. Most seasonal changes occurred when female tortoises were in vitellogenesis. According to O'Connor et al. (1994) and Rostal et al. (1994), female desert tortoises had higher levels of albumin, calcium, phosphorus, and cholesterol compared to males. Elevated calcium levels have been associated with vitellogenesis (Ho 1987). Taylor and Jacobson (1982) associated higher levels of cholesterol with vitellogenesis in female gopher tortoises *(Gopherus polyphemus)*. Higher levels of cholesterol and lipids were observed in gravid female Mediterranean tortoises *(Testudo graeca* and *T. hermanni)* in August (time of nesting) compared to males (Lawrence 1987). In addition, Palmer and Guillette (1990) found increased levels of estrogen during vitellogenesis in free-ranging gopher tortoises. Vitamin A and E levels, first reported by the AGFD health study, may also be influenced by vitellogenesis.

The observed peaks in cholesterol and triglycerides in May and September suggest that female Sonoran desert tortoises may be in vitellogenesis from September through May. We suspect vitellogenesis is continuous from September until May and not biphasic during the calendar year. Rostal et al.

(1994) found captive Mohave tortoises in vitellogenesis in spring and fall, as indicated by increased calcium levels and follicular growth. Unlike Mohave tortoises, Sonoran tortoises are monestrous, laying a single clutch of eggs each year (Murray et al. 1996; Averill-Murray et al., this vol., ch. 7).

Higher plasma levels of aspartate aminotransferase (an enzyme contained in skeletal muscle as well as other tissues) in male tortoises possibly resulted from increased activity and male aggression associated with mating. O'Connor et al. (1994) suggested an increase in this enzyme in male desert tortoises was associated with mating or fighting in the spring. Male tortoises have been observed fighting in the fall in the Sonoran Desert (Vaughan 1984).

Bacteria

Few studies have reported on desert tortoise bacteria (Fowler 1977; Snipes and Biberstein 1982; Jackson and Needham 1983; Jacobson et al. 1991). Most of our information on Sonoran tortoise bacteria comes from the AGFD health study. This study's objectives were to collect baseline data on aerobic bacteria found on nasal and cloacal epithelia of free-ranging desert tortoises; determine if differences occurred as a result of site, sex, season; and determine if we could differentiate between ill and healthy tortoises based on bacteria types.

In the AGFD health study, the cloacal and choanal (nasal) regions of captured tortoises were swabbed and the swabs stored in transtubes. The swabs were kept on ice and sent to the Arizona Veterinary Diagnostic Laboratory within twenty-four hours for bacterial isolation. In the laboratory, bacterial cultures were grown using MacConkey agar (gram-negative bacteria), selenite agar (*Salmonella* spp., *Shigella* spp.), Hektoen agar (*Salmonella* spp., *Shigella* spp.), and Campylobacter agar (*Campylobacter* spp.). Bacteria were classified as gram positive or gram negative and, when possible, identified to species.

Five species of bacteria were found in the nasal cavity (table 10.2), with two potential pathogens: *Pasteurella testudinis* and *Pseudomonas* spp. Interestingly, there were higher levels of *P. testudinis* in 1991 compared to all other years. In health studies of Mohave populations, tortoises with signs of upper respiratory tract disease (URTD) had significantly higher levels of *P. testudinis* in their nasal cavities than healthy tortoises. Jacobson et al. (1991)

TABLE 10.2. Bacteria isolated from the nasal cavity of Sonoran Desert tortoises from Little Shipp Wash and Harcuvar Mountains, Arizona, 1990–1991. Bacteria were collected with nasal flushes.

Organism	Number	n
Corynebacterium spp.	2	25
Pasteurella spp.	12	26
Pseudomonas spp.	5	6
Staphylococcus spp.[a]	4	20
Streptococcus spp.	1	6

[a] Harcuvar Mountain tortoise only.

found eleven species of bacteria in the respiratory tract (nasal cavity, trachea, lung) of ill and healthy desert tortoises. Most respiratory tract bacteria were found in the choanae; *P. testudinis* was only found in the nasal cavity (Jacobson et al. 1991). Christopher et al. (1993) found high heterophil counts in Mohave tortoises with heavy growth of *P. testudinis* in the nasal cavity. Snipes and Biberstein (1982) found *P. testudinis* in ill and healthy captive desert tortoises and reported the "bacterium appears commensal in healthy free-ranging desert tortoises."

The AGFD health study isolated seventeen cloacal bacteria, two of which were considered pathogens (*Pseudomonas* spp., *Salmonella* spp.; table 10.3). Jacobson (1987) found four bacterial pathogens (*Morganella* spp., *Pseudomonas* spp., *Proteus* spp., and *Providencia* spp.) accounted for 44 percent of the bacterial isolates, mostly from diseased snakes. *Salmonella* spp. has been frequently cultured from reptilian abscesses (Frye 1981; Marcus 1981).

Pasteurella testudinis, *Pseudomonas* spp., and *Salmonella* spp. are the least debatable potential pathogens found on epithelial swabs in the AGFD health study. *Citrobacter* spp., *Corynebacterium* spp., *Enterobacter-Klebsiella*, *Flavobacterium* spp., *Pasteurella* spp. (other than *P. testudinis*), *Proteus* spp., *Staphylococcus* spp., *Streptococcus* spp., and possibly *Campylobacter* spp. and *Shigella* spp. can, under certain circumstances, be pathogenic to chelonians (Jacobson et al. 1998).

The higher incidence of *P. testudinis*, *Pseudomonas* spp., and *Salmo-*

TABLE 10.3. Bacteria isolated from the cloacal cavity of Sonoran Desert tortoises from Little Shipp Wash and Harcuvar Mountains, Arizona, in 1990–1994. Bacteria were collected with transtube cloacal swabs.

Organism	Number Positive	n
Bacillus spp.	13	201
Campylobacter spp.	9	201
Citrobacter spp.	10	201
Coliforms	12	201
Corynebacterium spp.[a]	13	105
Diptheroids	39	201
Enterobacter-klebsiella	66	201
Escherichia coli	31	201
Lactobacillus spp.	18	201
Pasteurella spp.	7	201
Proteus spp.[a]	3	105
Pseudomonas spp.	58	201
Salmonella spp.	18	201
Shigella spp.	42	201
Staphylococcus spp.	151	201
Streptococcus spp.	17	201

[a] Harcuvar Mountain tortoise only.

nella spp. in ill tortoises demonstrates their usefulness in identifying individuals with underlying pathology. These pathogens appear to be truly opportunistic; that is, the replacement of normal, nonpathogenic flora (e.g., bacteria and fungi) on the nasal and cloacal cell surfaces by these potential pathogens is highly indicative of debility and suppression of the immune system resulting from underlying disease processes, including URTD. Epithelial colonization by these opportunists may lead to further pathology, debilitation, and death if the tortoise is unable to recover from the disease.

Diseases of the Sonoran Tortoise

Mycoplasmosis, or Upper Respiratory Tract Disease

The mycoplasmas (class Mollicutes, order Mycoplasmatales) are the smallest free-living organisms. Unlike bacteria, they have no cell wall but are surrounded by a plasma membrane. The class Mollicute contains more than forty pathogens of humans and other animals including the causative agents of three respiratory diseases formerly thought to be due to viruses. URTD in desert tortoises was originally attributed to the bacteria genus *Pasteurella*. A landmark clinical study in 1994 showed that the cause was really a mycoplasma, later named *Mycoplasma agassizii* (Brown et al. 1994).

There are two ways to diagnose tortoise URTD: (1) sample the blood plasma with an enzyme-linked immunosorbent assay (ELISA) for antibodies assembled by the tortoise to combat the disease; and (2) flush the nares and test the sample genetically for the mycoplasma. The latter is known as a polymerase chain reaction (PCR) test and is specific for the 16S ribosomal ribonucleic acid (rRNA) gene sequence of *M. agassizii*. PCR is a laboratory technique that can replicate RNA or DNA amino acid sequences into millions of copies in a very short time, and the sequence patterns can be compared electrophoretically. In the AGFD health study, tortoise plasma was analyzed by an ELISA as described by Schumacher et al. (1993) for two years (1993–1994), and tortoise nasal aspirate by a PCR test as described by Brown and Brown (1994) for one year (1994). All laboratory analyses were conducted at the University of Florida, where the causative agent of URTD was first identified.

Healthy tortoises were defined as animals with no clinical signs of URTD and negative ELISA titers for *M. agassizii*, and ill tortoises are those with clinical signs of URTD and/or positive ELISA titers for *M. agassizii*. Clinical signs of URTD are nasal discharge, ocular discharge, conjunctivitis, and palpebral edema (eyelid swelling; Jacobson et al. 1991); more than one of these signs indicates URTD. There were no observable signs of URTD at either Little Shipp Wash or the Harcuvar Mountains.

In the AGFD health study, three tortoises tested positive for *M. agassizii*, the causative agent of URTD. Harcuvar tortoise 208 had a positive ELISA and a positive PCR, Little Shipp tortoise 500 had a negative ELISA but a positive PCR, and Harcuvar tortoise 203 was identified as suspect ELISA and negative PCR. Conflicting ELISA and PCR results could be due to a poor nasal flush (false negative PCR); a new infection (false negative ELISA); or the tortoise not having developed an immune response, at least at the time it was tested. In addition, other strains of *Mycoplasma* live in tortoises, and cross-reactions with nonpathogenic strains could produce weak or suspect ELISA reactions.

The average duration of URTD in free-ranging tortoises is debatable. In the five years of another AGFD health study on Mohave tortoises, several free-ranging tortoises had signs of URTD for two years (Dickinson et al. 1995). Jacobson et al. (1991) reported URTD in free-ranging tortoises was chronic in nature and lasted up to a year before the tortoises eventually died. He also reported that ill tortoises in captivity "may survive for several years before succumbing to systematic disease." This does not always appear to be the case in free-ranging desert tortoises, and more long-term health studies may find tortoises with URTD living longer. Interestingly, Sonoran tortoises have a low incidence of overt *M. agassizii* infection (as determined by positive ELISA) compared to Mohave tortoises. Comparing the two AGFD health studies, eleven Mohave tortoises had positive ELISA tests in addition to signs of URTD, whereas only one Sonoran tortoise showed a definite positive ELISA and no Sonoran tortoise showed signs of URTD, in spite of the documented existence of *M. agassizii* (by PCR) in the Sonoran study plots.

Shell Disease

Cutaneous dyskeratosis, a shell disease of desert tortoises, was originally referred to as *shell necrosis*. This condition of the carapace, plastron, and thickened scales of the forelimbs of the tortoise was first described in the Mohave Desert near Riverside, California (Jacobson et al. 1994). Between 1982 and 1988, a population of Mohave tortoises on the Chuckwalla Bench experienced a mortality rate of 70 percent, a die-off associated with a shell disease. The disease actually existed earlier because photographs taken in 1979 reveal shells with the characteristic condition—affected portions are grayish white, dry, roughened, and flaky. The lesions appear to start at the scute seams and spread outward toward the middle of the scute. Histological findings of the Mohave tortoises suggested that either a nutritional deficiency or toxicosis (i.e., toxins such as arsenic, mercury, selenium) could be responsible for these shell lesions (Jacobson et al. 1994). Still, the exact cause of shell disease could not be determined from this study.

In Sonoran tortoises, shell disease has been monitored on the Sonoran Desert monitoring plots since 1992 (table 10.4; Roy C. Averill-Murray, unpubl. data). In the AGFD health study, detailed notes on shell disease were not taken until 1992, when word of shell necrosis reached researchers in Arizona. Still, we were able to identify individual tortoises with signs of shell disease as early as 1990 based on the field drawings of shells. Of the thirty-six individual tortoises studied from 1990 to 1995, only five tortoises from Little Shipp Wash showed signs of shell disease (Vanessa M. Dickinson, unpubl. data). All five individuals were females (estimated ages greater than thirty years) and were still living when the health study ended. Shell disease may not be a serious problem among Sonoran tortoises, but additional study is warranted regarding its cause in Mohave and Sonoran tortoises.

Parasites of Sonoran Tortoises

The majority (95 percent) of tortoises in the AGFD health study had pinworms (order Oxyurida). This is unremarkable, because pinworms are among the most common and numerous intestinal worms of lizards and turtles (Marcus 1981). The host can reinfect itself by inhaling or ingesting pinworm ova. Tortoises share burrows and have been documented eating soil and scat (Esque and Peters 1994), activities that contribute to continued reinfection. What is remarkable is that we did not find any other intestinal parasites using the fecal flotation technique. Glassman et al. (1979) found higher numbers of eosinophils in alligators *(Alligator mississippiensis)* with intestinal leeches. We found that tortoises with pinworms did not have increased numbers of eosinophils in their blood. In addition, external parasites such as ticks were extremely rare.

A fecal analysis study conducted in 1996 (Jeanine Baker, unpubl. data) of eleven Sonoran tortoises from the Meyer Ranch near Florence, Arizona, also found pinworm ova in all the samples, in addition to *Entamoeba* spp. (a protozoan). Interestingly, an unknown ova found in eight of the eleven fecal samples appears to be an oospore of a new species of fungus. The Chytridiomycete fungi found in the gut of herbivorous mammals have very strong cellulolytic ability and are very active in forage digestion, especially of the lignocellulose of cell walls (Li and Heath 1993). Currently, a chytrid

TABLE 10.4. Summary of cutaneous dyskeratosis (CD) at Sonoran Desert tortoise monitoring plots in Arizona. AD, adults ($\geq$ 180 mm carapace length); JV, juveniles (< 180 mm carapace length).

		Adults			Juveniles			Total		
Year	Plot	CD	Total	%	CD	Total	%	CD	Total	%
1992	Granite Hills	0	45	0	0	30	0	0	75	0
	Little Shipp Wash	14	76	18.4	0	12	0	14	88	15.9
	Eagletails	0	22	0	0	5	0	0	27	0
	Bonanza Wash	7	14	50.0	0	3	0	7	17	41.2
	Tortillas	3	49	6.1	0	3	0	3	52	5.8
	Mazatzals	2	46	4.4	0	5	0	2	51	3.9
1993	Granite Hills	1	55	1.8	0	40	0	1	95	1.1
	Little Shipp Wash	32	83	38.6	1	20	5.0	33	103	32.0
	Eagletails	1	23	4.4	0	14	0	1	37	2.7
	Harcuvars	7	44	15.9	0	2	0	7	46	15.2
	Eastern Bajada	30	43	69.8	0	3	0	30	46	65.2
1994	Granite Hills	15	60	25.0	8	49	16.3	23	109	21.1
	Little Shipp Wash	24	61	39.3	0	16	0	24	77	31.2
	Eagletails	4	28	14.3	0	19	0	4	47	8.5
	Harquahalas	12	17	70.6	0	2	0	12	19	63.2
1995	West Silver Bells	7	74	9.5	0	16	0	7	90	7.8
	San Pedro Valley	1	84	1.2	0	6	0	1	90	1.1
	Mazatzals	20	49	40.8	3	17	17.7	23	66	34.9
1996	Hualapai Foothills	5	37	13.5	0	2	0	5	39	12.8
	Tortillas	5	60	8.3	2	12	16.7	7	72	9.7
1997	Arrastras	1	13	7.7	0	1	0	1	14	7.1
	Bonanza Wash	5	10	50.0	0	3	0	5	13	38.5
	Eastern Bajada	28	43	65.1	1	2	50.0	29	45	64.4
	Harcuvars	1	50	2.0	0	4	0	1	54	1.9

fungus is being investigated by AGFD as a cause of mortality in declining leop-
ard frog *(Rana yavapaiensis)* populations. Chytrid oospores were found in
fecal pellets from healthy Sonoran tortoises. Additional research is needed
to establish the role of these fungi in the gut flora of the desert tortoise and
other herbivorous reptiles.

Discussion

So, are Sonoran tortoises healthy? In our opinion, the health of Sonoran tortoises is good. Disease is apparently not a major factor limiting
populations of this tortoise. The negative health conditions encountered (dehydration, starvation, liver and kidney disease) are products of below-average
rainfall. The bacteria and parasites found in the tortoises are either widespread *(Pseudomonas* sp.) or probably evolved with the tortoises *(M. agassizii,
P. testudinis,* and *Salmonella* sp.). At this time, there are no confirmed deaths
in free-ranging Sonoran tortoises due to *M. agassizii.* This is in contrast to
the Mohave tortoise populations, in which mycoplasmosis is a mortality factor. In the AGFD health studies, 55 percent of the Mohave tortoises (eleven
of twenty sampled) had either URTD or antibodies to *M. agassizii,* although
no Sonoran tortoise exhibited signs of URTD, and only 9 percent (2 of 22
sampled) showed any antibody response to *M. agassizii.* No external parasites were found on Sonoran tortoises during the AGFD health studies, although a tick *(Ornithodoros turicata)* from California desert tortoises may
carry spirochaetes of public health significance (Ryckman and Kohls 1962).

Sonoran tortoises may be faring better than their northern neighbors for several reasons. One primary difference is that Mohave tortoises
construct more substantial burrows in valley-bottom habitats, freeing them
from the sheltersite limitations of the rocky slope habitats of the Sonoran
tortoise. Tortoises in such high-density populations in relatively large, continuous areas are likely to share more diseases and parasites. Mohave tortoise
habitat is being lost and impacted much faster than Sonoran tortoise habitat
(Howland and Rorabaugh, ch. 14 of this volume).

Another important factor in the apparent health differences between
the Sonoran and Mohave tortoises is the climate. The Sonoran tortoise lives
in a biseasonal rainfall regime with opportunities to rehydrate during both

winter-spring and summer rains. Lowe (1990) suggested that Mohave tortoises may be stressed because they are at the northern limit of the tortoise range, where summers are hotter, winters are colder, and rainfall is limited to a single rainfall season. Peterson (1996b) reached similar conclusions that Mohave tortoises are on the edge physiologically and vulnerable to catastrophic population declines during severe droughts. Van Devender (ch. 2 of this volume) suggested that the Mohave tortoise is likely the most recently evolved North American tortoise and the only one to have lived in a winter-rainfall Mediterranean climate with regular droughts of eighteen months and occasional droughts of more than thirty months.

The continued good health of the Sonoran tortoise population is in human hands. If people can limit habitat destruction and the introduction of new diseases and parasites from reintroduced captive tortoises, Sonoran tortoises should persist indefinitely. Tortoise biologists now have normal reference values for hematological and biochemical tests that allow them to evaluate the health of free-ranging desert tortoises. Nasal and cloacal bacterial screens, however, are the simplest, least invasive, and least expensive way to assess health in a tortoise population and to identify individuals that are in some way debilitated. Bacterial screens would not identify the cause of debility, but could provide rapid comparisons of the relative health status of different populations of desert tortoises. For future physiological assessment of Sonoran tortoise populations, we recommend designing and implementing appropriate health monitoring programs and establishing a standard protocol to ensure the data is comparable.

ACKNOWLEDGMENTS

We thank assistants John Snider, Sue Trachy, and Diana Parmley; interns Alyssa Carnahan, Josh Hurst, Petra Lowe, and Laura Sychowski; and many volunteers for data collection. Financial support for this project was provided by the Bureau of Land Management, Arizona Game and Fish Department Heritage Fund, and the U.S. Fish and Wildlife Service. Tom Van Devender helped edit the manuscript.

LITERATURE CITED

Berry, K. H. 1988. Bureau of Land Management's techniques manual for collecting and ana-lyzing data on desert tortoise populations and habitats. Bureau of Land Management, Riverside, Calif.

Brown, D. R., and M. B. Brown. 1994. The 16S rRNA gene identifies *Mycoplasma agassizii* isolated from ill desert tortoises. *Proceedings of the Desert Tortoise Council Symposium* 1994:146.

Brown, D. R., I. M. Schumacher, P. A. Klein, K. Harris, T. Correll, and E. R. Jacobson. 1994. *Mycoplasma agassizii* causes upper respiratory tract disease in the desert tortoise. *Infectious Immunity* 62:4580–86.

Cable, D. R. 1975. Influence of precipitation on perennial grass production in the semidesert Southwest. *Ecology* 56:981–86.

Christopher, M. M., I. Wallis, K. A. Nagy, B. T. Henen, C. C. Peterson, B. Wilson, C. Meien-berger, and I. Girard. 1993. *Laboratory health profiles of free-ranging desert tortoises in California: Interpretation of physiological and pathological alterations, March 1992–October 1992*. Bureau of Land Management, Riverside, Calif.

Christopher, M. M., R. Brigmon, and E. R. Jacobson. 1994. Seasonal alterations in plasma B-hydroxybutyrate and related biochemical parameters in the desert tortoise *(Gopherus agassizii)*. *Comparative Biochemistry and Physiology* 108:303–10.

Christopher, M. M., K. A. Nagy, I. Wallis, J. K. Klaassen, and K. H. Berry. 1997. Labora-tory health profiles of desert tortoises in the Mojave Desert: A model for health status evaluation of chelonian populations. Pp. 76–82 *in* J. Van Abbema, ed. *Proceedings: Con-servation, restoration, and management of tortoises and turtles—An international conference.* New York Turtle and Tortoise Society, New York.

Christopher, M. M., K. H. Berry, I. R. Wallis, K. A. Nagy, B. T. Henen, and C. C. Peterson. 1999. Reference intervals and physiologic alterations in hematologic and biochemical values of free-ranging desert tortoises in the Mojave Desert. *Journal of Wildlife Diseases* 35:212–38.

Dickinson, V. M., T. Duck, C. R. Schwalbe, and J. L. Jarchow. 1995. *Health studies of free-ranging Mojave desert tortoises in Utah and Arizona*. Technical Report no. 21, Arizona Game and Fish Dept., Phoenix.

Dickinson, V. M., J. L. Jarchow, and M. H. Trueblood. 1996. *Health studies of free-ranging So-noran desert tortoises in Arizona*. Technical Report no. 24, Arizona Game and Fish Dept., Phoenix.

Esque, T. C., and E. L. Peters. 1994. Ingestion of bones, stones, and soil by desert tortoises. Pp. 105–11 *in* R. B. Bury and D. J. Germano, eds. *Biology of North American tortoises.* Fish and Wildlife Research no. 13, U.S. Dept. of the Interior National Biological Survey, Washington, D.C.

Fowler, M. E. 1977. Respiratory diseases in desert tortoises. Pp. 79–99 *in* M. E. Fowler, ed. *Annual proceedings of the American Association of Zoo Veterinarians.* Davis, Calif.

Frye, F. L. 1981. *Biomedical and surgical aspects of captive reptile husbandry.* Veterinary Medical Publications, Edwardsville, Kans.

Germano, D. J., R. B. Bury, T. C. Esque, T. H. Fritts, and P. A. Medica. 1994. Range and habitats of the desert tortoise. Pp. 73–84 *in* R. B. Bury and D. J. Germano, eds. *Biology of North American tortoises.* Fish and Wildlife Research no. 13, U.S. Dept. of the Interior National Biological Service, Washington, D.C.

Glassman, A. B., T. W. Holbrook, and C. E. Bennett. 1979. Correlation of leech infestation and eosinophilia in alligators. *Journal of Parasitology* 65:323–24.

Ho, S. 1987. Endocrinology of vitellogenesis. Pp. 145–69 *in* D. Norris and R. Jones, eds. *Hormones and reproduction in fishes, amphibians, and reptiles.* Plenum, New York.

Hoffman, R. G. 1971. *Establishing quality control and normal ranges in the clinical laboratory.* Exposition, New York.

Jackson, O. F., and J. R. Needham. 1983. Rhinitis and virus antibody titers in chelonians. *Journal of Small Animal Practitioners* 24:31–36.

Jacobson, E. R. 1987. Reptiles. *Veterinary Clinics of North America* 17:1203–25.

Jacobson, E. R., J. M. Gaskin, M. B. Brown, R. K. Harris, C. H. Gardiner, J. L. LaPointe, H. P. Adams, and C. Reggiardo. 1991. Chronic upper respiratory disease of free-ranging desert tortoises *(Xerobates agassizii). Journal of Wildlife Diseases* 27:296–316,

Jacobson, E. R., T. J. Wronski, J. Schumacher, C. Reggiardo, and K. H. Berry. 1994. Cutaneous dyskeratosis in free-ranging desert tortoises, *Gopherus agassizii,* in the Colorado Desert of Southern California. *Journal of Zoo and Wildlife Medicine* 25:68–81.

Jacobson, E. R., J. L. Behler, and J. L. Jarchow. 1998. Health assessment of chelonians and release into the wild. Pp. 232–42 *in* M. E. Fowler, and R. E. Miller, eds. *Zoo and wild animal medicine: Current therapy.* Vol. 4. W. B. Saunders Co., Philadelphia, Pa.

Lance, V., K. A. Vliet, and J. L. Bolaffi. 1985. Effect of mammalian luteinizing hormone-releasing hormone on plasma testosterone in male alligators, with observations on the nature of alligator hypothalamic gonadotropin releasing hormone. *General Comparative Endocrinology* 60:138–43.

Lawrence, K. 1987. Seasonal variation in blood biochemistry of long-term captive Mediterranean tortoises *(Testudo graeca* and *T. hermanni). Research in Veterinary Science* 43:379–83.

Li, J., and I. B. Heath. 1993. Chytridiomycetous gut fungi, oft overlooked contributors to herbivore digestion. *Canadian Journal of Microbiology* 3:1003–13.

Lowe, C. H. 1990. Are we killing the desert tortoise with love, science, and management? Pp. 84–102 *in* K. R. Beaman, F. Chaporaso, S. McKeown, and M. D. Graff, eds. *Proceedings of the first international symposium on turtles and tortoises: Conservation and captive husbandry.* Chapman University, Chapman, Calif.

Maixner, J. M., E. C. Ramsay, and L. H. Arp. 1987. Effects of feeding on serum uric acid in captive reptiles. *Journal of Zoo and Wildlife Medicine* 18:62–65.

Marcus, L. C. 1981. *Veterinary biology and medicine of captive amphibians and reptiles.* Lea and Febiger, Philadelphia, Pa.

Murray, R. C., C. R. Schwalbe, S. J. Bailey, S. P. Cuneo, and S. D. Hart. 1996. Reproduction in a population of the desert tortoise, *Gopherus agassizii,* in the Sonoran Desert. *Herpetological Natural History* 4:83–88.

Nagy, K. A., and P. A. Medica. 1986. Physiological ecology of desert tortoises in southern Nevada. *Herpetologica* 42:73–92.

O'Connor, M. P., J. S. Grumbles, R. H. George, L. C. Zimmerman, and J. R. Spotila. 1994.

Potential hematological and biochemical indicators of stress in free-ranging desert tortoises and captive tortoises exposed to a hydric stress gradient. *Herpetological Monograph* 8:5–26.

Palmer, B. D., and L. J. Guillette Jr. 1990. Morphological changes in the oviductal endometrium during the reproductive cycle of the tortoise, *Gopherus polyphemus*. *Journal of Morphology* 204:323–33.

Peterson, C. C. 1996a. Anhomeostasis: Seasonal water and solute relations in two populations of tortoise *(Gopherus agassizii)* during chronic drought. *Physiological Zoology* 69:1324–58.

———. 1996b. Ecological energetics of the desert tortoise *(Gopherus agassizii)*: Effects of rainfall and drought. *Ecology* 77:1831–44.

Rosskopf, W. J. 1982. Normal hemogram and blood chemistry values for California desert tortoises. *Veterinary Medicine/Small Animal Clinicians* 1:85–87.

Rostal, D. D., V. A. Lance, J. S. Grumbles, and A. C. Alberts. 1994. Seasonal reproductive cycle of the desert tortoise *(Gopherus agassizii)* in the eastern Mojave Desert. *Herpetological Monograph* 8:72–82.

Ryckman, R. E., and G. M. Kohls. 1962. The desert tortoise, *Gopherus agassizii*, in California. *Journal of Parasitology* 48:502–3.

Schumacher, I. M., M. B. Brown, E. R. Jacobson, B. R. Collins, and P. A. Klein. 1993. Detection of antibodies to a pathogenic mycoplasma in desert tortoises *(Gopherus agassizii)* with upper respiratory tract disease. *Journal of Clinical Microbiology* 31:1454–60.

Snipes, K. P., and E. L. Biberstein. 1982. *Pasteurella testudinis* sp. nov.: A parasite of desert tortoises. *International Journal of Systematic Bacteriology* 32:201–10.

Taylor, R. W., and E. R. Jacobson. 1982. Hematology and serum chemistry of the gopher tortoise, *Gopherus polyphemus*. *Comparative Biochemical Physiology* 72:425–28.

Turner, R. M., and D. E. Brown. 1982. Sonoran desertscrub. *Desert Plants* 4:121–81.

U.S. Fish and Wildlife Service. 1990. Endangered and threatened wildlife and plants; determination of threatened status for the Mojave population of the desert tortoise. *Federal Register* 55:12178–91.

Vaughan, S. L. 1984. Home range and habitat use of the desert tortoise, *Gopherus agassizii*, in the Picacho Mountains, Pinal Co., Arizona. M.S. thesis, Arizona State University, Tempe.

Woodbury, A. M., and R. Hardy. 1948. Studies of the desert tortoise, *Gopherus agassizii*. *Ecological Monographs* 18:146–200.

CHAPTER 11

Growth of Desert Tortoises

Implications for Conservation and Management

DAVID J. GERMANO, F. HARVEY POUGH, DAVID J. MORAFKA, ELLEN M. SMITH, AND MICHAEL J. DEMLONG

Desert tortoises *(Gopherus agassizii)* are slow-growing and long-lived animals. Newly hatched tortoises require fifteen to twenty years to reach sexual maturity, and for the first five or six years their small size (fig. 11.1) leaves them vulnerable to many predators. Adult desert tortoises are large enough to be safe from most natural predators, and individuals that survive to age 20 have a good chance of living for another twenty years or more. This long life expectancy is a critical element in the life history of desert tortoises, because the annual reproductive potential of female desert tortoises is low and mortality of juveniles before they reach sexual maturity is high. In the Sonoran Desert, tortoises lay a single clutch per year of one to twelve eggs, with an average of about five (Averill-Murray et al., this vol., ch. 6). These factors are offset by a reproductive period that may last twenty years or more. Thus, adult longevity must compensate for low reproductive rates, high juvenile mortality relative to adults, and slow maturation to ensure that the lifetime reproductive success of adult desert tortoises is sufficient to maintain tortoise populations.

Conservation plans for desert tortoises must include an understanding of the normal rates of growth of free-ranging tortoises and of the ecological, behavioral, and nutritional requirements of both juveniles and adults. This chapter focuses on the importance of growth rates in the context of life history and conservation of tortoises by reviewing natural growth rates and experimental manipulations of growth in captivity.

Figure 11.1. A neonate Sonoran desert tortoise feeding on filaree *(Erodium cicutarium)* near Vail, Rincon Mountains, Pima County, Arizona. (Photograph by Cecil Schwalbe)

Growth of Free-Ranging Tortoises in the Sonoran Desert

In the Sonoran Desert the desert tortoise lives in habitats that range from the extremely arid lower Colorado River Valley in California and Arizona to the relatively mesic Plains of Sonora in Sonora, Mexico (Germano et al. 1994). The habitat in the north intergrades with the Mohave Desert and in the south into foothills thornscrub and tropical deciduous forest (Germano et al. 1994). Growth of tortoises can be influenced by proximate causes (environmental conditions; Andrews 1982) and evolutionary causes (a genetic component to growth). Like other tortoises, those of the Sonoran Desert carry a history of past growth on their shells in the form of growth rings. Growth rings can be used to age many individuals in a population and to assess past variability in growth (Landers et al. 1982; Zug 1991; Germano 1992). Germano (1998) and Germano and Bury (1998) reviewed the use of scute rings to estimate age.

In this section we present data and results of an analysis by one of us (David J. Germano) on growth of free-ranging desert tortoises from the Sonoran Desert. From these data, equations of growth for male and female tortoises were constructed and age at first reproduction of female tortoises and longevity of adults were estimated.

Growth Measurements

Using tree calipers, whole-shell measurements were recorded to the nearest 1 mm from live and preserved tortoises from the Sonoran Desert of Arizona and Mexico. Shell measurements included straight-line carapace length (CL) taken along the shell from nuchal to supracaudal, maximum plastron length, and greatest shell height. In addition to shell measurements, the individual's age was determined using counts of scute layers from the carapace and plastron. Scute layers (rings) have been found to match age (annuli) in desert tortoises until about twenty-five years of age (Germano 1988, 1992); some tortoises could be classified only as older than twenty-five years because scute rings were worn and the animal was large, indicative of a tortoise that is no longer depositing large, countable rings.

Growth curves for desert tortoises were constructed from CLs using the Richards' (1959) growth model. This growth curve is a general mathematical model of growth that does not make assumptions about the shape of the curve. Curves for males and females were constructed using data for juveniles added to each sex. The Richards' growth model estimates three parameters using CL and age data: M, the shape of the growth curve; K, the growth constant; and I, the point at which curve inflection begins. The model uses the general formula:

$$CL = \textit{asymptotic size}[1 + (M - 1)e^{-K(age - I)}]^{[1/(1 - M)]}$$

to solve for CL at various ages. The CL at which the curve reaches its asymptotic size was preset in the equations to match the upper 10 percent of the samples (upper decile CL; Germano 1994a). Comparisons of growth of male and female tortoises were made for mean and upper decile CLs of adults and calculated CL as determined from the growth equations.

These calculations show that male and female Sonoran desert tortoises grow at similar rates (fig. 11.2). Both sexes require almost seven years

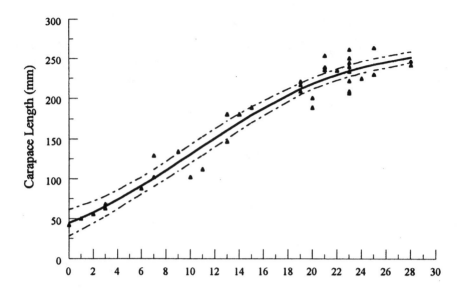

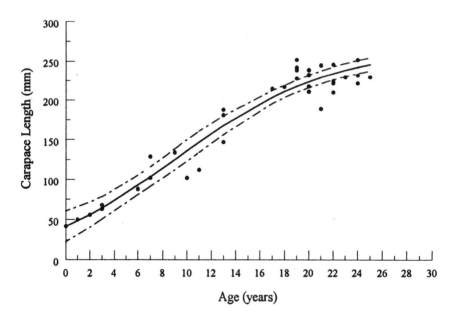

Figure 11.2. Growth curves of free-ranging male (top) and female (bottom) tortoises in the Sonoran Desert. The dashed curves represent 95 percent confidence intervals.

TABLE 11.1. Mean and upper decile carapace lengths of male and female Sonoran desert tortoises. Calculated carapace lengths (95% confidence interval) at various ages were determined from growth equations for each sex (see text).

	Male	Female
MEAN CARAPACE LENGTH (MM)	263.9	258.1
UPPER DECILE CARAPACE LENGTH (MM)	270.8	269.8

	Calculated Carapace Length (mm)	
Age (years)	Male	Female
o (Hatchling)	44.7 (28.1–61.3)	47.7 (22.8–60.6)
5	82.3 (71.4–93.8)	82.8 (70.7–96.1)
10	131.1 (119.3–142.8)	136.3 (122.9–149.7)
15	180.3 (170.6–190.0)	187.3 (175.9–196.1)
20	218.5 (211.5–225.6)	224.1 (216.1–232.1)
25	242.9 (235.5–250.4)	246.3 (237.9–254.7)

to reach 100 mm CL, and females require almost seventeen years and males almost eighteen years to reach 200 mm. The growth of the Sonoran desert tortoise is similar to that of the Texas tortoise (*Gopherus berlandieri*; the smallest tortoise in North America) for their first twenty years, but desert tortoises ultimately attain a much larger size (Germano 1994a). Both sexes reach about 270 mm maximum CL (table 11.1), which is similar to the largest mean size of desert tortoises from the western Mohave Desert and tropical deciduous forest (Germano 1994a). The Richards' growth equations for each sex are:

$$\text{male CL} = 270.8 \times [1 + 0.861e^{-0.145(T - 10.06)}]^{-1.161}$$

and

$$\text{female CL} = 269.8 \times [1 + 0.709e^{-0.149(T - 9.13)}]^{-1.410}$$

where CL is carapace length in millimeters and T is age in years. The model fits the data well with a coefficient of determination of 0.945 for males and 0.933 for females. Growth analyses of three populations of desert tortoises in Arizona (Murray and Klug 1996) gave similar parameter values to what we found using tortoises throughout the Sonoran Desert.

Based on the above equations, males are estimated to hatch at about 45 mm CL and females at about 42 mm CL (table 11.1). These estimates correspond to measured sizes of hatchling desert tortoises (or more properly, neonates, as young tortoises of known age less than one year should be designated; Morafka 1994), thus showing that the fitted equation estimates size well. Although asymptotic sizes are virtually the same, females grow slightly faster than males early on so that by age 10, females average 136 mm CL and males average 131 mm CL; at age 15 females average 187 mm and males 180 mm CL; and by age 20 females average 224 mm and males 218 mm (table 11.1).

Age at Maturity

Because studies of Sonoran tortoises have not been extensive enough to give a minimum reproductive size, age at maturity for Sonoran Desert females was estimated using a published value for the minimum size of Mohave Desert females with eggs, 190 mm CL (Turner et al. 1986). Age at this size was determined in two ways. We ascertained age at maturity from scute annuli for those females that had reached or exceeded 190 mm CL as determined by the annuli length measurement. In this way, age was determined for each female individually. Age at maturity was estimated to be 15.7 years by using the mean of measurements of scute annuli for each individual. The age at which 190 mm CL is reached in the Richards' growth equation was also calculated, and age at maturity determined from the equation for female growth yields a similar value of 15.3 years.

The smallest size at which female tortoises have been found to carry eggs in the Sonoran Desert is 220 mm CL (Maricopa and Mazatzal Mountains in Arizona; Averill-Murray et al., ch. 6 of this volume). Using the growth equation for female tortoises, age at maturity of a female at 220 mm

CL is 19.3 years (fig. 11.2). Although future studies involving more animals may find smaller reproductive females, the age at maturity of female tortoises in the Sonoran Desert appears to be greater than that of other regions within the range of the desert tortoise (Germano 1994a, 1994b).

Longevity

The oldest age attained by desert tortoises in the Sonoran Desert estimated from the Richards' growth equation was 62.2 years for females and 64.4 years for males. Longevity of each sex was assumed to be the age at which calculated CLs came within 0.1 mm of the asymptotic size. This estimate of maximum age is considerably older than an estimate of thirty-five years that was based on accumulated microscopic layers formed at the edge of costal scutes, which gives a minimal age estimate (Germano 1992). The estimate from the growth equation may be inaccurate because it relies on size as the estimate of age, and size is poorly correlated with age for many species of turtles (Tucker and Moll 1997). Turtles live so long, however, that estimates rather than direct observation may provide the best information we have.

The Biology of Juvenile Desert Tortoises

Because small turtles are so difficult to find, we know much less about the ecology and behavior of juveniles than we do about adults for most species of turtles. Indeed, juvenile desert tortoises have historically been among the least-studied terrestrial vertebrates of the Sonoran Desert, but not because they are rare. During hatching in September and October (Averill-Murray et al., ch. 6 of this volume) and emergence in early spring, neonates are probably the most abundant age class in a tortoise population. Nonetheless, juvenile tortoises are easy to overlook. Young tortoises do not have the conspicuous silhouettes on slopes, washes, and roads that make adult tortoises so recognizable, and the lighter weight and smaller size of juveniles makes their scat and footprints less distinctive and more ephemeral than those of adults. Additionally, hatchlings may spend a great deal of time hiding in vegetation and burrows. Bolson tortoise *(Gopherus flavomarginatus)* neonates can use rodent burrows or other preformed shelters as their first refuges after leaving the nest (Tom 1994), and free-ranging desert tortoises, at least in some areas, may use only preformed burrows as their first shelters (David J. Morafka, pers. obs.). Unlike the burrows of adult Mohave desert tortoises, the juveniles' shelters have no unique features that identify them as tortoise burrows, and they are easily overlooked in population surveys and other ecological studies. In the typical rocky ridgetop habitat of the Sonoran desert tortoise in Arizona, hatchlings are very difficult to observe (Averill-Murray et al., ch. 6 of this volume).

In this account, we will combine information about Sonoran tortoises with extrapolations from studies of Mohave tortoises. Given the substantial ecological and genetic differences between these geographical forms, however, our conclusions are tentative. Newly hatched tortoises are only 1–2 percent of the weight of an adult tortoise (25–50 g for Mohave hatchlings, slightly less for Sonoran desert tortoises). Juvenile tortoises in the Mohave population differ from adults in the time of their activity periods. At Fort Irwin in the western Mohave Desert, juveniles commence activity in early January, and in some years surface activity is greater during February than in April and May, which are usually viewed as peak activity months for diurnal reptiles in the Mohave Desert. These small tortoises are active not only early in the year, but early in the day as well, often at body temperatures as low as 17°C (Hillard 1996). By May, juvenile tortoises in the Mohave Desert begin activity as early 5:00 or 6:00 A.M., and may retreat to their burrows by 9:30 A.M., just when other diurnal reptiles and the herpetologists who study them are emerging.

The timing of daily and seasonal activity periods of juvenile tortoises may have important implications for their diet and water balance. Early spring activity in the Mohave and Sonoran Deserts gives young tortoises access to plants that would not be available later in the year, when many plants have grown too tall for juvenile tortoises to graze on (see Van Devender et al., fig. 8.2 of this volume). Furthermore, early morning activity during the spring gives juvenile tortoises access to water in the form of dew. In the Sonoran Desert, where the summer monsoon rains stimulate growth in perennials and germination of a different set of annuals and provide pools of drinking water, summer activity is more important for juvenile tortoises (Averill-Murray et al., ch. 7 of this volume).

Virtually all young tortoises appear to hatch with considerable yolk reserves. Gopher tortoises *(Gopherus polyphemus)* in Florida hatch with enough residual yolk to sustain the hatchling for several months (Linley and Mushinsky 1994). Desert tortoises apparently retain enough stored energy to live four to six months without feeding (Nagy et al. 1997; Lance and Morafka 2001). Hibernation of hatchlings is important for the yolk nutrients to remain relatively undepleted and available for use in the critical first spring (Jarchow et al., ch. 12 of this volume). In spite of the energy stores of hatchling tortoises, their survivorship to adulthood is probably low. We know little about the actual annual survival of young desert tortoises, in general, and even less about young tortoises in the Sonoran Desert, and survival probably varies substantially from year to year depending on the weather. Recruitment of tortoises into a population is probably relatively high when several wet years follow each other and low in periods of drought. Studies in the Mohave Desert suggest annual survival of 47 percent for neonates but annual survival as high as 77–80 percent for individuals less than 100 mm CL (Turner et al. 1987a).

Manipulating Growth Rates of Juvenile Desert Tortoises

The slow growth rate and delayed maturity of juvenile tortoises in the Sonoran Desert combined with their initial abundance and their relatively high mortality from predation compared to adults suggest that protecting young tortoises until they have grown beyond their most vulnerable period might be a useful technique for conservation. This practice, which is often called head-starting, appears to have several potential advantages: Newly hatched tortoises are often available in large numbers, without human intervention most of these hatchlings will be eaten by predators before they reach maturity, and juvenile tortoises grow rapidly with careful husbandry. Thus, newly hatched tortoises appear to represent a large reservoir of potential adults that could be released when they have passed the most vulnerable stage of their life history.

Opponents of these programs, in contrast, point out that head-starting focuses conservation efforts on the very individuals that are least likely to survive to reproduce. Head-starting has been attempted for several species of sea turtles, but no programs have clearly succeeded, and in some instances head-starting appears to have accomplished nothing more than subsidizing predators (Mrosovsky 1983; Frazer 1992). Four problems have been identified in head-start programs for sea turtles: (1) failure to continue protection until the hatchery cohort reaches reproductive age; (2) failure to measure the effectiveness of head-starting by comparing the survivorship of hatchery- or nursery-produced tortoises to survivorship of local natural control groups; (3) failure to produce anatomically and behaviorally competent young, ready for independent living and reproduction in a natural environment; and (4) releasing hatchery-generated cohorts of young turtles in high localized concentrations that may surpass local carrying capacities in terms of food and shelter or subsidize predation when local carnivores focus their feeding on concentrations of vulnerable young.

These problems could apply to head-start programs for desert tortoises as well. We propose that studies of juvenile tortoises, including methods of head-starting, are relevant to conservation provided that these programs avoid the pitfalls that have been identified. In the following sections we describe two approaches to head-starting desert tortoises, a low-effort program conducted in the Mohave Desert in California and a high-effort program conducted at the Phoenix Zoo. Both of these projects are still in their early stages, and the information is presented as progress reports, not as finished studies.

Growth Rates of Tortoises in Field Enclosures

Since 1990 one of us (David J. Morafka) has established two predator-resistant hatchery enclosures in good tortoise habitat at Fort Irwin, San Bernardino County, California. These units, called Fort Irwin Study Site (FISS) units, were established to define the ecological niche of juvenile desert tortoises and to develop effective head-start methods for restocking existing populations. The enclosures were designed to overcome some of the problems classically inherent in sea turtle head-start efforts. This strategy is devised to raise young tortoises on natural diets in natural settings with which they are genetically compatible. Protection has been continued for up to seven years, until the tortoises' shells become large and hard enough to resist most predators.

FISS I and FISS II have been described in detail elsewhere (Morafka et al. 1997). They are located on a 2 percent slope on northeast-facing sandy hillsides at Fort Irwin in Mohave desertscrub dominated by creosotebush *(Larrea divaricata)*, white bursage *(Ambrosia dumosa)*, rabbit thorn *(Lycium pallidum)*, Mormon tea *(Ephedra californica)*, big galleta grass *(Pleuraphis rigida)*, and introduced annual grasses *(Bromus rubens, Schismus barbatus)*. Local densities of desert tortoises average about twenty adults per square kilometer.

Studies in FISS I

FISS I measures 60 × 60 m and was constructed by hand to minimize disturbance to soil crust and vegetation within its confines. The entire unit is fenced and roofed with chicken wire and is further protected by hardware cloth buried to a depth of 0.76 m and extending 0.6 m aboveground. The unit has two entrance doors and an automated weather station.

FISS I was intended to concentrate young tortoises for ecological study and to raise them under natural conditions before releasing them into the local population. Each spring gravid female desert tortoises were temporarily moved into FISS I from the surrounding 5 km. The females were radiographed repeatedly, allowed to nest naturally, and then returned to their original burrows.

Hatching success was 90 percent for most nests, and the mixed-age group included fifty to one hundred neonates and other juveniles. Densities of juvenile tortoises within FISS I at peak use were nearly five hundred times greater than the maximum densities estimated for natural populations (up to 344 individuals/ha in FISS I compared to 0.75 individuals/ha at the Desert Tortoise Natural Area in Kern County, California). No water or food was provided; the tortoises ate plants growing in the enclosure and drank when rain fell. Growth rates for 1993, measured by CL and weight, averaged 8.7 percent (± 3.9 SD) and 21.7 percent (± 16.5), respectively, for the mixed pool of one- to three-year-old individuals (Morafka et al. 1997).

Estimates of survivorship of hatchling and juvenile tortoises require assumptions that can introduce sources of error. In a study of free-ranging Mohave desert tortoises, Turner et al. (1987b) subtracted the number of carcasses they found from the estimated number of tortoises present at the beginning of the period. This method assumes that all dead tortoises are found and it will overestimate survivorship if some carcasses are missed. In contrast, the studies in FISS I counted surface-active tortoises. This method assumes that any tortoise that was not seen had died and will underestimate survivorship if some tortoises remain in their burrows during the entire census period. However, ongoing quarterly surveys usually correct underestimates or confirm previous counts within three to twelve months.

Annual survivorship of wild juvenile tortoises estimated by counting dead tortoises was 70 percent during a 3-year period and 61 percent during a different 15.5-month period (see Turner et al. 1987b, tables 7 and 8, respectively). Turner et al. (1987a) also provided information on the number of marked tortoises recaptured over an 8-year period and used the modified Jolly-Seber method to generate estimates of annual survivorship. Annual survivorships estimated with these models were 77–80 percent. Annual survivorship in FISS I ranged from 50 percent to 100 percent for different years and different cohorts of tortoises (see Morafka et al. 1997, table 2). The average annual survivorship in FISS I was 87.5 percent. Statistical comparisons of the rates of survivorship in the different studies would not be meaningful because assumptions and methods were so different. However, the average annual rate of survivorship in FISS I appears to be somewhat higher than that reported for free-living tortoises by Turner et al. (1987a, 1987b).

In May 1994 nine individuals (three- and four-year-olds) were released into the surrounding habitat (Spangenberg 1997). The release method was designed to duplicate the conditions that free-ranging tortoises of that age would encounter. The nine tortoises were placed individually near abandoned rodent burrows within 200 m of FISS I. Nine months later radiotelemetry and visual tracking confirmed that six of the tortoises (66.7 percent) were still in the vicinity. (The fates of the other three tortoises are unknown.) Growth rates since 1994 of the six released tortoises have generally exceeded those of tortoises that were kept at high densities in FISS I.

Studies in FISS II

FISS II is like FISS I in most details including its location, but it is slightly larger and protected more rigorously from human foot traffic, leaving the soil crust more intact and a greater variety of annual forage available. The density of tortoises in FISS II was generally kept below twenty-five individuals per hectare.

A different release strategy was employed when the first set of eight two- to five-year-old individuals were released in spring 1995. Radio-transmitters were placed on the animals while they were still inside a partitioned southern quarter of FISS II. Holes were cut in the lateral fencing at ground level that allowed young tortoises to leave the exclosure. (The holes were too small for large predators.) Within a month, five of the tortoises had been killed by avian predators. Common ravens *(Corvus corax)* perched directly over the apertures, apparently waiting for juvenile tortoises to emerge.

We conclude that juvenile survivorship inside the pens may have been higher than that of robust wild populations in relatively undisturbed areas. However, there were many assumptions necessary for the estimates of survivorship of the wild tortoises, so the accuracy of these estimates is unknown. Densities of juveniles within FISS I were higher than those of naturally occurring populations by more than two orders of magnitude, and the availability of forage plants, growth rates, physiology, and behavior may all have been affected by these high densities. The attempt to allow voluntary dispersal by juveniles from FISS II was frustrated by ravens that learned to seek emerging tortoises—in effect we had created a tortoise vending machine for these predators. Three- and four-year-old tortoises released from FISS I had an annual survivorship comparable to wild counterparts, and they grew normally and appeared to function as free-ranging, independently living individuals.

Growth of Tortoises in Captivity

Captive rearing programs and public education can play important roles in the conservation of some species of tortoises (Klemens 1989), and a method of head-starting hatchling tortoises that produces behaviorally and physiologically competent adults in manageable periods can contribute to both these activities. In captivity desert tortoises can grow to adult size in as little as two years when they are fed year-round and prevented from hibernating (Jackson et al. 1976), but this rapid growth is usually accompanied by deformation of the carapace. In addition, it is not clear whether tortoises that have grown rapidly to adult size are behaviorally and physiologically mature. That is, are they young adults or just big babies?

A pelleted tortoise diet developed by Olav T. Oftedal and Mary E. Allen of the Smithsonian Institution (U.S. National Zoo) produced rapid growth with little or no deformation of the carapace (E. M. Smith, O. T. Oftedal, F. H. Pough, M. E. Allen, T. Christopher, K. Bardeen, B. Henen, and D. C. Rostal, unpubl. data). In that study, young, wild-caught tortoises from the Mohave Desert (1 to 5 years old at the beginning of the study) grew to adult size in 2.5 to 4.5 years when they were fed this diet and prevented from hibernating. The collaborative study at the Phoenix Zoo by three of us (Pough, Smith, and Demlong) described here refines and extends the work of Oftedal and Allen by using tortoises of known age and origin and adding hibernation and sibling group as variables.

Diet and Growth

The experimental design included two variables, diet and hibernation, yielding four combinations of experimental treatments (fig. 11.3). Tortoises were fed either a salad composed of chopped mixed produce or a high-protein, medium-fiber 1/8-inch pelleted diet (table 11.2). The salad was selected as the control treatment because it is the standard diet for herbivorous reptiles at the Phoenix Zoo and the intent of this study was to compare a modification of husbandry methods (the pelleted diet) to the standard diet.

Newly hatched desert tortoises, believed to be from Sonoran tortoise parents, were obtained during October and November 1995 from hobbyists in Phoenix and Tucson, Arizona. In some cases the exact date of hatching was known, but more often we knew only the month. All tortoises were less than two months old when they entered the study. During a thirty-day quarantine period in the zoo's animal care center, each tortoise was weighed, measured, and individually marked by gluing a numbered plastic tag to the center of the fourth vertebral scute (placed to avoid inhibiting growth around the margins of scutes).

The experimental design (two experimental variables) required four individuals from each sibling group (i.e., from the same clutch of eggs). It was not always possible to get exactly four neonates from a clutch, however, and some groups were represented by more or fewer individuals. All of the

Diet

		Salad	Pellets
Winter Conditions	Hibernation	Sibling 1	Sibling 2
	No Hibernation	Sibling 3	Sibling 4

Figure 11.3. Experimental design illustrating the distribution of four individuals from a single sibling group (clutch of eggs) among the four combinations of diet and winter conditions in the study.

tortoises were kept warm and fed during their first winter (1995–1996), and were moved to outdoor pens (approximately 3 × 6 m) in the Suzan L. Biehler Tortoise Conservation Center in March 1996. The logistics of feeding required that the two diet groups be kept in separate pens, so the groups were switched every month to minimize the effects of differences in microclimate between pens.

All tortoises were fed salad initially, and the pelleted group was shifted gradually to its manufactured diet. Ground pellets were sprinkled on the salad and the proportion of pellets was progressively increased over a two-month period. We had anticipated that some tortoises would refuse to eat pellets, but that was not the case — every tortoise assigned to the pelleted food group made the transition. The condition of the tortoises was checked daily, and they were fed and given access to water. We weighed each tortoise every two weeks, and at monthly intervals we measured carapace and plastron lengths, carapace width, and shell height.

The hibernation treatment was started in the second winter (1996–1997). In November half of the tortoises in each diet group were randomly chosen and placed in hibernation for nine weeks while the remaining tortoises were kept warm and fed. Food was withheld from the hibernation group for two weeks, and then the tortoises were put in cardboard cartons packed with bermuda hay and moved to a cool area (approximately 10°C).

TABLE 11.2. Composition of diets. Phoenix Zoo salad diet: chopped produce, 10–20 percent by volume (bell pepper, carrot, celery, cucumber, green beans, zucchini). Chopped greens, 80–90 percent by volume (chard, kale, mustard greens, spinach, turnip greens). Composition based on Haytowitz and Matthews (1984). Weighted average based on 10 and 20 percent by volume. Zeigler Brothers Medium Fiber Tortoise Diet (no. 5365021800): alfalfa, cane molasses, corn, isolated soy, oat hulls, soy oil, soybean meal, wheat. Vitamins and minerals: ascorbic acid, avian vitamins and minerals, calcium phosphate, DL methionine, L lysine, limestone, iodized salt. Protein, lipid, and fiber values from company literature; others from Arizona Veterinary Diagnostic Lab analyses.

	Composition (% dry mass)					
	Protein	Lipid	Fiber	Calcium	Phosphorus	Ca:P Ratio
Phoenix Zoo Salad Diet						
Produce Chopped	15.6	2.3	11.5	0.39	0.37	1.22:1
Greens Weighted	25.2	3.4	10.4	1.2	0.5	2.5:1
Average	23.3–24.2	3.2–3.3	10.5–10.6	1.04–1.10	0.48–0.50	2.2:1–2.4:1
Zeigler Brothers Diet	21.2	5.6	14.4	1	0.18	5.6:1

Every two weeks the hibernating tortoises were moved to a warm room (approximately 25°C) overnight, removed from the hay, and soaked in lukewarm water to prevent dehydration. After nine weeks the tortoises were moved back to the warm room and feeding was resumed.

A difference in the growth rates of the tortoises on the two diets became apparent in the fifth month of the study and increased steadily. In May 1997 (when the animals were about eighteen months old) tortoises on the pelleted diet had an average weight of 184 g and an average CL of 95 mm, compared to 123 g and 84 mm for tortoises fed salad (fig. 11.4). Both of these differences were highly significant ($P < 0.002$ for both). Tortoises on the pelleted diet showed no evidence of abnormal shell growth.

Hibernating tortoises lost an average of 6 g during the nine weeks of hibernation, whereas the nonhibernating animals gained an average of 5.5 g during that period. The hibernating tortoises quickly regained weight when

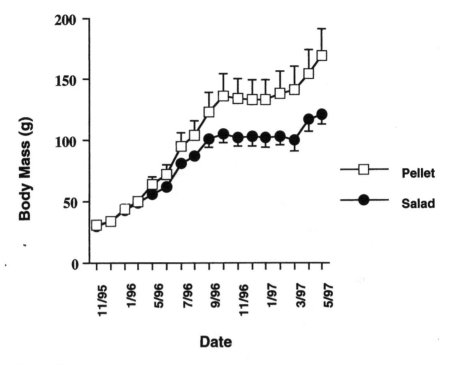

Figure 11.4. Comparison of growth rates of tortoises on the pelleted and salad diets (monthly mean values ± one standard error).

they resumed eating, and by May 1997 the weights of hibernating and non-hibernating tortoises did not differ (averages of 155 versus 152 g, respectively, $P > 0.94$).

Sibling group had a profound effect on growth rate (fig. 11.5). By May 1997 tortoises in the fastest-growing group had increased their initial mass approximately seven-fold, whereas those in the slowest group had only tripled their initial body mass. There was marginally significant interaction between diet and sibling group ($P = 0.05$).

Both the salad and pelleted diets had higher protein concentrations than most natural foods of desert tortoises, and this is probably one factor contributing to the rapid growth of both groups of tortoises in this study compared to growth rates of free-ranging hatchlings (see Baer 1994). The salad and the experimental diet had gross protein contents above 20 percent of dry mass, whereas the gross protein content of plants eaten by tortoises

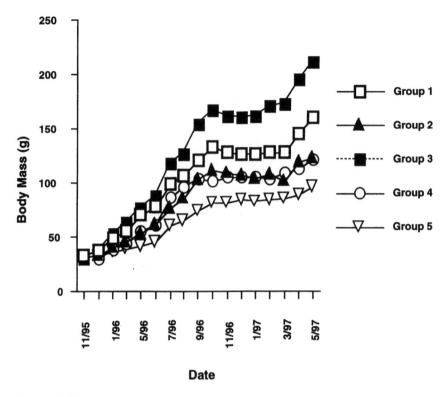

Figure 11.5. Comparison of growth rates of tortoises in the five sibling groups (monthly mean values).

in Nevada was substantially lower—2.5 percent and 2.7 percent of dry mass for two forbs *(Erodium cicutarium, Malacothrix glabrata)* and 0.8 percent and 0.9 percent for two grasses *(Oryzopsis [Achnatherum] hymenoides, Schismus barbatus;* Meienberger et al. 1993; Nagy et al. 1998). Somewhat higher nitrogen contents were reported for food plants eaten by piebald chuckwallas *(Sauromalus varius)*, but gross protein content was below 10 percent in nine of seventeen food items and only two of seventeen items had protein concentrations above 20 percent (Lawler et al. 1994).

Gross nitrogen content provides only an estimate of nutritional quality, because the apparent digestibility of nitrogen varies among food plants (Nagy et al. 1998). Apparent digestibility was 79 percent for desert dandelion *(Malacothrix glabrata)*, 72 percent for filaree *(Erodium cicutarium)*, and 7.2 percent for Indian ricegrass *(Achnatherum hymenoides)*. For Mediterranean grass *(S. barbatus)*, the apparent digestibility of nitrogen was −6.9 percent, which means that tortoises excreted more nitrogen in their urine and feces than they took in by eating this grass.

The calcium content of both salad and pellets (approximately 1 percent of dry weight) was lower than the average of twelve food items eaten by chuckwallas (3.6 percent). The phosphorus content of pellets (0.18 percent) was approximately the same as the average of chuckwalla food items (0.20 percent), and lower than that of salad (approximately 0.5 percent). The calcium:phosphorus ratios of both the salad (from 2.2:1 to 2.4:1) and pellets (5.6:1) are lower than the average calcium:phosphorus ratios of plants eaten by chuckwallas (35.6:1).

Educational Component

Making the public aware of challenges to desert tortoise survival can help change attitudes toward land use and ultimately may help reduce desert tortoise mortality related to human activity. We hope that such changes will increase adult survivorship and therefore allow tortoises to achieve their natural longevity. Accordingly, education has been an important component of this study from the beginning.

A newsletter kept donors of tortoises informed of the progress of the study, and individual and group tours of the Suzan L. Biehler Tortoise Conservation Center were provided for interested members of the community. With the help of undergraduate students from Arizona State University West, we worked with classes from four schools in Phoenix. The program included two days of classroom work at each school and a trip to the Phoenix Zoo during which students weighed and measured the tortoises. The students analyzed the data and wrote reports that included graphic and statistical analyses of growth.

Desert tortoises exemplify the problems of habitat loss and human-induced mortality that affect most species of vertebrate animals in North American deserts. The relationship between the life history of desert tortoises and their current problems of survival is easy to explain in the classroom. While the students were at the zoo, this lesson was extended to other species of tortoises and the potential use of head-starting became apparent. The same message was transmitted to the public by including the desert tortoise project in the behind-the-scenes tour at the zoo, features in magazine and newspaper articles (Brooks 1997), and a program on public television (KAET Phoenix).

Summary and Conclusions

Free-ranging Sonoran desert tortoises grow slowly. Under field conditions females desert tortoises are estimated to reach sexual maturity at about sixteen years, assuming 190 mm is the CL at maturity. This is a slower growth rate than that seen in female tortoises from the western Mohave Desert, but faster than those in the eastern Mohave (Germano 1994a). Longevity is at least thirty-five years and may be as long as seventy to ninety years. Female tortoises are thought to produce eggs for most of their adult lives.

Growth rates can potentially be increased by head-starting hatchling tortoises in captivity. Tortoises fed the high-protein, medium-fiber pelleted diet used in the Phoenix Zoo study had reached an average CL of 95 mm after eighteen months, whereas a free-ranging hatchling would require an estimated five years to reach that length. Tortoises fed salad also showed more rapid growth than expected for free-ranging tortoises, and their average CL at eighteen months (84 mm) corresponds to an estimated age of about four years for a free-ranging tortoise.

The Phoenix Zoo study represents a labor-intensive husbandry regime in which the tortoises were fed daily and provided access to water. In contrast, tortoises in the FISS enclosures did not receive supplemental food or water, and these tortoises grew at approximately the same rate as free-ranging Mohave desert tortoises. Tortoises from the FISS enclosures that were released into the field survived at the same rate as wild tortoises. The ability of the tortoises from the Phoenix Zoo study to adjust to field conditions has not been tested. Neither study has yet produced mature tortoises, so we cannot yet evaluate the reproductive status of head-started tortoises.

The utility of head-starting programs for turtles is an open question (Frazer 1997). In the long term we accomplish nothing by adding turtles to a shrinking population unless we can also address the factors that are causing the population to shrink. In fact, in Arizona it is not legal to release tortoises, including hatchlings and juveniles, into the wild (Howland and Rorabaugh, ch. 14 of this volume). In addition, many species of turtles probably have a life history that is based on high mortality of eggs and juveniles and long reproductive lives for individuals that reach adulthood. Head-starting alters this selective regime, and the long-term effect of that change on the biology of turtles is unknown.

Frazer's (1997) critique, while valid for the species and project he examined, may not apply to all chelonians. In a short-term context, head-starting can be a useful technique for management. For example, Galapagos tortoises were so close to extinction on Española that all of the tortoises known to live on the island were captured and brought to the Charles Darwin Research Station between 1963 and 1974 (Cayot and Morillo 1997). Captive-bred hatchlings have been repatriated to Española since 1975, and the first live hatchlings produced on Española were found in 1991. With continued protection of adults, the Española population of tortoises is expected to become self-sustaining. Management techniques that yield high growth rates of hatchlings and produce ecologically and physiologically competent adults can be used to reestablish populations in situations like Española, where the problems that originally threatened the wild populations of tortoises can be addressed.

Furthermore, hatcheries and nurseries provide opportunities to study large numbers of tortoise eggs, neonates, and juveniles under conditions in which physical and biological variables can be controlled and growth and behavior can be studied. Juvenile desert tortoises are rarely encountered in the field (e.g., Morafka 1994; Morafka et al. 1997; Averill-Murray et al., chs. 6 and 7 of this volume), and hatchery studies can provide information about age-specific characters that can be used to make decisions about habitat preservation and tortoise management.

ACKNOWLEDGMENTS

D. J. Morafka thanks Southern California Edison Company and its environmental scientist, J. Palmer, for substantial contractual support, advice, and encouragement in initiating studies of neonate desert tortoises. Morafka is also grateful to M. Quillman, cultural and natural resources manager in the Directorate of Public Works, Fort Irwin, for sustaining the FISS facility and its research over the last decade.

LITERATURE CITED

Andrews, R. M. 1982. Patterns of growth in reptiles. Pp. 273–320 *in* D. C. Gans and F. H. Pough, eds. *Biology of the Reptilia.* Vol. 13, *Physiology.* Academic Press, London.

Baer, D. J. 1994. The nutrition of herbivorous reptiles. Pp. 83–90 *in* J. B. Murphy, K. Adler, and J. T. Collins, eds. *Captive management and conservation of amphibians and reptiles.* Society for the Study of Amphibians and Reptiles, Ithaca, N.Y.

Brooks, D. 1997. Speeding tortoises. *Arizona State University Research* 11:34–37.

Cayot, L., and G. Morillo. 1997. Rearing and repatriation of Galapagos tortoises: *Geochelone nigra hoodensis*, a case study. Pp. 178–83 *in* J. Van Abbema, ed. *Proceedings: Conservation, restoration, and management of tortoises and turtles—An international conference.* New York Turtle and Tortoise Society, New York.

Frazer, N. B. 1992. Sea turtle conservation and halfway technology. *Conservation Biology* 6:179–84.

———. 1997. Turtle conservation and halfway technology: What is the problem? Pp. 422–25 *in* J. Van Abbema, ed. *Proceedings: Conservation, restoration, and management of tortoises and turtles—An international conference.* New York Turtle and Tortoise Society, New York.

Germano, D. J. 1988. Age and growth histories of desert tortoises using scute annuli. *Copeia* 1988:914–20.

———. 1992. Longevity and age-size relationships of populations of desert tortoises. *Copeia* 1992:367–74.

———. 1994a. Growth and age at maturity of North American tortoises in relation to regional climates. *Canadian Journal of Zoology* 72:918–31.

———. 1994b. Comparative life histories of North American tortoises. Pp. 175–85 *in* R. B. Bury and D. J. Germano, eds. *Biology of North American tortoises.* Fish and Wildlife Research no. 13, U.S. Dept. of the Interior National Biological Survey, Washington, D.C.

———. 1998. Scutes and age determination of desert tortoises revisited. *Copeia* 1998:482–84.

Germano, D. J., and R. B. Bury. 1998. Age determination in turtles: Evidence of annual deposition of scute rings. *Chelonian Conservation and Biology* 3:123–32.

Germano, D. J., R. B. Bury, T. C. Esque, T. C. Fritts, and P. A. Medica. 1994. Range and habitats of the desert tortoise *(Gopherus agassizii)*. Pp. 73–84 *in* R. B. Bury and D. J. Germano, eds. *Biology of North American tortoises.* Fish and Wildlife Research no. 13, U.S. Dept. of the Interior National Biological Survey, Washington, D.C.

Haytowitz, D. B., and R. H. Matthews. 1984. *Composition of foods: Vegetables and vegetable products.* U.S. Dept. of Agriculture, Agriculture Handbook no. 8-11, Human Nutrition Information Service, Washington, D.C.

Hillard, S. 1996. The importance of the thermal environment to juvenile desert tortoises. M.S. thesis, Colorado State University, Fort Collins.

Jackson, C. G., Jr., J. A. Trotter, T. H. Trotter, and M. W. Trotter. 1976. Accelerated growth rate and early maturity in *Gopherus agassizii. Herpetologica* 32:139–45.

Klemens, M. 1989. The methodology of conservation. Pp. 1–4 *in* I. R. Swingland and M. W. Klemens, eds. *The conservation biology of tortoises.* Occasional Paper no. 5, IUCN Species Survival Commission, Gland, Switzerland.

Lance, V. A., and D. J. Morafka. 2001. Postnatal lecithotroph: A new age class in the ontogeny of reptiles. *Herpetological Monographs* 15:123–33.

Landers, J. L., W. A. McCrae, and J. A. Garner. 1982. Growth and maturity of the gopher tortoise in southwestern Georgia. *Bulletin of the Florida State Museum (Biological Science)* 27:81–110.

Lawler, H. E., T. R. Van Devender, and J. L. Jarchow. 1994. Ecological and nutritional management of the endangered piebald chuckwalla *(Sauromalus varius)* in captivity. Pp. 333–41 *in* J. B. Murphy, K. Adler, and J. T. Collins, eds. *Captive management and conservation of amphibians and reptiles.* Society for the Study of Amphibians and Reptiles, Ithaca, N.Y.

Linley, T. A., and H. R. Mushinsky. 1994. Organic composition and energy content of eggs and hatchlings of the gopher tortoise. Pp. 113–28 *in* R. B. Bury and D. J. Germano, eds. *Biology of North American tortoises.* Fish and Wildlife Research no. 13, U.S. Dept. of the Interior National Biological Survey, Washington, D.C.

Meienberger, C., I. Wallis, and K. A. Nagy. 1993. Food intake rate and body mass influence transit time and digestibility in the desert tortoise *(Xerobates agassizii)*. *Physiological Zoology* 66:847–62.

Morafka, D. J. 1994. Neonates: Missing links in the life histories of North American tortoises. Pp. 161–73 *in* R. B. Bury and D. J. Germano, eds. *Biology of North American tortoises.* Fish and Wildlife Research no. 13, U.S. Dept. of the Interior National Biological Survey, Washington, D.C.

Morafka, D. J., K. H. Berry, and E. K. Spangenberg. 1997. Predator-proof field enclosures for enhancing hatching success and survivorship of juvenile tortoises: A critical evaluation. Pp. 147–65 *in* J. Van Abbema, ed. *Proceedings: Conservation, restoration, and management of tortoises and turtles — An international conference.* New York Turtle and Tortoise Society, New York.

Mrosovsky, N. 1983. *Conserving sea turtles.* British Herpetological Society, London.

Murray, R. C., and C. M. Klug. 1996. Preliminary data analysis from three desert tortoise long-term monitoring plots in Arizona: Sheltersite use and growth. *Proceedings of the Desert Tortoise Council Symposium* 1996:10–17.

Nagy, K. A., D. J. Morafka, and R. A. Yates. 1997. Young desert tortoise survival: Energy, water, and food requirements in the field. *Chelonian Conservation and Biology* 2:396–404.

Nagy, K. A., B. T. Henen, and D. B. Vyas. 1998. Nutritional quality of native and introduced food plants of wild desert tortoises. *Journal of Herpetology* 32:260–67.

Richards, F. J. 1959. A flexible growth function for empirical use. *Journal of Experimental Botany* 10:290–300.

Spangenberg, K. E. 1997. Movement and dispersal orientation of neonatal and juvenile desert tortoises. *In Abstracts of the third world congress of herpetology.* 2–10 August 1997, Prague, Czech Republic.

Tom, J. 1994. Microhabitats and use of burrows of Bolson tortoise hatchlings. Pp. 139–46 *in* R. B. Bury and D. J. Germano, eds. *Biology of North American tortoises.* Fish and Wildlife Research no. 13, U.S. Dept. of the Interior National Biological Survey, Washington, D.C.

Tucker, J. K., and D. Moll. 1997. Growth, reproduction, and survivorship in the red-eared turtle, *Trachemys scripta elegans*, in Illinois, with conservation implications. *Chelonian Conservation and Biology* 2:352–57.

Turner, F. B., P. Hayden, B. L. Burge, and J. B. Roberson. 1986. Egg production by the desert tortoise *(Gopherus agassizii)* in California. *Herpetologica* 42:93–104.

Turner, F. B., K. H. Berry, D. C. Randall, and G. C. White. 1987a. *Population ecology of the desert tortoise at Goffs, California, 1983–1986.* Report to Southern California Edison Company, Rosemead.

Turner, F. B., K. H. Berry, B. L. Burge, P. Hayden, L. Nicholson, and J. Bickett. 1987b. Population ecology of the desert tortoise at Goffs, San Bernardino County, California. *Proceedings of the Desert Tortoise Council Symposium* 1984:68–82.

Zug, G. R. 1991. *Age determination in turtles.* Herpetological Circular no. 20, Society for the Study of Amphibians and Reptiles, Ithaca, N.Y.

CHAPTER 12

Care and Diet of Captive Sonoran Desert Tortoises

JAMES L. JARCHOW, HOWARD E. LAWLER,

THOMAS R. VAN DEVENDER, AND CRAIG S. IVANYI

Desert tortoises *(Gopherus agassizii)* have been kept as pets in Arizona for many decades. It is not surprising that so many people find these desert denizens so appealing. Tortoises are highly personable and often appear to interact with people and other animals around them. Unfortunately, these positive qualities have become yet another threat to their existence, especially near rapidly growing urban areas.

In general, we do not believe that desert tortoises should be kept as pets, if by "pet," one means an animal that is frequently handled. Tortoises fare best when handled or disturbed as little as possible; much enjoyment and understanding of the desert tortoise can be gained simply by observing their natural behavior. However, so many tortoises are presently held in captivity in Arizona that it would be impossible and impractical to strictly prohibit possession of them. Wild desert tortoises have been protected by state law since 1987 due to concerns about the potential for decline of their populations (Howland and Rorabaugh, ch. 14 of this volume). Tortoises may not be collected from or released into the wild, imported, or exported from Arizona without specific permission from the Arizona Game and Fish Department. Tortoises obtained from a captive source, however, may be kept, with one tortoise per family member allowed under Arizona wildlife regulations.

Tortoise Husbandry

Diet

The desert tortoise is herbivorous, feeding mostly on grasses, leafy plants, and flowers. Bjorndal (1987) established that the gopher tortoise *(Gopherus polyphemus)* is an efficient hindgut fermenter that relies on anaerobic microflora to degrade cellulose in plants. Presumably, the digestive processes of the desert tortoise are similar. These difficult-to-digest plant fibers are a valuable source of metabolic energy that augments energy derived from more easily digested plant tissues. In this regard, the tortoise's digestive physiology approximates that of a horse or rabbit, and feeding principles are similar to the extent that plant fiber is a fundamentally important nutrient. As in the horse and rabbit, a segment of the tortoise hindgut is enlarged and modified to accumulate a large volume of fibrous plant material. Conditions are maintained in the hindgut that promote the growth of microbes that ferment the cellulose in plant cell walls. This fermentation releases protein and other nutrients from plant cells and results in the production of volatile fatty acids, which provide a significant energy source for the host herbivore (Bjorndal 1979). To maintain the digestive microbial flora and to discourage the growth of competing bacteria, which may be opportunistic pathogens, cellulose must be consistently ingested to provide a suitable nutrient substrate.

Nutrient analyses of the most frequently consumed plant species reported from four studies of free-ranging desert tortoises (Burge and Bradley 1976; Hansen et al. 1976; Coombs 1977) reveal overall dry matter content of crude protein ranging from 9.0 to 15.4 percent, lipids (i.e., fats, ether extract) below 3.0 percent, crude fiber from 18.7 to 34.0 percent, carbohydrate (nitrogen-free extract) from 45.5 to 52.3 percent, and calcium-phosphorus ratios of 3.2:1 to 5.8:1 (Jarchow 1987). In the absence of controlled studies of specific nutrient requirements of desert tortoises, these ranges have been successfully used as dietary guidelines for captive animals for the past two decades. Horticultural forages that approach or meet these criteria include alfalfa (Medicago sativa), Bermuda grass (*Cynodon dactylon*), clover (*Trifolium* spp.), dichondra (*Dichondra* sp.), and ryegrass (*Lolium* spp.; see table 12.1). Seed of these plants are available at local health-food markets or nurseries. Leaves of grape and mulberry are also suitable forage. Clinical assessment of tortoises provided this diet over a period of twenty years or longer has indicated the diet is adequate and complete, that is, vitamin and mineral supplementation is unnecessary.

Desert tortoises should have access to growing plants so that they may feed freely, and foraging areas should be sufficient to sustain daily grazing. Landscaping the tortoise area can be an opportunity to create an interesting and colorful area for people as well. A variety of native perennial

TABLE 12.1. Landscape plants eaten by desert tortoises that are commercially available as plants or seeds (most annuals; John F. Wiens, unpubl. data). Growth forms: CT, cactus; GA, annual grass; GP, perennial grass; HA, annual herb; HP, perennial herb; SH, shrub; SS, subshrub; TR, tree; VH, herbaceous vine; VW, woody vine. Plant parts eaten: FL, flower; FR, fruit; VG, vegetative material (leaves, stems). An asterisk indicates a native plant in the range of desert tortoise.

Common Name	Species	Plant Parts
TREES AND SHRUBS		
Baja fairy duster	*Calliandra californica* (SH)	FL
Chuparrosa	*Justicia californica** (SH)	FL, VG
Creosotebush	*Larrea divaricata** (SH)	FL (VG)
Fern acacia	*Acacia angustissima** (SH)	FL, VG
Sweet acacia	*Acacia farnesiana** (TR)	VG
Desert honeysuckle	*Anisacanthus thurberi** (SH)	FL, VG
Desert willow	*Chilopsis linearis** (TR)	FL (VG)
Mexican bird-of-paradise	*Caesalpinia mexicana* (SH, TR)	VG
Mulberry	*Morus alba* (TR)	VG
Ocotillo	*Fouquieria splendens** (SH)	FL
Yellow trumper flower	*Tecoma stans** (SH)	FL
SUBSHRUBS AND WOODY VINES		
Catclaw vine	*Macfadyena unguis-cati** (VW)	VG
Chinese lantern	*Hibiscus schizopetalus* (SS)	FL
Desert rose mallow	*Hibiscus coulteri** (SS)	FL, VG
Globe mallow	*Sphaeralcea ambigua** (SS)	FL, FR, VG
Globe mallow	*Sphaeralcea emoryi** (SS)	FL, FR, VG
Indian mallow	*Abutilon palmeri** (SS)	FL, VG
Rock rose mallow	*Hibiscus denudatus** (SS)	FL, VG
Desert rock pea	*Lotus rigidus** (SS)	FL
Fairy duster	*Calliandra eriophylla** (SS)	FL, VG
Gallinitas	*Callaeum macropterum** (VW)	FL
Grape (cultivated)	*Vitis* spp.	VG
Jasmine	*Jasminum sambac* (SS)	FL
Queen's wreath	*Antigonon leptopus** (VW)	VG
Smooth bouvardia	*Bouvardia ternifolia* (SS)	FL, VG
CACTI		
Hedgehog cacti	*Echinocereus* spp.* (CT)	FL, FR
Pincushion fishhook cacti	*Mammillaria* spp.* (CT)	FL, FR
Prickly pear	*Opuntia engelmannii** (CT)	FR, VG
Tuna prickly pear	*Opuntia ficus-indica* (CT)	FR
Saguaro	*Carnegiea gigantea** (CT)	FR

TABLE 12.1. Continued

Common Name	Species	Plant Parts
HERBACEOUS PERENNIALS		
Alfalfa	*Medicago sativa* (HP)	FL, FR, VG
Arizona cottontop	*Digitaria californica*• (GP)	VG
Bamboo muhly	*Muhlenbergia dumosa* (GP)	VG
Bush muhly	*Muhlenbergia porteri*• (GP)	VG
Beardtongue	*Penstemon pseudospectabilis*• (HP)	FL
Beardtongue	*Penstemon superbus*• (HP)	FL
Parry penstemon	*Penstemon parryi*• (HP)	FL, VG
Bermuda grass	*Cynodon dactylon* (GP)	VG
Black grama	*Bouteloua eriopoda*• (GP)	VG
Blue grama	*Bouteloua gracilis* (GP)	VG
Sideoats grama	*Bouteloua curtipendula*• (GP)	VG
Curly mesquite	*Hilaria belangeri*• (GP)	FL, VG
Desert senna	*Senna covesii*• (HP)	FL, VG
Evening primrose	*Calylophus hartwegii* (HP)	FL, VG
Evening primroses	*Oenothera* spp. (HP)	FL, VG
Purple threeawn	*Aristida purpurea*• (GP)	VG
Rama del toro	*Dicliptera resupinata*• (HP)	FL, VG
Snapdragon vine	*Maurandya antirrhiniflora*• (VH)	FL, FR, VG
Thurber dyssodia	*Thymophylla pentachaeta*• (HP)	FL, VG
Windmills	*Allionia incarnata*• (HP)	FL, FR, VG
Yellow morning glory	*Merremia aurea* (VH)	FL, VG
ANNUALS (CULTIVARS)		
Canteloupe	*Cucumis pepo* (VH)	FL, VG
Corn	*Zea mays* (GA)	VG
Hollyhock	*Alcea rosea* (HA)	FL, VG
Okra	*Abel moschatus* (HA-HP)	FL, VG
Pumpkin	*Cucurbita* sp. (VH)	(FL), VG
Radish	*Raphanus sativus* (HA)	VG
Tepary bean	*Phaseolus acutifolius* (VH)	FL, VG
Watermelon	*Citrullus lanatus* (VH)	VG
Zucchini squash	*Cucumis pepo* var. *melopepo*	VG
ANNUALS (OTHERS)		
Desert Canterbury bells	*Phacelia campanularia* (HA-SP)	FR, VG
Lupine	*Lupinus sparsiflorus*• (HA-SP)	FL, VG
Owl's clover	*Castilleja exserta*• (HA-SP)	FL, VG
Summer poppy	*Kallstroemia grandiflora*• (HA-SU)	(FL), VG

grasses, herbs, and shrubs are available from commercial sources in Arizona (table 12.1), and tortoise custodians are encouraged to plant natural forage species wherever possible. An adult desert tortoise should be provided with a grass area (Bermuda grass is fine, but nonflowering sterile hybrids are preferred) of at least 5.6–9.3 m^2 and a habitat area with alfalfa, clover, dichondra, and/or grape vines (*Vitis* spp.) measuring 2.8–5.6 m^2. Alternatively, the grass area may be planted with commercially available native grasses such as Arizona cottontop *(Digitaria californica)*, curly mesquite *(Hilaria belangeri)*, bush muhly *(Muhlenbergia porteri)*, dropseeds (*Sporobolus* spp.), gramas (*Bouteloua* spp.), or purple threeawn *(Aristida purpurea)*. The habitat could be planted with beardtongues (*Penstemon* spp.), desert willow *(Chilopsis linearis)*, legumes (e.g., *Calliandra californica, C. eriophylla, Senna covesii*), mallows (*Abutilon* spp., *Hibiscus* spp., *Sphaeralcea* spp.), prickly pear cacti (*Opuntia* spp.; avoid long-needled chollas), primroses (*Calylophus* spp., *Oenothera* spp.), *rama del toro (Dicliptera resupinata)*, and windmills (trailing four o'clock, *Allionia incarnata*). During summer months, roadside weeds including chinchweed *(Pectis papposa)*, grasses, horse purslane *(Trianthema portulacastrum)*, pigweed *(Amaranthus palmeri)*, spiderlings (*Boerhavia* spp.), spurges (*Euphorbia* spp.), summer poppy *(Kallstroemia grandiflora)*, verdolagas *(Portulaca oleracea, P. umbraticola)*, and windmills *(Allionia incarnata)* can be offered to augment the diet.

It is advisable to provide as much variety as possible because tortoises display individual variation in food preferences, eat various parts of the plants (e.g., flowers, fruits, leaves, and stems), nutrient content may vary seasonally, and tortoises may become conditioned to eating a limited number of forages if that is all that is offered. Although tortoises may not be active in the spring (Averill-Murray et al., ch. 7 of this volume), they readily consume the dried remains of spring annuals during the summer rainy season (Van Devender et al., ch. 8 of this volume). Leaving skeletons of such commercially available spring wildflowers as lupines *(Lupinus sparsiflorus)* and owl's clover *(Castilleja exserta)* and volunteer annuals including the native caterpillar weed *(Phacelia distans)*, combbur *(Pectocarya recurvata)*, fiddleneck *(Amsinckia intermedia, A. tessellata)*, Indian wheat *(Plantago ovata)*, nievitas (e.g., *Cryptantha barbigera*), peppergrass *(Lepidium lasiocarpum)*, and wild carrot *(Daucus pusillus)* and the introduced filaree *(Erodium cicutarium)* in the habitat area for later harvest might be beneficial.

Neonates and juvenile tortoises (*Gopherus* spp.) exhibit a preference for less-fibrous, higher-protein diets (Adest et al. 1989), a pattern typical of herbivores. Consequently, these age groups can be expected to feed more heavily on tender herbs than grasses, although we have observed captive neonates feeding heavily on less-fibrous, higher-protein new growth of previously grazed grass.

Troyer (1984) demonstrated the importance of soil ingestion to the development of the microbial fermentation system of neonatal iguanas *(Iguana iguana)*, another herbivorous hindgut fermenter. The ability to ferment plants in the hindgut depends on the acquisition of cellulose-degrading bacteria from the environment and the successful colonization of the tortoise's large intestine by those bacteria. Furnishing juvenile desert tortoises with a topsoil substrate is preferable to the common but risky practice of offering dried scat from adult tortoises as a source of beneficial microbes. In the latter case, pathogens may be transferred as well. Sand should be avoided due to its propensity to become impacted in the intestine if ingested.

In general, feeding commercial produce is discouraged. A review of twenty-nine varieties of produce indicated deficient levels of crude fiber and usually suboptimal calcium:phosphorus ratios compared to natural forages (Jarchow 1987). Additionally, the high moisture content of produce can dilute nutrients (Fowler 1976). Some dark green leafy vegetables contain sufficient crude protein and calcium:phosphorus ratios to justify their inclusion as a dietary supplement for adults and as a temporary diet for neonates; these include beet greens, collards, dandelion greens, kale, mustard greens, parsley, Swiss chard, and turnip greens.

Some tortoises will accept alfalfa, Bermuda grass, or timothy *(Phleum pratense)* hay and combinations thereof as dietary supplements. All of these hay varieties provide major nutrients within guidelines from natural forages, but palatability and acceptance are problematic. Pelletized and other commercially prepared diets often pose problems with palatability and are largely unproven as being nutritionally adequate; these diets are, in general, not recommended. A noteworthy effort that may prove to be an exception is the diet discussed by Germano et al. (ch. 11 of this volume). Other foods that have commonly been included in captive desert tortoise diets such as apples, avocado, bananas, cat and dog food, corn, dairy products, lettuce, melons, peaches, and primate biscuit fall egregiously outside of dietary guidelines from natural forages and should be avoided entirely.

Water

In their desert environment, tortoises are dependent on the moisture in their food and the relative humidity of their shelter to maintain hydration for most of the year. In the Sonoran Desert, tortoises readily drink and rehydrate during summer and winter rains. Often, captive tortoise shelters do not provide sufficient relative humidity to deter evaporative water loss through the skin and respiratory system. To mitigate water loss, the captive tortoise should be provided water to drink at least once a week.

In our experience, the most effective method to induce drinking is to simulate rain. This can be accomplished by using a sprinkler, letting a hose run so puddles form around the tortoise, or soaking the tortoise in a tub of shallow water. In this case, the water should be clean, room temperature, and should be chin level. As the tortoise drinks it may completely submerse its head in water, behavior that might unduly alarm its owner. Once the tortoise is through drinking or if it shows no interest in drinking after thirty minutes, it may be removed from the water. Pools, ponds, and standing water should be avoided. Not only is there risk of drowning, but pathogens found in standing water may cause infections.

Physical Environment

In Arizona, desert tortoises should be kept outdoors all year whenever possible. Temperature, light, and climatic change stimulate activity, foraging, and behavior patterns that are generally conducive to good health. In our experience tortoises kept inside, even under full spectrum artificial lighting, exhibit a far higher incidence of metabolic bone disease than those kept outdoors.

In southern Arizona, most captive desert tortoises are maintained in walled backyards, where they generally do quite well (fig. 12.1). The minimum enclosure area for a single adult desert tortoise should encompass at least 18 m^2. This area is large enough for a single male or up to three females.

Figure 12.1. An example of a backyard tortoise habitat in Tucson, Arizona. This enclosure features a diverse habitat with grass and herbaceous foraging areas, a shade tree, and secure parimeter walls. A mound of soil (at least 20 cm deep) over the shelter would improve insulation. (Photograph by Jim Jarchow)

The enclosure should have solid walls, constructed of materials such as concrete block, adobe, or tightly fitted wooden slats set in a concrete footer. Wire fencing should be avoided due to the hazards of head, neck, and limb injuries. The enclosure should be at least 0.6 m high. Desert tortoises climb well, so the shelter, den, or other interior structures should be at least 0.3 m from the enclosure perimeter. Tortoises will readily escape and may travel considerable distances.

The enclosure should be landscaped with grass and herbaceous feeding areas (see Diet above) and shrubs or trees to provide shade. Sonoran desert tortoises inhabit rocky hillsides, and the provision of hilly, rocky terrain in the enclosure will promote natural activity and may shorten acclimation periods for relocated captives.

At least two dens should be provided for each tortoise. A den can be constructed aboveground using concrete blocks and a three-quarter-inch plywood roof. The structure should be covered by at least 20 cm of soil for insulation. One den should have an eastern or southern exposure for cool seasons and the other a northern exposure for summer. In veterinary practice, far more tortoises are presented for hyperthermia (overheating) in the summer than are presented for freeze-related injury in the winter. A shaded, well-insulated summer shelter is critical to maintaining a healthy tortoise year-round. All dens should be well drained. Persistent dampness in the den may predispose the tortoise to dermatitis and respiratory infections. Shelter-sites should be far enough from feeding areas so that watering practices do not contribute to den dampness.

The tortoise enclosure should be free of hazards, many of which are not so obvious. If the tortoise is housed in the backyard, it must be excluded from swimming pool and decorative pond areas as well as steep drop-offs. Tortoises should not have access to garages, storage sheds, or other areas where antifreeze, fertilizer, pesticides, solvents, or other toxic materials are kept. Dogs, no matter how trustworthy, should not be left unattended with the tortoise. Dog-inflicted injuries are the most common form of trauma in captive tortoises brought to veterinarians. Enclosures housing hatchlings and juveniles up to 70 mm carapace length should be covered with poultry netting or a similar mesh material to keep out predatory birds, rodents, and house cats. Unless the enclosure is very large (i.e., 140 m^2 or more) and many sheltersites and foraging areas are available, male desert tortoises should be maintained individually to prevent injury from male–male agonistic behavior and overzealous courtship of females. Desert tortoises should not be housed with other species of chelonians due to the risks of interspecific parasite and disease transmission and injurious behavioral interaction. Ants pose a serious threat, especially to juvenile tortoises. Bait poison granules (Amdro, available from American Cyanamid Co., Parsippany, New Jersey) may be used to control ants outside the enclosure. If used inside the enclosure, all tortoises should be removed until the pellets are gone. The tortoise area should be free of string, wire, nails, plastic debris, cigarette butts, and other refuse that could entangle or be ingested by the tortoise. Although most tortoise care manuals warn of the danger of poisonous plants, clinical intoxication of captive tortoises from this source appears very rare. In twenty-five years of tortoise veterinary practice, the only cases of plant intoxication observed by one of us (James L. Jarchow) were the result of ingested toadstools. This is in spite of the fact that tortoises are commonly kept in yards with oleander

(Nerium oleander), sacred datura *(Datura wrightii)*, and a host of other toxic plants. Finally, if the tortoise enclosure has a gate, it should have a lock or at least a secure latch. Escaped captive tortoises are frequently found on city and suburban streets and are often lost or injured as they wander through unfamiliar areas. Additionally, escaped and released captives may have negative impacts on free-ranging tortoise populations, including disturbing or displacing resident tortoises and introducing disease. The increased incidence of mycoplasmosis (upper respiratory tract disease) at desert tortoise release sites has been documented in the Mohave Desert (Jacobson et al. 1991). Exotic diseases that may be introduced by captive tortoises into wild populations have the potential to cause catastrophic declines, a problem that has only recently been addressed (Jacobson et al. 1999).

Cold-Weather Care

As the weather turns cool in the fall, captive tortoises prepare to hibernate. Their appetites decrease and they become less active. The tortoises have nutrient and water reserves built up and should easily survive through the winter hibernation if they have eaten well during the warm months. If tortoises retire to their shelters, there is very little attention required during winter months.

If freezing or subfreezing temperatures are forecast for your area, crumpled newspaper or a blanket may be placed over or in the shelter entrance to further insulate the tortoise from freezing. This cover should be removed the next morning to allow air circulation into the den. The tortoise may emerge during winter rains to drink, but should otherwise remain torpid until emergence in the spring. Tortoises that bask or frequently emerge from their shelters during the winter are often diagnosed with pathological processes, malnutrition, or severe injuries; these tortoises should be examined by a veterinarian.

Desert tortoises introduced to a new enclosure in late summer or fall may exhibit hyperactive, pacing behavior—often along enclosure walls. Although this activity is associated with acclimation to a new environment and may be seen at other times of the year, it is especially detrimental during the weeks prior to hibernation; nutrient and water stores may be depleted without the opportunity to recoup losses or repair tissue injury. These tortoises should be presented to a veterinarian for evaluation, rehydration, and other appropriate therapy prior to hibernation.

If the tortoise does not accept its winter shelter, or if an adequate den is not available, artificial hibernation quarters may be provided indoors. A cool, dry protected area such as a garage, storeroom, or unheated closet provides an adequate environment. Place the tortoise in a cardboard box, wooden crate, or similar container that is large enough to allow the animal to freely turn around. Each tortoise should be housed individually in its own container. The floor should be covered with several layers of newspaper or straw. Then cover the box with porous fabric, such as a blanket or towel, to maintain darkness but allow ventilation. A temperature range of 2–20°C should be maintained for the duration of hibernation.

Dehydration is a significant risk during hibernation for tortoises kept indoors, but can be avoided if the tortoise is offered water periodically. This is accomplished by soaking the adult tortoise in shallow (chin-level), room-temperature water for thirty to sixty minutes once every four weeks. Juvenile and neonatal tortoises should be soaked once every two to three weeks because their proportionate rates of evaporative water loss are much greater. When the tortoise has finished drinking, or if it attempts to climb out of the soaking container, it may be blotted dry and returned to its hibernation box. Tortoises hibernating indoors will often drink more readily if they are placed outside and provided with puddles from a garden hose on warm (15.6°C or higher), sunny winter days.

Tortoises hibernating outdoors generally maintain hydration without assistance, provided there are winter rains. Exceptionally dry winters may necessitate rain simulation with a sprinkler once every four to six weeks, but excessive dampness should be avoided. Tortoises may be observed when watering to see that no health problems are developing, otherwise they should not be exposed to light or other disturbances during hibernation.

Desert tortoises typically emerge from hibernation on warm days from late February through April; those that hibernated indoors may be placed outside at this time, as long as adequate shelters are provided. Water should be provided upon emergence from hibernation.

Tortoises requiring medical treatment or nutritional support during the winter should not hibernate, and special care is required. The adult tortoise should be housed in an indoor enclosure measuring at least 1 m². The enclosure should be maintained at a daytime temperature between 27 and 30°C. This can be achieved by placing a light above the enclosure and installing a thermometer inside, close to the floor. A 150-watt infrared flood lamp works well for this purpose; different wattages may be tried until the desired temperature is achieved. Provide food daily and soak the tortoise in shallow, room-temperature water to stimulate drinking at least three times weekly. Take the tortoise outside whenever the sun is shining and temperatures are above 21°C. Frequent exposure to sunlight is beneficial to tortoises in rehabilitation and will usually stimulate their appetite, although shade must always be available. Maintain a normal daily photoperiod and thermal cycle by turning off the light at sunset. Leaving the light on at night may prove stressful. If the tortoise's health is in question, it should be examined by a qualified veterinarian to determine if hibernation is advisable. Likewise, debilitated tortoises that have been kept warm and cared for during the winter should be examined by a veterinarian before they are placed outdoors in the spring.

Desert Tortoise Reproduction

Arizona Wildlife Law and Adoption Programs

Desert tortoises reproduce readily in captivity under proper conditions. This often presents a dilemma for the tortoise custodian in Arizona. Under Arizona Game and Fish Department wildlife regulations, the hatchlings may only be kept for up to twenty-four months, at which time they must be disposed of by gift or as directed by the department. Captive tortoises may not be released into the wild, imported, or exported under any circumstances, without authorization by the department.

For these reasons, tortoise adoption programs have been established to help facilitate transfer of the increasing number of surplus or unwanted captives to custodians who are sufficiently informed and committed to providing an appropriate home for them. Two desert tortoise adoption programs sanctioned by the Arizona Game and Fish Department are conducted by the Arizona-Sonora Desert Museum and the Arizona Game and Fish Department's Wildlife Center at Adobe Mountain. From spring 1995 through fall 1998 these two programs placed 535 captive tortoises with private individuals. Although these tortoises are owned by the State of Arizona, people may become custodians to ensure the welfare and longevity of existing captives. Because tortoises are so long-lived (fifty years or more), tortoise custodians may be undertaking a lifetime endeavor.

Since its inception in the early 1980s, the Arizona-Sonora Desert Museum Tortoise Adoption Program has placed hundreds of tortoises with members of the public. The program was created to provide an outlet for unwanted captive tortoises in the Tucson area by facilitating their transfer to willing and informed custodians. Prospective custodians are carefully screened and educated on all aspects of desert tortoise husbandry and the current status of wild tortoises. These individuals must meet rigid guidelines that address the tortoise's physical environment and diet. Additionally, potential hazards are outlined, including swimming pools or ponds, harmful chemicals of any kind, toxic plants, or additional pets (including other types of tortoises or turtles), as well as excessive human handling. Individuals must demonstrate that they will always operate in the best interest of their tortoise(s) and prevent intentional or accidental injury, disease, or death.

Backyard breeding of desert tortoises is discouraged because disposition of young tortoises is often difficult and time-consuming. Custodians must consider whether they have sufficient space and resources to meet the requirements of hatchling and young tortoises and whether they have the means to legally and ethically dispose of surplus offspring. Most custodians are not equipped to provide for the needs of these young animals. Thus, a surplus of juvenile tortoises almost *always* exists. For these reasons, the museum adoption program strongly discourages backyard breeding of desert tortoises and will not place sexual pairs, or opposite sex animals that complete a pair, with tortoise custodians.

Tortoise adoption programs can have a positive effect on tortoise conservation in several ways. By educating the public regarding the risks to wild tortoises of releasing captive animals and providing an outlet for surplus and unwanted captives, adoption programs have a direct effect on the preservation of wild desert tortoise populations. A natural by-product of this interaction with the public is that an increasing number of people are made aware of the plight of tortoises and are more likely to take an active role in their conservation.

Distinguishing the Sexes

The sex of adult tortoises can be determined by several criteria. First examine the plastron (lower shell). The posterior portion is concave in mature males; this depression enables the male to fit the carapace (upper shell) of the female during coitus. The plastron is flat in mature females. It is important to note that immature tortoises of either sex have flat plastra. Therefore age, which is best determined by size in the absence of an actual life history, is important in determining sex. Tortoises are generally mature sexually when the carapace straight-line measurement reaches 200 mm. The rear margin of the carapace typically flares out in females and is nearly vertical in males. Males have proportionally longer tails and well developed glands on the ventrolateral surfaces of their chins.

Courtship and Mating

Male tortoises court females throughout their activity season during the summer months. The male nods his head at the female as he approaches, often circling her and biting at her forelimbs before attempting to copulate. The female will often ignore the male. She may refuse his advances or become receptive as courtship behavior ensues. Captive female desert tortoises often appear to be very selective in accepting a mate; males, however, are typically attracted to any female they encounter and may court even sub-adult individuals. Copulation can be very brief or may continue for hours. The male tortoise characteristically extends his head and neck and vocalizes in the form of grunts, groans, and/or wheezes during copulation. Both males and females may mate with several individuals in the course of a year.

Eggs and Incubation

In the Sonoran desert tortoise, eggs are typically laid in a single clutch in June or July prior to or at the beginning of the summer rainy season (Averill-Murray et al., ch. 7 of this volume). Because viable sperm are retained by the female, she is capable of laying fertile eggs for several years from a single mating. In such cases, the number of fertile eggs per clutch may diminish with time. The clutch size is generally six to twelve eggs.

The female normally digs a vase-shaped nest hole about as deep as her carapace length in which to deposit the eggs. When appropriate nesting substrate is not available, eggs laid on the surface desiccate and usually die. The female fills in the nest hole after oviposition. Females can be quite secretive in their nesting activities, and the custodian may be unaware of the eggs until they hatch. There is evidence some females may defend the nest site for some time against potential predators, although they do not care for the offspring (Barrett and Humphrey 1986).

The incubation period ranges from 80 to 120 days. The hatchling cracks the eggshell with a temporary protrusion on the upper jaw called the egg tooth, which is lost soon after hatching. Alternating between periods of activity and rest, the hatchling emerges from the egg and digs its way to the surface. At this time, the tortoise is smaller than a silver dollar and the shell is quite pliable.

In most cases, it is best to leave the eggs in the nest to hatch. However, if their location is at risk and the custodian wants to attempt to incubate the eggs, the following procedure is recommended: Mark the top of each egg with a soft graphite pencil before moving them from the nest, and be careful not to turn them. The slightest rotation of the egg can result in the embryo's death. Carefully remove each egg from the nest and place it in an incubator in the exact position in which it was laid.

Reptile egg incubators are commercially available (Lyon Electric Co., Chula Vista, California; Randall Burkey Co., Boerne, Texas), or one can be constructed using a polystyrene (Styrofoam) box; a heating pad or incandescent light for heat; 10 cm of slightly moistened vermiculite, sand, or soil; and a piece of glass to cover the top of the box. A small opening must be provided for ventilation. A thermometer should be used to calibrate the incubation temperature and kept in the incubator to monitor temperature throughout incubation. Appropriate temperature and moisture are the most important parameters for successful incubation. The temperature should be maintained at 29–33°C. The light wattage required to maintain this temperature will vary according to the size of the incubator and the ambient temperature in the room where the incubator is kept. The substrate should be slightly moist but not wet, with a relative humidity of about 60 percent inside the incubator. A slight amount of condensation should always be evident on the cover glass. If excessive condensation forms, resulting in water dripping onto the eggs, wipe the glass clean until condensation is reduced to a light fog. Add water as needed by dripping along the edges of the sand, but *never* directly on the eggs. It is best to set up the incubator and calibrate the temperature and humidity before the eggs arrive.

The Care of Hatchlings

Never remove hatching tortoises from their eggs. They generally require two to four days to fully absorb the yolk before breaking through the shell. The hatchlings may emerge with the yolk sac still attached to the plastron. This yolk is an important food source that provides nutrients for many months after hatching and should not be disturbed. Individual hatchlings can be placed in a margarine container lined with clean wax paper to protect the yolk until it is absorbed.

Once the yolk is absorbed, hatchlings may be placed immediately in outside enclosures provided with adequate sheltersites, herbaceous foraging areas, and shade plants. Hatchlings are easy prey for many native birds and mammals as well as domestic dogs and cats, so outside enclosures must be covered with poultry netting or similar material. If outdoor facilities are not available, neonates may be temporarily housed indoors in plastic boxes, terraria, or similar containers. Because they can easily tip over onto their backs, usually by climbing against the wall or over siblings, only one hatchling per container is recommended. Desert topsoil should be used as a substrate, and a hiding box or similar shelter should be provided. A water dish suitable for drinking and soaking may be kept in the container if the water is changed daily, or the tortoise may be soaked twice weekly. Recommended food items—native herbs, new grass growth, alfalfa (cuttings from growing plants, not sprouts), beet greens, clover, collards, dandelion greens, dichondra, kale, mustard greens, parsley, and turnip greens—should be offered daily. Fruits and lettuce, for reasons previously addressed, should be avoided entirely. A daytime temperature range of 27–30 °C should be maintained (an overhead incandescent lamp is effective) and a full spectrum fluorescent light (e.g., Reptisun, Zoo Med Laboratories, Inc., San Luis Obispo, California) employed, although the latter may prove to be a meager substitute for direct sunlight. A recently developed incandescent lamp, Active UV Heat (T-REX Products, Inc., Chula Vista, California), has been tested at the Arizona-

Sonora Desert Museum for several years, and it appears to be a reasonable substitute for direct sunlight. Lights should be left on for eight hours a day and the container allowed to cool to room temperature at night.

Indoor housing of hatchlings, as described above, should be continued only from the time of hatching until mid-November; then hatchlings should be placed in a suitable outdoor enclosure or allowed to hibernate indoors. A gradual cooling period should commence about 1 November, which entails removing the incandescent lamp and moving the container to a cooler area of the house. Food should be withheld at this time and not offered again until emergence from hibernation, when all hatchlings should be moved to an outdoor enclosure. Hibernation the first winter affords considerable benefit to hatchlings, even those that hatch only a few days prior to hibernation, in that yolk nutrients remain relatively undepleted well into the following spring, when the neonates are actively exploring a rigorous environment, foraging, and acquiring a microbial fermentation system.

Signs of Health Problems

Custodians are encouraged to weigh their desert tortoises once a month (except during hibernation). Abrupt or sustained weight loss may be an early indication of pathological processes. Healthy desert tortoises typically exhibit less than an 8 percent reduction in body weight during hibernation (Jacobson et al. 1999). Unprecedented or unexpected activity changes, including torpor and restlessness (the latter especially in nesting females), are also associated with a variety of metabolic and pathological problems. Loss of coordination, labored breathing (often with mouth open, neck extended, and accompanying forelimb movements), regurgitation, and straining to defecate are obvious signs of potentially critical situations.

Changes in physical appearance, including generalized swelling, eyelid swelling, closed eyelids, ocular discharge or opacity, nasal discharge, encrusted nares, salivation (which often appears frothy), sunken or protruding eyes, a bony appearance of the head or limbs, a soft or pliant shell, prolapsed tissue from the vent, and asymmetry of the head or limbs are important signs to recognize. Changes in gait such as lameness and dragging the shell while walking are also noteworthy.

Each of these situations warrants veterinary attention. Modern diagnostic and therapeutic techniques are usually very effective provided the tortoise is presented to a qualified veterinarian before irreversible pathological change occurs. Several veterinarians in Arizona volunteer their services to desert tortoise adoption programs. In the event a tortoise custodian cannot afford medical treatment, the tortoise may be relinquished to an adoption agency and medical attention provided.

Injured and ill tortoises should not be exposed to temperature extremes or handled unnecessarily. To minimize further stress, the tortoise should be placed in a cardboard box, kept dark, and maintained at 25–29°C until it is brought to the veterinarian.

Overheating

Tortoises that are overturned or trapped in sunny areas during warm months quickly become overheated. Signs of overheating include heavy salivation resulting in an accumulation of frothy saliva around the mouth, open-mouth breathing, and voiding urine. The tortoise should be moved to a shaded area and placed on the ground right side up; cool water should be poured continuously over its carapace until heavy breathing subsides and the tortoise appears normal. A veterinarian should be called for consultation. Tortoises may suffer long-term effects from an acute episode of overheating, but debility is often preventable with early medical intervention.

Drowning

Tortoises that fall into swimming pools, ponds, or are trapped in flooded enclosures typically engorge themselves with water and sink to the bottom. Unfortunately, the vertical walls of these structures preclude the animals from simply walking up the bank and out of the water. These tortoises are often found totally submerged and motionless. In such an event, the tortoise should be removed from the water and held with the head pointed down and the rear of the carapace straight up. Hold the mouth open to drain water from the stomach. Once the flow of water out of the mouth has ceased, place the tortoise upright, resting on its plastron. Hold the forelimbs flexed at a right angle at the elbow so that they are in a normal anatomical position, then push the upper forelimbs straight into the body cavity, directly toward the hind limbs. Water may be observed gushing from the mouth or nares. Grasp the forelimbs by the elbows, then pull them straight forward out of the body cavity. Air may be heard as it flows into the lungs. Repeat these steps until there is no further water discharge and the tortoise has resumed breathing on its own. The tortoise should then be taken to a veterinarian for further evaluation and treatment. Again, tortoises should never have access to swimming pools or ponds.

Ant Bites and Stings

Desert tortoises, especially juveniles, are frequently attacked by ants. The tortoise is usually found frantically attempting to escape the attacking ants and scratching at exposed areas of vulnerable skin where ants are attached. The ants should be removed with forceps or tweezers. A paste of baking soda and water may then be applied to skin surfaces to reduce the effects of formic acid. A veterinarian should be consulted for follow-up treatment.

Abrasions, Lacerations, and Ulcerations

Female tortoises may exhibit abrasions from the bites of overzealous, amorous males on the anterior surfaces of their forelimbs. Tortoises may also experience abrasions, usually on limbs or shell, from canine bite wounds or repeated contact with abrasive surfaces in their environment. These wounds may be effectively treated by the topical application of a noncaustic antiseptic such as povidone-iodine solution (Betadine, Purdue Frederick Co., Norwalk, Connecticut). The antiseptic should be applied daily until the redness has subsided. Ointments should be avoided; repeated use of ointments or oils on the integument may result in sloughing of epithelial cells and scarring.

Deeper wounds that expose bone, muscle, or connective tissue should be treated by a veterinarian. If the wound is large enough and deep enough to allow fly larvae to gain access, they usually will. Untreated small wounds frequently degrade into large, maggot-infested necrotic wounds, and chronic debility may result.

Constricting Bands

Tortoises occasionally become entangled with wire, string, fishing line, or thin fibrous vines. The only evidence may be swelling of the distal portion of a limb or head and neck. The affected appendage should be extended sufficiently to allow examination of the proximal (inward) extent of the swelling. If a constricting band is located, it should be carefully cut and removed. If swelling persists longer than two hours or if the skin has been lacerated, the tortoise should be examined by a veterinarian.

Shell Fractures

Fractures of the shell may result from den collapse, canine bite wounds, vehicular trauma, falling, vandalism, or being stepped on. There are several first aid procedures that will enhance tissue viability and improve prognosis for healing. First, plant material, hair, dirt, and other debris should be carefully removed from the traumatized area with forceps or clean hands. Care should be taken to avoid touching exposed viscera with fingers. Next, the wound should be thoroughly irrigated with a sterile saline solution (soft contact-lens wetting solutions are acceptable). The tortoise should be held with the wound facing down during irrigation so that debris is washed out of the wound and not carried deeper into the wound. Under no circumstances should hydrogen peroxide, water, or other liquids be used for irrigation. If sterile saline is not available, cover the wound with sterile gauze or a clean cloth. Then porous first-aid, duct, or masking tape should be wrapped around the shell to hold the bandage securely in place. If any shell fragments have been detached, they should be gathered up and taken to a veterinarian with the bandaged tortoise. Veterinary shell repair of recent fractures usually entails the application of fiberglass material and epoxy or acrylic using sterile techniques and should not be attempted by the layman.

ZOONOSES (DISEASES COMMUNICABLE TO HUMANS). *Salmonella* and *Shigella*, bacteria of considerable public health significance, have been isolated from cloacal cultures of free-ranging Sonoran desert tortoises (Dickinson et al. 1996). In James L. Jarchow's veterinary experience, *Salmonella* often appears on fecal culture of stressed or debilitated tortoises but may be shed by apparently healthy tortoises as well. These organisms are not often pathogenic for their host and appear to be ubiquitous in chelonian populations (Cooper 1981). For these reasons, precautions must be taken to prevent human exposure to fecal-borne pathogens. First, wash hands thoroughly after contact with tortoise feces. Second, keep tortoises and their soaking tubs, water dishes, and indoor containers out of food-preparation areas. Third, remove feces from tortoise enclosures on a regular basis and dispose of wastes in sealed plastic bags. Finally, young children and immunocompromised individuals should not come into contact with tortoise feces or handle tortoises with feces on their exterior surfaces; thorough hand washing should follow the handling of any tortoise.

Discussion and Conclusions

Historically, captive desert tortoises have been subjected to a variety of feeding practices and environmental conditions with varying degrees of success. Too often, captive reptiles, including desert tortoises, are cared for using guidelines derived from anecdotal information gleaned from popular literature, Internet websites, and well-meaning "experienced" individuals. Because these animals may survive many years under less-than-ideal conditions, inappropriate husbandry practices may not be readily apparent even to the experienced keeper or veterinarian. Criteria such as achievement of longevity, growth and reproductive rates, morphology, and behavior typical of healthy free-ranging tortoises reflect relatively unincumbered physiology and should be used to evaluate husbandry practices. The goal is to effectively provide those elements of the tortoise's habitat essential to maintain health and vigor (and thereby minimize the stress of captivity); and the appearance and function of healthy wild tortoises should provide the model.

In the future, feeding trial studies with decades-long evaluation of test subjects will more precisely define nutritional requirements. Until then, the logical approach to providing a complete diet is to duplicate that of robust, wild individuals. As more feeding observation studies and dietary analyses of scat of wild desert tortoises are conducted, age, sex, and regional and seasonal variations in nutrient consumption will become apparent, thus providing direction for further refinement of feeding practices.

Likewise, observations of microhabitat selection and behavior of wild tortoises should guide enclosure design and husbandry practices. Ready acceptance of the new captive environment and natural but moderate climatic cycles promote normal activity and reduce the animal's period of acclimation. Clearly, the best preparation for the prospective desert tortoise custodian is to become familiar with the animal's natural history through credible literature. In addition, advice should be sought from well-informed adoption program representatives or tortoise biologists, and a qualified veterinarian should be located before emergencies arise.

Living in the heart of the species' geographical range is a key to success in desert tortoise husbandry. If the tortoise is kept outdoors, fed a diet equivalent to that of wild tortoises, provided a well-designed and secure habitat, and afforded adequate protection from injury and climatic extremes, it should remain healthy for decades.

The captive population of desert tortoises provides not only enjoyment to their custodians but, more importantly, opportunities for education of the public and increased awareness of the species among those who may never see a desert tortoise in nature. Thus, the captive population may play an important role in mustering public support for conservation of their wild relatives.

ACKNOWLEDGMENTS

The tremendous efforts of the late volunteer Tortoise Adoption Coordinators Betty Vance and Kathy McNaughton were largely responsible for the early development and success of the Arizona-Sonora Desert Museum program. We thank Sandra Cate for information on the Arizona Game and Fish Department's tortoise adoption program at the Adobe Mountain Wildlife Center. The Herpetology staff, especially Janice Perry, and volunteer Tortoise Adoption Program Coordinator Cindy Wicker provided information about the tortoise adoption program at the Arizona-Sonora Desert Museum. John Wiens shared the results of his feeding trials of native and horticultural plants in his Tucson yard, and provided information about their commercial availability. Linda Jarchow processed the manuscript.

LITERATURE CITED

Adest, G. A., G. Aguirre, D. J. Morafka, and J. L. Jarchow. 1989. Bolson tortoise *(Gopherus flavomarginatus)* conservation. 1. Life history. *Vida Silvestre Neotropical* 2:7–13.

Barrett, S. L., and J. H. Humphrey. 1986. Agonistic interactions between *Gopherus agassizii* (Testudinidae) and *Heloderma suspectum* (Helodermatidae). *Southwestern Naturalist* 31:261–63.

Bjorndal, K. A. 1979. Cellulose digestion and volatile fatty acid production in the green turtle, *Chelonia mydas*. *Comparative Biochemical Physiology* 63A:127–33.

———. 1987. Digestive efficiency in a temperate herbivorous reptile, *Gopherus polyphemus*. *Copeia* 1987:714–20.

Burge, B. L., and W. B. Bradley. 1976. Population density, structure and feeding habits of the desert tortoise, *Gopherus agassizii*, in a low desert study area in southern Nevada. *Proceedings of the Desert Tortoise Council Symposium* 1976:51–74.

Coombs, E. M. 1977. Status of the desert tortoise, *Gopherus agassizii*, in the state of Utah. *Proceedings of the Desert Tortoise Council Symposium* 1977:95–101.

Cooper, J. E. 1981. Bacteria. Pp. 187–88 *in* J. E. Cooper and O. F. Jackson, eds. *Diseases of the Reptilia*. Vol. 1. Academic Press, London.

Dickinson, V. M., J. L. Jarchow, and M. H. Trueblood. 1996. *Health studies of free-ranging Sonoran desert tortoises in Arizona*. Technical Report no. 24, Arizona Game and Fish Dept., Phoenix.

Fowler, M. E. 1976. Respiratory disease in captive tortoises. *Proceedings of the Desert Tortoise Council Symposium* 1976:89–98.

Hansen, R. M., M. K. Johnson, and T. R. Van Devender. 1976. Foods of the desert tortoise, *Gopherus agassizii*, in Arizona and Utah. *Herpetologica* 32:247–51.

Jacobson, E. R., J. M. Gaskin, M. B. Brown, R. K. Harris, C. H. Gardiner, J. L. LaPointe, H. P. Adams, and C. Reggiardo. 1991. Chronic upper respiratory disease of free-ranging desert tortoises *(Xerobates agassizii)*. *Journal of Wildlife Disease* 27:296–316.

Jacobson, E. R., J. L. Behler, and J. L. Jarchow. 1999. Health assessment of chelonians and release into the wild. Pp. 232–42 *in* M. E. Fowler and R. E. Miller, eds. *Zoo and wild animal medicine: Current therapy*. Vol. 4. W. B. Saunders, Philadelphia, Pa.

Jarchow, J. L. 1987. Veterinary management of the desert tortoise, *Gopherus agassizii*, at the Arizona-Sonora Desert Museum: A rational approach to diet. *Proceedings of the Desert Tortoise Council Symposium* 1984:83–94.

Troyer, K. 1984. Behavioral acquisition of the hindgut fermentation system by hatchling *Iguana iguana*. *Behavioral Ecology and Sociobiology* 14:189–93.

CHAPTER 13

Fire Ecology of the Sonoran Desert Tortoise

TODD C. ESQUE, ALBERTO BÚRQUEZ M., CECIL R.
SCHWALBE, THOMAS R. VAN DEVENDER, PAMELA J.
ANNING, AND MICHELLE J. NIJHUIS

Desert wildfires are dramatic phenomena with the potential to kill desert tortoises and to reduce the biotic and structural diversity in their habitats. Although the first published observations of fires in desert tortoise habitats were made decades ago in the Mohave Desert (Woodbury and Hardy 1948), only recently has this threat to tortoise habitats in the Sonoran Desert been acknowledged by researchers, land managers, and the public. Wildfires cause disturbances that can affect how desert tortoises make their living in a variety of ways. In this chapter, we discuss how introduced exotic plants predispose tortoise habitats to wildfires and the short- and long-term effects of increased fire frequency, magnitude, and intensity on desert tortoise populations.

Desert Wildfires: Why Now?

Wildfires and related habitat changes have recently become widespread in desert tortoise habitats throughout parts of the Sonoran Desert and elsewhere (Rogers 1985, 1986; Brown and Minnich 1986; Loftin 1987; Schmid and Rogers 1988; Búrquez M. et al. 1996). In the Arizona Upland subdivision, abundant, finely textured fuels from introduced annual and perennial plants after winters of high precipitation (Steenbergh and Lowe 1977; Rogers and Steele 1980; Wilson et al. 1995; Todd C. Esque, unpubl. data) have fueled increasing numbers of fires (Narog et al. 1995). In the pre-Colombian Southwest, abundant fine fuels were not present in desert environments. Indeed, fire was not an important ecological process in the evolution of desert plants in dry tropical forest and thornscrub in the Oligocene and Miocene (about 30–8 million years ago) or the formation of the Sonoran Desert by the late Miocene (8 million years ago; Van Devender 1995). Today, high temperatures, low relative humidity, and gusty winds during the dry summer months from late May into July just prior to the monsoon season in the Sonoran Desert, often called the arid foresummer, can create the ideal conditions for the spread of fires.

Exotic plants are those that are introduced either purposefully or accidentally into new habitats primarily by the agency of humans or in association with their livestock and cultivated plants. The pseudoriparian habitats along highways are important dispersal corridors. Species such as the annual grass red brome *(Bromus rubens)* from the Mediterranean region and the perennial buffelgrass *(Pennisetum ciliare)* from Africa can be extremely abundant, essentially filling the interstices between perennial desert plants and creating unbroken fuel beds of finely textured fuels (fig. 13.1). In the Sonoran Desert and more tropical communities to the south in Sonora, introduced perennial plants are the primary carriers of fire. Buffelgrass is a remarkable grass that grows into an almost-woody subshrub and accumulates flammable materials over the course of several years, in effect unlinking fire frequency from annual climatic variability and increasing the fire intensity. After the fire, buffelgrass rapidly resprouts from the root crown as a simple bunch grass, but one with a huge root system.

Although fires often start naturally by lightning strikes (Bahre 1995; Bureau of Land Management, unpubl. data), the frequencies of ignition have increased substantially due to normal urban activities, such as burning trash, parking vehicles on dry grass, throwing cigarettes out of vehicles, and igniting fireworks, and by accident while camping, hunting, and hiking in the backcountry (Brussard et al. 1994; Búrquez M. et al. 1996). In the central Sonoran Desert, fire is used as a tool to maintain the vigor of buffelgrass (Búrquez M. et al. 1996). Certainly, lightning struck in desert habitats prior to the introduction of exotic plants, but those fires were not expected to have carried far from the stricken plants due to lack of fuels in most years. Although native annual grasses such as annual six-weeks needle grama *(Bouteloua aristidoides)* in summer or six-weeks fescue *(Vulpia octoflora)* in spring can form dense stands between shrubs in some years, their dried, slender stems are poor fuel. In central Sonora, short-lived perennials such as Rothrock grama *(Bouteloua rothrockii)* may become locally dense in Sonoran desertscrub after an exceptionally wet year or series of years, creating a buildup of fuel. Brown's (1982) Sonora Savannah Grassland was probably based on

Figure 13.1. Fires resulting from accumu-
lations of fine fuel from introduced grasses
are a serious threat to desert tortoises in
the Sonoran Desert. Above: Tortoise tunnel
in dense, dried red brome (winter annual)
in August 1993 in the Black Mountains,
Mohave County, Arizona. (Photograph by
Cecil Schwalbe) Insert is a close-up of red brome. (Photograph by Thomas Van
Devender) Right: Close-up of buffelgrass (perennial) near Alamos, Sonora. Note
the coarse, dense foliage. (Photograph by Thomas Van Devender)

photographs from such times, thus exaggerating the importance of grass in
these communities. False grama *(zacate borreguero, Cathestecum brevifolium)*
is a dwarf-tufted, stoloniferous perennial that can form lush turfs 2.3 cm
tall in open areas during the summer monsoonal rains. Even then, the accumulated fuel buildup may not be enough to carry fires. During and after
El Niño events with exceptional rains, native plants may have accumulated
enough fuel to support lightning-caused fires. Fires started in front of early
summer storms that generated gusty winds would have been highly direc-
tional (Arthur W. Bailey, pers. comm. 1996). The dramatic increases in fire

frequency fueled by introduced species may also have changed the intensity and behavior of wildfires. Today fires may start during relatively calm conditions by human activities and spread out slowly in all directions on an abundant bed of dried exotic plants, rather than the historic condition of intense heat passing across the landscape quickly for only seconds or minutes. Because today's fires may move slowly, they can cause more severe damage to plants. Conditions within desert wildfires vary considerably depending on fuel loads, fuel geometry, and environmental conditions (Whelan 1995). Air temperatures measured near various burning plants in one fire in Arizona Upland habitat ranged between 140 and 440 °C (Patten and Cave 1984). With the introduced grass fuel to carry fire and ignite any accumulated woody materials, desert wildfires can become extremely intense, with temperatures exceeding 1,000 °C (Arthur W. Bailey, pers. comm. 1996) and flame lengths exceeding 10 m (Todd C. Esque, pers. obs.). Such extreme temperatures boil the moisture out of live cactus. These fires may almost entirely incinerate aboveground vegetation over large areas.

Interactions between introduced plants and increased fire ignitions result in a dynamic feedback mechanism known as the "grass/fire cycle" (D'Antonio and Vitousek 1992). The grass/fire cycle is found when intro-' duced plant species invade habitats, thus increasing the production of fuels and changing fuel types such that fires ignite more frequently than they did prior to invasions. Once the habitats burn, they are increasingly prone to burning again (D'Antonio and Vitousek 1992). With each successive burn- ing event, woody and/or long-lived species of plants decline in distribution and abundance in the affected areas, whereas exotic grasses increase in abun- dance. The grass/fire cycle is observed to be fueled by annual or peren- nial grasses in the Sonoran Desert. Annual grasses such as red brome are known to fuel fires in the Sonoran Desert at least as far south as Tucson, Arizona. Introduced perennial grasses including Lehmann's lovegrass *(Era- grostis lehmanniana)* are known to fuel fires in desert grassland south of Tucson; Natal grass *(Melinus [Rhynchelytrum] repens)* in desert grassland south of Nogales, Sonora; and buffelgrass in desertscrub and thornscrub in central Sonora.

The Importance of Desert Wildfires

The importance of fires in desert habitats has only recently been acknowledged in the literature (Humphrey 1974; Cave and Patten 1984; Swantek et al. 1999). This lack of attention is partly due to incomplete fire records. Even though a great deal of the Sonoran Desert in the United States is administered for the public by state and federal agencies, wildfire records in desert habitats were not standardized and centrally stored until approximately 1980. Conventional methods of analyzing fire history from tree rings (e.g., in forested habitats) are unsuccessful in most desert habitats due to the multiple stems of the few trees that grow there and their ability to drought-prune major branches and clone. Loftin (1987) did not consider fires to be important landscape phenomena in deserts, but others have de- scribed the increase in frequency and distribution of fires (Brown and Min- nich 1986; Rogers 1986; Rogers and Vint 1987; Schmid and Rogers 1988). However, the severe impacts of fires in the Sonoran Desert are increasingly visible in the plant communities themselves. For example, foothills paloverde *(Parkinsonia microphylla)* and saguaro *(Carnegiea gigantea)*, dominants in the Arizona Upland desertscrub, are readily killed by fire (Steenbergh and Lowe 1977; McLaughlin and Bowers 1982; Rogers 1985; Wilson et al. 1995; Todd C. Esque, unpubl. data).

Both community structural dominants and introduced species represent small percentages of local floras, however. Herbaceous plants are the most speciose life forms in most temperate and tropical floras (Hammel 1990). In the Tucson Mountains, where paloverde-saguaro desertscrub is the dominant vegetation type, 76 percent of the flora are herbs compared to such structural dominants as shrubs (9 percent), subshrubs (7 percent), succulents (6 percent), and trees (2 percent). The first and third most speciose genera *(Euphorbia*, sixteen species; *Bouteloua*, ten species; Rondeau et al. 1996) in the flora are strictly herbs. The effect of fire on the diversity of herbaceous plants has not yet been addressed in the Sonoran Desert. Only 13 percent of the flora are nonnative species; most of these are innocuous and have little impact on the flora or vegetation. Thus, a few aggressive exotic species bearing a new ecological process (fire) have a dramatic impact on a relatively small but very important part of the flora.

In large natural systems that evolved with fire, such as the Great Victoria Desert of south-central Australia, wildfires increase the diversity of plant and animal assemblages and are considered to be beneficial (Pianka 1992). The ecosystem is large enough to support a mosaic of many large patches at different stages of recovery from disturbance. In contrast, tortoise habitats in the Sonoran Desert are not fire-adapted communities, and the dominant trees, shrubs, and succulents do not quickly recover from disturbances or recolonize from adjacent areas. Introduced species alter habitat in two very different ways: modification and conversion. Modification refers to subtle changes in community composition or cover quantity and quality. Examples of this are the introduction of Lehmann's lovegrass into desert grassland in the southwestern United States (McClaran and Anable 1992) and Natal grass in northern and eastern Sonora (Van Devender et al. 1997). After introduction of these species, the vegetation is still desert grassland.

Conversion is a much more severe ecological impact in which one community is replaced by another. In tortoise habitat in the Sonoran Desert, introduced species and fires result in conversion from desertscrub to desert grassland, although in three very different situations. At the upper elevation limits (1,300–1,400 m) of Arizona Upland desertscrub (Turner et al. 1995), fires fueled by such winter-spring Mediterranean annuals as red brome, wild barley *(Hordeum murinum)*, and wild oats *(Avena fatua)* may result in the lowering of the desert grassland–Arizona Upland ecotone (Van Devender et al. 1997; fig. 13.2). The fire season fueled by Mediterranean annual grasses begins in the arid foresummer in May. Two terrible examples were the Basin Fire near Saguaro Lake and the Río Fire near Scottsdale northeast of Phoenix in 1995.

At lower elevations in the Arizona Upland, desertscrub is converted to a barren landscape with few plants other than the Mediterranean weeds, as can be seen near the junction of Interstate 17 and Arizona Highway 74 north of Phoenix, a process that may be exacerbated by grazing. Similar large-scale habitat changes were first observed when fires fueled by cheatgrass *(Bromus tectorum)* devastated desertscrub communities in the Great Basin (Klemmedson and Smith 1964; Callison et al. 1985; Young et al. 1987; Melgoza and Nowak 1991). Great Basin desertscrub, which is dominated by big sagebrush *(Artemisia tridentata)*, blackbrush *(Coleogyne ramosissima)*, other woody shrubs, and perennial grasses, is being converted to exotic annual grasslands (Young et al. 1987; Billings 1990). Intervals between fires decreased from approximately fifty years to less than twenty years (Wright et al. 1979). In the western and central Mohave Desert, Arabian grass *(Schismus arabicus)* and Mediterranean grass *(S. barbatus)* are important introduced plants because of the fires that they fuel (Brooks 1998). However, these species appear not to be generally problematic for tortoises in the Sonoran Desert because they typically grow on relatively flat valley surfaces below the tortoise habitats on rocky slopes in the Arizona Upland (Averill-Murray et al., ch. 7 of this volume).

The third type of conversion is from Arizona Upland desertscrub in Arizona and Plains of Sonora desertscrub and foothills thornscrub in central Sonora to a subtropical Africanized savanna (Búrquez M. et al. 1996, 1999). In Arizona, conversion has been fueled by red brome, but in Sonora buffelgrass is the primary agent of change. Buffelgrass, a shrubby perennial grass originally from Kenya, grows in dense monocultures, accumulates great amounts of fuel, and burns easily — even when green. Buffelgrass grows widely in eastern and southern Africa. (The name *buffelgrass* is a South

Figure 13.2. Arizona Upland Sonoran desertscrub between Florence and Oracle Junction, Pinal County, Arizona. Top: Natural habitat; note the dominant triangleleaf bursage *(Ambrosia deltoidea)* subshrubs. Bottom: Burned habitat in which the bursages were completely incinerated. This photograph was taken about one month after the fire. (Photographs by Thomas Van Devender)

Figure 13.3. Buffelgrass invaded Sonoran desert tortoise habitat near El Batamote, 47 km north of Hermosillo, Sonora. This site at 480-m elevation in the Plains of Sonora subdivision of the Sonoran Desert with desertscrub on the flats and foothills thornscrub on emergent desert ranges was surveyed for tortoises by Fritts and Jennings (1994) and Treviño et al. (1994). Left: View to the southwest showing pastures cleared and planted with buffelgrass. Note the *chorizos*,

African variant of "buffalograss," not to be confused with the buffalograss *[Buchloë dactyloides]* of the southern Great Plains in the United States.)

The buffelgrass fire season begins at the end of the summer rainy season in late September and early October and extends until the summer rains begin the following July, although winter rains and cool-season growth will reduce the number of fires. Buffelgrass fires result in temperatures so hot that the soil is scorched and the bedrock cracked. A very sad example is the granite ridges near El Batamote, 47 km north of Hermosillo on Mexico Highway 15 (fig. 13.3A). In this case the fire was so intense that the varnished surfaces on the granite boulders exfoliated and erosion has accelerated (fig. 13.3B). Tortoises were surveyed here by Fritts and Jennings (1994) and Treviño et al. (1994). Since about 1990, buffelgrass has been spreading like a fire-driven cancer from an adjacent pasture upslope into tortoise habitat

the sausage-like walls of rock cleared from the pastures, and foothills thornscrub on the emergent hills. (Photograph by Cecil Schwalbe) Right: Boulder area in September 1997 with accelerated erosion and exfoliation of the granite due to intense buffelgrass fires. Ironwood *(Olneya tesota)*, torote *(Bursera fagaroides, B. laxiflora)*, and tree morning glory *(palo santo, Ipomoea arborescens)* trees formerly grew on the knob. (Photograph by Thomas Van Devender)

(Bury et al., ch. 5 of this volume). Near the highway, the foothills thornscrub has been completely decimated. Keystone trees such as ironwood *(Olneya tesota)* and torote *(Bursera fagaroides)* were not only killed but completely incinerated. Eventually buffelgrass will likely cover the entire ridge as it now covers the flats, thus eliminating most of the tortoises in this area.

Furthermore, many parts of the Sonoran Desert can no longer be considered natural communities due to the intense management of open spaces; fragmentation by agricultural fields, livestock fences, trails, roads, and urbanization (Murray and Dickinson 1996; Búrquez M. et al. 1996); and ecosystem changes due to encroachment by introduced plants. In the Sonoran Desert, areas that formerly supported one hundred to more than three hundred plant species per square kilometer (McLaughlin and Bowers 1982; Felger 1990; Turner et al. 1995) may be reduced to supporting only a handful of species over the course of several years. The losses of diversity can be even greater in areas where tortoises inhabit foothills thornscrub and tropical deciduous forest (Búrquez M. et al. 1996). Fires often result in dramatic changes in plant species composition and physical structure in tortoise habitats due to loss of dominant plant species. The seeds or vegetative components of many native species do not possess adaptations to avoid mortality due to fires. Native plants may also have difficulty establishing after fires due to physical changes in soil properties (nutrient and organic matter content), surface temperatures, or competition with introduced plant species. Perhaps even more importantly, the structure of these habitats may be dramatically altered. For example, areas previously supporting a structurally diverse community with a wide array of trees, shrubs, subshrubs, succulents, and many herbaceous species may be reduced to herb-dominated disturbance communities.

Thus, in the Sonoran and Mohave Deserts maintenance of natural communities demands that fires be excluded. If allowed to burn, they could consume considerable desert tortoise habitat and likely kill numbers of tortoises.

Fire Effects on Desert Tortoises

Some animals that evolved in disturbance-prone communities may benefit from habitat changes due to fires (Mushinsky and Gibson 1991; Pianka 1992), whereas other species, such as the desert tortoise, are predicted to decline (Goldingay et al. 1996; Esque et al., in review). More subtle long-term affects on tortoise populations are expected from fire-mediated changes adversely affecting the viability of fire survivors. Fire may cause both injury and death to individual desert tortoises. Many studies have described or quantified responses of small animals to fires (Cook 1959; Howard et al. 1959; Kahn 1960; Greenburg et al. 1994; McCoy et al. 1994; Trainor and Woinarski 1994; Masters 1996), but the mechanisms for changes in animal populations remain unresolved, and this issue has not been approached experimentally.

In their seminal work on desert tortoise ecology in the Mohave Desert, Woodbury and Hardy (1948) stated briefly that tortoises died from exposure to fires in the Mohave Desert. Only recently has the fire mortality

TABLE 13.1. Results of standardized surveys of burned areas to determine presence/absence of mortalities of desert tortoises in Arizona Upland subdivision of the Sonoran Desert of Arizona and Mohave Desert of Utah.

Site	Date	Area Burned (ha)	Area Surveyed (ha)	Search Time (h)	Number Live	Number Dead
SONORAN DESERT SITES						
Pusch Ridge	20 Sept 1987	1,750	15	41	1	7
Skyline	6 June 1992	202	40	21	1	0
Rock Peak	20 April 1993	117	35	8	1	2
Mother's Day	8 May 1994	486	138	112	6	7 (5)[a]
MOHAVE DESERT SITES						
Mill Creek	20 June 1993	895	41	46	3	0
Bulldog	4 Aug 1993	1,079	23	11	0	0

[a] Number in parentheses is the actual number of tortoises determined to have died as a result of the fire.

of tortoises been quantified in standardized surveys after fires in the Mohave and Sonoran Deserts (Esque et al., in review). Here, we report the results of postfire surveys at five locations in the northern Sonoran Desert of Arizona. In Sonora, fire mortality of tortoises has not been systematically studied, but due to the extreme temperatures of buffelgrass fires the mortality of tortoises and many other small animals surely must be high (fig. 13.3).

Surveys to determine the effect of fires on desert tortoises were conducted at four Sonoran and two Mohave sites (Esque et al., in review). Dead tortoises were not found at all of the sites surveyed (table 13.1). The amount of damage to burned individuals varied, but in many cases tissue damage was severe enough to incinerate limbs or portions of the tortoises' shells (fig. 13.4). One desert tortoise severely burned in a Mohave Desert fire was determined to have eventually died from secondary injuries related to heat and smoke exposure due to massive infection and changes in body chemistry (Homer et al. 1994). Although live tortoises were also found at all of these sites, each survey resulted in more dead than live tortoises. We did not observe any fire-related injuries to live tortoises at these sites, but such injuries have been observed elsewhere in Arizona (Cecil R. Schwalbe and Roy C. Averill-Murray, unpubl. data).

We speculate that losses of adult female desert tortoises in fires may

Figure 13.4. A charred Sonoran desert tortoise in the foothills of the Bradshaw Mountains, Yavapai County, Arizona, on 29 July 1991. The burned epidermal scute will slough off, beginning the regeneration process. (Photograph by Cecil Schwalbe)

be greater than males because peak fire season during the arid foresummer (June, July, and August; Swantek et al. 1999; P. J. Swantek, pers. comm. 1997) coincides with the egg-laying of desert tortoises (Averill-Murray et al., ch. 7 of this volume). Whereas male tortoises often remain dormant in shelters during the arid foresummer, females are more active during this period.

A Case Study: The Mother's Day Fire in Saguaro National Park

The Mother's Day Fire started on 8 May 1994 in the Rincon Management Unit of Saguaro National Park. The area is in Arizona Upland desertscrub on the western base of the Rincon Mountains in Pima County, Arizona. The fire was ignited by faulty brakes on a car parked on a roadside and primarily fueled by red brome. Initial surveys in mid-June yielded 5 dead tortoises (table 13.1). With the 34 live adult tortoises found during a population survey of the area one year after the fire (Wirt et al. 1997), there were a minimum of 39 adult desert tortoises inhabiting the area prior to the fire. The estimated mean population density for the sites of 44.8 tortoises/km^2 falls within the range of population estimates for Arizona Upland habitats (Averill-Murray et al., ch. 6 of this volume). Considering the possibility of undetected tortoises, the mortality due to fire would have ranged from 4.3 percent, which is twice as high as expected background mortality in healthy populations (Averill-Murray et al., ch. 6 of this volume), to 12.8 percent based on the observed number of animals. A 12 percent population loss in a long-lived population of animals such as the desert tortoise is considered to be catastrophic (Brussard et al. 1994). Because the Mother's Day Fire was only of moderate size and intensity, it is easy to surmise greater mortality during larger, hotter fires in the Arizona Upland and especially during the buffelgrass-fueled fires in the central Sonoran Desert. The numbers of dead tortoises found after fires may seem small, but increased mortality and recurrent fires pose a serious threat to desert tortoise populations.

Long-Term Mortality due to Wildfire

We have shown that not all of the tortoises in a burned area are killed outright. How do the tortoises that survive fires fare in modified habitats in the years following fires? Habitat changes that affect the thermal environment can be important ecological events to reptiles (Heatwole and Taylor 1987), especially desert tortoises. Habitat changes can range from subtle to quite dramatic depending on prefire vegetation type, fire intensity, and whether repeated fires have burned in the same location. Sometimes previously burned areas are not recognizable to the casual observer unless repeated observations have been made over time.

In some cases, however, fire-induced changes have dramatic effects on plant communities (McLaughlin and Bowers 1982; Loftin 1987; Wilson et al. 1994). For example, preliminary analyses of the loss of saguaros at the Mother's Day Fire indicated greater than 20 percent mortality as a result of fire (Schwalbe et al. 1999). Changes in plant community composition and structure may cause alteration in the availability of thermal shelter, forage plants, and incubation sites for eggs. Habitat structural changes, especially the loss of large shrubs and white-throated packrat *(Neotoma albigula)* houses, decrease the number and quality of microhabitats with deep shade, potentially inhibit movements across large open areas, and increase vulnerability to predators (Berry and Turner 1986). Monocultures of buffelgrass may become virtually impassable barriers to tortoises. It also seems likely that dense stands of exotic annual grasses could be a problem for very small tortoises.

Normally, predation on desert tortoises is greatest in the smallest size classes because adults in excess of 200 mm carapace length are too large for most natural predators in their habitats (Berry and Turner 1986; Brussard et al. 1994). However, smaller tortoises are susceptible to predation by a host of predators, including, but not limited to, badgers *(Taxidea taxus)*, common ravens *(Corvus corax*; Boarman and Berry 1995), coyotes *(Canis latrans)*, foxes *(Urocyon cinereoargenteus, Vulpes macrotus)*, and golden eagles *(Aquila chrysaetos*; Brussard et al. 1994). Some instances of predation are described in references found in Grover and DeFalco (1995), yet the majority of these are anecdotal accounts with little value for determining population outcomes.

Availability of Food Plants

Desert tortoises are herbivorous and almost entirely dependent on plants for food (Luckenbach 1982; Nagy and Medica 1986; Jennings 1993; Esque 1994; Avery 1995; Van Devender et al., ch. 8 of this volume). Desert fires promote the domination of exotic plants in desert habitats and a reduction in plant species richness. However, is the loss of variety in diets a net loss or gain for the desert tortoise? Desert tortoises in two populations in the northeastern Mohave Desert were observed eating a majority of introduced plants in their diets, in consecutive years, with no outwardly visible health problems (Esque 1994). During years when the same tortoises had a greater variety of plants available, however, their diets were more varied. When compared with like growth forms (i.e., grasses to grasses and herbs to herbs), the nutritional content of native desert annuals do not differ significantly from introduced desert annuals (Nagy et al. 1998). Although it is clear that fires can cause a dramatic shift in plant species availability, there is as yet no consensus on how these changes affect desert tortoises in the long term. We suspect, however, that the lack of diversity in diets may have a net negative result over time.

Changes in the Thermal Environment

Desert tortoises are ectotherms and cannot generate heat by internal metabolic processes or cool themselves evaporatively (e.g., perspiring or panting), such as many other animals do. Instead, tortoises depend entirely on behavioral thermoregulation of ambient environmental temperatures to regulate their body temperatures (Averill-Murray et al., ch. 7 of this volume). By moving to cooler or warmer patches in their habitats, tortoises adjust their ambient and internal temperatures. These thermal microhabitats (Larmuth 1984; Zimmerman et al. 1994; Hillard 1996) are important resources to reptiles and can have a direct influence on their fitness (Magnuson et al. 1979; Grant 1990).

Different sources of thermal energy in desert habitats include direct solar radiation, conductance of thermal energy through the ground surface, reflected and/or reradiated thermal energy from the sky and ground, and convective thermal energy from the air. Desert tortoises may thermoregulate behaviorally by basking in the sun or placing their bodies on warm surfaces (e.g., rocks and soil surfaces) to increase body temperatures. Alternatively, tortoises use caves and burrows, surface structures (e.g., rocks and cliffs), or large plants for protection from both hot and cold temperature extremes. The animals can travel reliably from point to point among known resources such as watering sites (Medica et al. 1980), feeding sites (e.g., patches of curly mesquite, *Hilaria belangeri*; Peter Holm, pers. comm. 1999), or sheltersites in caves or under woody plants (Todd C. Esque, pers. obs.). Chelazi and Calzolai (1985) demonstrated wild Hermann's tortoises *(Testudo hermanni)* that were familiar with their environments thermoregulated more efficiently than naive tortoises experimentally moved into unfamiliar territories. In undisturbed Arizona Upland desertscrub, the great diversity of plant growth forms as well as packrat houses, rocky outcrops, boulders, and arroyos provide many shade opportunities readily used by tortoises, many of which are lost in severely burned habitats. Surface temperatures where desert tortoises live often exceed 40°C. Because body temperatures in excess of 40°C are lethal for desert tortoises after only a few minutes, increased exposure to extreme temperatures is a serious problem.

Fires can alter the physiognomy of desert tortoise habitats significantly. Areas converted to buffelgrass, either for agriculture or by invasion, lose most of the cover of trees and large shrubs after the fires. Standing crop biomass in converted areas could be up to five times less than areas with native vegetation (e.g., 3,000–4,000 kg/ha in buffelgrass versus 20,000 kg/ha in natural xeroriparian vegetation; Búrquez M. et al. 1996). After fires, sheltered sites for tortoises disappear, and there are marked changes in the thermal regime. As more soil is exposed, the boundary layer becomes thinner and temperatures increase. Buffelgrass rapidly regrows into dense stands, buffering temperatures that may then be lower than in more open, unaltered desertscrub. In central Sonora, desertscrub and thornscrub transformed to buffelgrass savannas can stretch for several kilometers without major interruptions. Eventually, the cleared areas can become so large that they are almost certainly unsurmountable barriers to tortoise dispersal. We do not know enough about the cues tortoises use to orient themselves, but if plants were used as cues for orientation, then it could be difficult for tortoises to find long-used resources such as watering, feeding, and cover sites.

We postulate that desert tortoise habitats become relatively inhospitable to tortoises after severe fires. This trend becomes more pronounced as habitats degrade toward exotic annual monocultures. The worst-case scenario would be a habitat relatively void of shade sources other than grasses. It seems likely that some measurable qualitative changes occur in habitats that are highly modified by fires. Future desert tortoise research in the Sonoran and Mohave Deserts should determine the extent to which tortoise behavior is altered by habitat alterations and whether this alters the vigor or fitness of individual tortoise, thus affecting the health of populations.

ACKNOWLEDGMENTS

We thank Brenda Carbajal, Cecily Westphal, Sandy Mosolf, and Carol Wakely for administrative support. Research technical assistance was provided by Cristina Jones and Bridgett Watts. Bill Halvorson provided valuable logistical and technical support. Field assistants at Saguaro National Park included Jeff Clarke, Lesley DeFalco, Sara Eckert, Dustin Haines,

Ryan Schwarz, and numerus Student Conservation Association volunteers and student interns. We are grateful for the support of Frank Hayes, Mark Holden, Natasha Kline, and Meg Weesner at Saguaro National Park. Funds for Alberto Búrquez's research were provided by DGAPA-UNAM project IN212894. Steven P. McLaughlin read and commented on the manuscript.

LITERATURE CITED

Avery, H. W. 1995. *Digestive performance of the desert tortoise* (Gopherus agassizii) *fed native and non-native desert vegetation.* Report to California Dept. of Parks and Recreation, Off-highway Motor Vehicle Recreation Division, Sacramento.

Bahre, C. J. 1995. Human impacts on the grasslands of southeastern Arizona. Pp. 249–52 *in* M. P. McClaran and T. R. Van Devender, eds. *The desert grassland.* University of Arizona Press, Tucson.

Berry, K. H., and F. B. Turner. 1986. Spring activities and habits of juvenile tortoises *(Gopherus agassizii)* in California. *Copeia* 1986:1010–12.

Billings, W. D. 1990. *Bromus tectorum*, a biotic cause of ecosystem impoverishment in the Great Basin. Pp. 301–22 *in* G. M. Goodwell, ed. *The Earth in transition: Patterns and processes of biotic impoverishment.* Cambridge University Press, Cambridge.

Boarman, W. I., and K. H. Berry. 1995. Common ravens in the Southwestern U.S. Pp. 73–75 *in* E. T. LaRoe, G. S. Farris, C. E. Puckett, P. D. Doran, and M. J. Mac, eds. *Our living resources: A report to the nation on the distribution, abundance, and health of U.S. plants, animals, and ecosystems.* U.S. Dept. of the Interior National Biological Service, Washington, D.C.

Brooks, M. L. 1998. Ecology of a biological invasion: Alien annual plants in the Mojave Desert. Ph.D. diss., University of California, Riverside.

Brown, D. E. 1982. 144.3 Sonoran savanna grassland. *Desert Plants* 4:137–41.

Brown, D. E., and R. A. Minnich. 1986. Fire and changes in creosote bush scrub of the western Sonoran Desert, California. *American Midland Naturalist* 116:411–22.

Brussard, P. F., K. H. Berry, M. E. Gilpin, E. R. Jacobson, D. J. Morafka, C. R. Schwalbe, C. R. Tracy, and F. C. Vasek. 1994. *Desert tortoise (Mojave population) recovery plan.* Region 1 of the U.S. Fish and Wildlife Service, Portland, Ore.

Búrquez M., A., A. Martínez Y., M. Miller, K. Rojas, M. A. Quintana, and D. Yetman. 1996. Mexican grasslands and the changing arid lands of Mexico: An overview and a case study in northwestern Mexico. Pp. 21–32 *in* B. Tellman, D. M. Finch, C. Edminster, and R. Hamre, eds. *The future of arid grasslands: Identifying issues, seeking solutions.* Proceedings RMRS-P-3, U.S. Forest Service, Rocky Mountain Station, Fort Collins, Colo.

Búrquez M., A., A. Martínez Y., R. S. Felger, and D. Yetman. 1999. Vegetation and habitat diversity at the southern desert edge of the Sonora Desert. Pp. 36–67 *in* R. H. Robichaux, ed. *Ecology of Sonoran Desert plants and plant communities.* University of Arizona Press, Tucson.

Callison, J., J. D. Brotherson, and J. E. Bowns. 1985. The effects of fire on the blackbrush

(Coleogyne ramosissima) community of southwestern Utah. *Journal of Range Management* 38:535–38.

Cave, G. H., and D. T. Patten. 1984. Short-term vegetation responses to fire in the Upper Sonoran Desert. *Journal of Range Management* 37:491–95.

Chelazi, G., and R. Calzolai. 1985. Thermal benefits from familiarity with the environment in a reptile. *Oecologia* 68:557–58.

Cook, S. F., Jr. 1959. The effects of fire on a population of small rodents. *Ecology* 40:102–8.

D'Antonio, C. M., and P. M. Vitousek. 1992. Biological invasions by exotic grasses, the grass/fire cycle, and global change. *Annual Review of Ecology and Systematics* 23:63–78.

Esque, T. C. 1994. Diet and diet selection of the desert tortoise *(Gopherus agassizii)* in the northeastern Mojave Desert. M.S. thesis, Colorado State University, Fort Collins.

Esque, T. C., C. R. Schwalbe, L. A. DeFalco, R. B. Duncan, and T. J. Hughes. In review. Effects of desert wildfires on desert tortoise *(Gopherus agassizii)* and other small vertebrates. *The Southwestern Naturalist.*

Felger, R. S. 1990. *Non-native plants of Organ Pipe Cactus National Monument, Arizona.* Technical Report no. 31, Cooperative National Park Resources Studies Unit, University of Arizona, Tucson.

Fritts, T. H., and R. D. Jennings. 1994. Distribution, habitat use and status of the desert tortoise in Mexico. Pp. 49–56 *in* R. B. Bury and D. J. Germano, eds. *Biology of North American tortoises.* Fish and Wildlife Research no. 13, U.S. Dept. of the Interior National Biological Survey, Washington, D.C.

Goldingay, R., G. Daly, and F. Lemckert. 1996. Assessing the impacts of logging on reptiles and frogs in the montane forests of southern New South Wales. *Wildlife Research* 23:495–510.

Grant, B. W. 1990. Trade-offs in activity time and physiological performance for thermoregulating desert lizards, *Sceloporus merriami. Ecology* 71:2323–33.

Greenberg, C. H., D. G. Neary, and L. D. Harris. 1994. Effect of high-intensity wildfire and silvicultural treatments on reptile communities in sand-pine scrub. *Conservation Biology* 8:1047–57.

Grover, M. C., and L. A. DeFalco. 1995. *Desert Tortoise* (Gopherus agassizii): *Status-of-knowledge outline with references.* General Technical Report no. INT-316, U.S. Forest Service Intermountain Research Station, Ogden, Utah.

Hammel, B. 1990. The distribution of diversity among families, genera, and habitat types in the La Selva Flora. Pp. 75–85 *in* A. H. Gentry, ed. *Four Neotropical rainforests.* Yale University Press, New Haven, Conn.

Heatwole, H. F., and J. Taylor. 1987. *Ecology of lizards.* Surrey Beatty & Sons, Chipping Norton, New South Wales.

Hillard, S. 1996. The importance of the thermal environment to juvenile desert tortoises. M.S. thesis, Colorado State University, Fort Collins.

Homer, B. L., E. R. Jacobson, M. M. Christopher, and M. B. Brown. 1994. Physiolgical and morphological effects of burn injury in a desert tortoise, *Gopherus agassizii. Proceedings of the Desert Tortoise Council Symposium* 1994:156.

Howard, W. E., R. L. Fenner, and H. E. Childs Jr. 1959. Wildlife survival in brush burns. *Journal of Range Management* 12:230–45.

Humphrey, R. R. 1974. Fire in the deserts and desert grassland of North America. Pp. 365–400 *in* T. T. Kozlowski and C. E. Ahlgren, eds. *Fire and ecosystems*. Academic Press, New York.

Jennings, W. B. 1993. Foraging of the desert tortoise *(Gopherus agassizii)* in the western Mojave Desert. M.S. thesis, University of Texas, Austin.

Kahn, W. C. 1960. Observations on the effect of a burn on a population of *Sceloporus occidentalis. Ecology* 41:358–59.

Klemmedson, J. O., and J. G. Smith. 1964. Cheatgrass *(Bromus tectorum* L.). *Botanical Review* 30:226–62.

Larmuth, L. 1984. Microclimates. Pp. 57–66 *in* J. L. Cloudsley-Thompson, ed. *The Sahara Desert*. Pergamon, Oxford, U.K.

Loftin, S. R. 1987. Postfire dynamics of a Sonoran Desert ecosystem. M.S. thesis, Arizona State University, Tempe.

Luckenbach, R. 1982. Ecology and management of the desert tortoise *(Gopherus agassizii)* in California. Pp. 1–37 *in* R. B. Bury, ed. *Biology of North American tortoises*. Fish and Wildlife Research no. 12, U.S. Fish and Wildlife Service, Washington, D.C.

Magnuson, J. J., L. B. Crowder, and P. A. Medvick. 1979. Temperature as an ecological resource. *American Zoologist* 19:331–43.

Masters, P. 1996. The effects of fire-driven succession on reptiles in *Spinifex* grasslands at Uluru National Park, Northern Territory. *Wildlife Research* 23:39–48.

McClaran, M. P., and M. E. Anable. 1992. Spread of introduced Lehmann lovegrass along a grazing intensity gradient. *Journal of Applied Ecology* 29:92–98.

McCoy, E. D., H. R. Mushinsky, and D. S. Wilson. 1994. Growth and sexual dimorphism of *Gopherus polyphemus* in central Florida. *Herpetologica* 50:119–28.

McLaughlin, S. P., and J. E. Bowers. 1982. Effects of wildfire on a Sonoran Desert plant community. *Ecology* 63:246–48.

Medica, P. A., R. B. Bury, and R. A. Luckenbach. 1980. Drinking and construction of water catchments by the desert tortoise, *Gopherus agassizii*, in the Mojave Desert. *Herpetologica* 36:301–4.

Melgoza, G., and R. S. Nowak. 1991. Competition between cheatgrass and two native species after fire: Implications from observations and measurements of root distribution. *Journal of Range Management* 44:37–33.

Murray, R. C., and V. Dickinson. 1996. *Management plan for the Sonoran Desert population of the desert tortoise in Arizona*. Arizona Game and Fish Dept., Phoenix.

Mushinsky, H. R., and D. J. Gibson. 1991. The influence of fire periodicity on habitat structure. Pp. 237–59 *in* S. S. Bell, E. D. McCoy, and H. R. Mushinsky, eds. *Habitat structure: The physical arrangement of objects in space*. Chapman and Hall, New York.

Nagy, K. A., and P. A. Medica. 1986. Physiological ecology of desert tortoises in southern Nevada. *Herpetologica* 42:73–92.

Nagy, K. A., B. T. Henen, and D. B. Vyas. 1998. Nutritional quality of native and introduced food plants of wild desert tortoises. *Herpetologica* 42:260–67.

Narog, G. M., A. L. Koonce, R. C. Wilson, and B. M. Corcoran. 1995. Burning in Arizona's giant cactus community. Pp. 175–76 *in* D. R. Weise and R. E. Martin, technical coordinators. *Proceedings of the Biswell Symposium: Fire issues and solutions in urban interface and wildland ecosystems.* General Technical Report no. PSW-GTR-158, U.S. Forest Service, Walnut Creek, Calif.

Patten, D. T., and G. H. Cave. 1984. Temperatures and physical characteristics of a controlled burn in the upper Sonoran Desert. *Journal of Range Management* 37:277–80.

Pianka, E. 1992. Disturbance, spatial heterogeneity, and biotic diversity: Fire succession and arid Australia. *Research and Exploration* 8:352–71.

Rogers, G. F. 1985. Mortality of burned *Cereus giganteus. Ecology* 66:630–32.

———. 1986. Comparison of fire occurrence in desert and nondesert vegetation in Tonto National Forest, Arizona. *Madroño* 33:278–83.

Rogers, G. F., and J. Steele. 1980. *Sonoran Desert fire ecology.* General Technical Report no. RM81, U.S. Dept. of Agriculture Forest Service, Rocky Mountain Range Experimental Station, Fort Collins, Colo.

Rogers, G. F., and M. K. Vint. 1987. Winter precipitation and fire in the Sonoran Desert. *Journal of Arid Environments* 13:47–52.

Rondeau, R., T. R. Van Devender, C. D. Bertelsen, P. Jenkins, R. K. Wilson, and M. A. Dimmitt. 1996. Annotated flora of the Tucson Mountains, Pima County, Arizona. *Desert Plants* 12:3–46.

Schmid, M. K., and G. F. Rogers. 1988. Trends in fire occurrence in the Arizona Upland Subdivision of the Sonoran Desert, 1955–1983. *Southwestern Naturalist* 33:437–44.

Schwalbe, C. R., T. C. Esque, M. J. Nijhuis, D. F. Haines, and P. J. Swantek. 1999. Effects of fire on Arizona Upland desertscrub at Saguaro National Park. Pp. 107–9 *in* L. Benson and B. Gebow, eds. *Proceedings of a century of parks in southern Arizona.* Second Conference on Research and Resource Management in Southern Arizona National Parks, U.S. National Park Service Southern Office and U.S. Geological Survey Sonoran Desert Field Station, University of Arizona, Tucson.

Steenbergh, W. F., and C. H. Lowe. 1977. *Ecology of the saguaro. II. Reproduction, germination, establishment, growth, and survival of the young plant.* Scientific Monograph Series Vol. 8. U.S. National Park Service, Washington, D.C.

Swantek, P. J., W. L. Halvordon, and C. R. Schwalbe. 1999. GIS database to analyze fire history in southern Arizona and beyond: An example from Saguaro National Park. Cooperative Park Studies Unit Technical Report no. 61, University of Arizona and the U.S. Geological Survey, Tucson.

Trainor, C. R., and J. V. C. Woinarski. 1994. Responses of lizards to 3 experimental fires in the savanna forests of Kakadu National Park. *Wildlife Research* 21:131–48.

Treviño, M. A., M. E. Haro, S. L. Barrett, and C. R. Schwalbe. 1994. Preliminary desert tortoise surveys in central Sonora, México. *Proceedings of the Desert Tortoise Council Symposium* 1987–1991:379–88.

Turner, R. M., J. E. Bowers, and T. L. Burgess. 1995. *Sonoran Desert plants: An ecological atlas.* University of Arizona Press, Tucson.

Van Devender, T. R. 1995. Desert grassland history: Changing climates, evolution, biogeography and community dynamics. Pp. 68–99 *in* M. P. McClaran and T. R. Van Devender, eds. *The desert grassland.* The University of Arizona Press, Tucson.

Van Devender, T. R., R. S. Felger, and A. Búrquez M. 1997. Exotic plants in Sonora, Mexico. Pp. 10–15 *in* M. Kelly, E. Wagner, and P. Warner, eds. *Proceedings of the California Exotic Pest Plant Council symposium.* Vol. 3. California Exotic Pest Plant Council, Concord, Calif.

Whelan, R. J. 1995. *The ecology of fire*. Cambridge Studies in Ecology. Cambridge University Press, New York.

Wilson, R. C., M. G. Narog, A. L. Koonce, and B. M. Corcoran. 1994. *Postfire regeneration in Arizona's giant saguaro shrub community*. General Technical Report no. RM-GTR-264, U.S. Forest Service, Tucson, Ariz.

————. 1995. Impact of wildfire on saguaro distribution patterns. *San Bernardino County Museum Association Quarterly* 42:46–47.

Wirt, E. B., B. E. Martin, and S. F. Hale. 1997. Population density estimate for the Mother's Day Burn Study Plot. Draft Report to Saguaro National Park, Tucson, Ariz.

Woodbury, A. M., and R. Hardy. 1948. Studies of the desert tortoise, *Gopherus agassizii*. *Ecological Monographs* 18:146–200.

Wright, H. A., L. F. Neuenschwander, and C. M. Britton. 1979. *The role and use of fire in sagebrush-grass and pinyon-juniper plant communities: A state-of-the-art review*. General Technical Report no. INT-58, U.S. Forest Service Intermountain Forest and Range Experiment Station, Ogden, Utah.

Young, J. A., R. A. Evans, R. E. Eckert Jr., and B. L. Kay. 1987. Cheatgrass. *Rangelands* 9:266–70.

Zimmerman, L. C., M. P. O'Connor, S. J. Bulova, J. R. Spotila, S. J. Kemp, and C. J. Salice. 1994. Thermal ecology of desert tortoises in the eastern Mojave Desert: Seasonal patterns of operative and body temperatures, and microhabitat utilization. *Herpetological Monographs* 8:45–59.

CHAPTER 14

Conservation and Protection of the Desert Tortoise in Arizona

JEFFREY M. HOWLAND AND JAMES C. RORABAUGH

The life history and ecology of desert tortoises *(Gopherus agassizii)* leave them vulnerable to various human impacts (U.S. Fish and Wildlife Service [USFWS] 1994). The need for conservation and protection has long been recognized (e.g., Woodbury and Hardy 1948). We begin this chapter with an overview of threats to Sonoran desert tortoises and a discussion of why tortoises are sensitive to these threats. Although our understanding of Sonoran tortoise biology has increased in the past decade, we must rely, to some extent, on information from the federally listed and more thoroughly studied Mohave population. We then examine the roles of government agencies in conserving desert tortoises, summarize historical and current conservation and protection efforts, and make recommendations for future research, monitoring, conservation planning, and management of the Sonoran desert tortoise in Arizona.

Threats

Desert tortoise populations have been impacted by fragmentation, degradation, and loss of habitat and by elevated mortality rates caused by drought, shooting, road mortality, predation by dogs and ravens (especially near human settlements and roads), disease, livestock trampling, and collecting, among other causes (USFWS 1994). Release of captive tortoises may result in disruption of wild populations or introduction of diseases.

Some of these impacts may be more prevalent in the Mohave Desert than in the Sonoran Desert. Shooting has been documented in the Sonoran Desert (Shields et al. 1990; Roy C. Averill-Murray, pers. comm. 1998), but may be more common in the Mohave Desert (Berry 1986a). Raven predation, which is locally common in the Mohave (Boarman 1997), is not known to be significant for Sonoran tortoises. Collection of tortoises for food is probably limited in Arizona, although it was practiced historically by Native Americans (Nabhan, ch. 15 of this volume) and Mexican Americans (Martin 1996), and is known to occur in California (USFWS 1994), Sonora, and Sinaloa (Bury et al., ch. 5 of this volume).

Differences in habitat preference render many of the threats to Mohave desert tortoises less important, or at least less direct, for Sonoran tortoises. Mohave tortoises typically use *bajadas* and valley floors, whereas Sonoran tortoises prefer rocky slopes and incised washes in bajadas. Activities that degrade or destroy habitat, such as urbanization, agricultural development, livestock grazing, military activities, use of off-highway vehicles, and construction of highways, canals, and utility lines, are concentrated in flats, leaving rocky slopes relatively intact.

Habitat Fragmentation, Degradation, and Loss

Habitat impacts are widely recognized as the leading threat to global biodiversity (Ehrlich 1988; Wilson 1992). As is the case for other species, fragmentation, degradation, and outright loss of habitat have substantially reduced desert tortoise populations, at least in local areas. Factors affecting Sonoran tortoise habitat include roads, mines, quarries, urban and agricultural development, microwave and radio facilities, and livestock grazing.

Canals, major highways, and urban and agricultural development have caused habitat loss, but their most detrimental effect may be as barriers to movement within or among montane populations (figs. 14.1, 14.2). We do not know how much tortoises moved historically or to what extent this movement has been reduced. Observations of tortoises at substantial distances from rocky slopes (e.g., Woodman et al. 1995) and long-distance movements of radio-telemetered animals across habitat atypical of Sonoran tortoises (Roy C. Averill-Murray, pers. comm. 1999) suggest that such movement occurs. As in classical island biogeography, frequency of movement among populations is likely a function of distance between populations (more migration with shorter distance), physical size of habitat areas (more immigrants into larger areas), size of source populations (more emigrants from larger, denser source populations), and characteristics of intervening areas that determine how readily they can be crossed.

Figure 14.1. Where urban development extends to the base of desert tortoise habitat, as here at Squaw Peak in Phoenix, tortoise populations are increasingly subject to illegal collection, road mortality, and predation by pets. Illegally released tortoises or escaped animals may spread disease to wild populations. Any natural movement of tortoises among montane populations is precluded by urban or other development in the intervening valleys. (Photograph by Jim Rorabaugh)

Human-made barriers to movement may intensify genetic isolation of populations and reduce the probability that isolated habitat islands will be recolonized after local extinction. With the exception of apparently released captives, tortoises appear to be scarce or absent from most suitable habitat in Phoenix-area mountain parks and reserves (Brian K. Sullivan, pers. comm. 1998), where they were found historically. Collection, vandalism, predation by pets, and road mortality cause local extinction, and an urban sea precludes recolonization from potential source populations. Living among rocks protects Sonoran tortoises from many threats, but constrains populations to mountain ranges, where suitable habitat is limited and patchy. Thus, most

Figure 14.2. Although the development shown here (Central Arizona Project in the distance and Interstate 10 in the foreground) is relatively minor compared to downtown Phoenix (fig. 14.1), the effect on movements between populations is the same. The canal and freeway have effectively isolated tortoise populations on Picacho Peak, from which this picture was taken, from the main body of the Picacho Mountains. (Photograph by Roy Averill-Murray)

Sonoran tortoise populations are small and isolated (Johnson et al. 1990; see map in Germano et al. 1994:77). Human impacts exacerbate population isolation that may naturally be near the limits at which normal metapopulation dynamics break down.

Population viability analyses conducted for the Mohave population (USFWS 1994; no such analysis exists for the Sonoran population) suggest that a population of 20,000 to 60,000 tortoises is required to assure a 50 percent probability of persistence for five hundred years. If Sonoran populations average 7.7 individuals/km^2 (a figure near the low end of observed population densities presumed to be viable; see Arizona Interagency Desert Tortoise Team [AIDTT] 1996a; Averill-Murray et al., ch. 6 of this volume), then continuous habitat of about 2,590–7,770 km^2 is needed to sustain a population under the scenario above (assuming no immigration from neighboring populations). Perhaps the only appropriate habitat of this size in Arizona is the area south of Kingman and north of the Bill Williams River. The Mohave analyses assumed mean population growth of −1.5 percent per year (declining population), based on data from thirteen Mohave sites (appendix C in USFWS 1994). Data from Sonoran Desert study plots do not suggest widespread population decline, so extrapolations from Mohave data may be inappropriate. Furthermore, for long-term conservation planning, it is not reasonable to assume a constant and sustained decline, especially if conservation actions are implemented to allow recovery. For stable (includes moderately fluctuating) populations, smaller (perhaps much smaller) population size and therefore a smaller area of continuous habitat should ensure persistence. Due to the naturally small size and isolation of most Arizona populations, occasional exchange of individuals may be important for recolonization and prevention of genetic deterioration (Frankel and Soule 1981; Ralls and Ballou 1983). The smaller the population, the more important occasional immigration is likely to be.

Because Sonoran tortoises spend proportionately more time on rocky slopes than do livestock, which concentrate use on flats, ridge tops, and drainage bottoms, trampling and alteration of forage availability may be less severe for Sonoran than for Mohave desert tortoises, whose preferred habitat is often more heavily used by livestock. Off-highway vehicles are also more prevalent in the Mohave Desert than in the Sonoran, perhaps due to more open vegetation and proximity to the coastal southern California megalopolis. Mines exist in Sonoran tortoise habitat, but their impacts, although long-lasting (including trapping of tortoises in abandoned mine shafts and pits; Shields et al. 1990), are typically localized. Microwave facilities, power lines, fiber optic cables, and quarries affect relatively small areas.

Desert tortoise habitat is degraded by nonnative plants such as Mediterranean or split grass *(Schismus barbatus)*, red brome *(Bromus rubens)*, and buffelgrass *(Pennisetum ciliare)*. Some nonnative plants have low nutritional value (Avery 1994; but see McArthur et al. 1994; Nagy et al. 1998) but, more importantly, frequency of fire in Sonoran desertscrub increases with nonnative plant invasions. Wildfire in Sonoran desertscrub is almost always associated with dense growth of nonnative plants because they carry fires more readily than natives (Esque et al., ch. 13 of this volume). Large areas of Sonoran desertscrub have burned in recent years, causing direct mortality of tortoises and leaving habitat that is of questionable value. In portions of the Pakoon Basin, north of the Grand Canyon in Mohave County, Arizona, repeated burning has caused conversion of Mohave desertscrub to a non-native annual grassland dominated by red brome that may no longer support desert tortoises at viable population densities (Timothy A. Duck, pers. comm. 1999).

Human-Induced Mortality and Take

The life history of desert tortoises makes them particularly sensitive to elevated levels of mortality, especially of adult animals. We begin this section with a brief review of desert tortoise life-history characteristics that explain this sensitivity.

The small size and soft shell of young desert tortoises leave them more vulnerable than adults to climatic extremes and predation. It may take ten to fifteen years for a desert tortoise to reach its mature size of about 180–200 mm (Germano et al., ch. 11 of this volume). Adult tortoises have few natural enemies, their size provides a buffer against extremes in temperature so they do not overheat or freeze as easily as small tortoises, and they are able to store internally large reserves of water and energy. Annual adult survival rates are often 95–98 percent (Luckenbach 1982; USFWS 1994; Howland and Klug 1996). A tortoise may live decades (Germano 1992), perhaps a century, and female Sonoran tortoises produce one clutch of eggs during most years of their mature life. Recruitment of juvenile tortoises into adult age classes is low, however (Turner et al. 1987; Germano 1994). For long-term population persistence, the lifetime reproductive effort of an adult female must yield, on average, at least two offspring that reach maturity, one of each sex (assuming a 50:50 sex ratio).

Tortoises prosper in wet years: They feed, grow, reproduce, and survive better than in dry periods (Peterson 1994; Henen et al. 1998; Roy C. Averill-Murray, pers. comm. 1998). With high rainfall, young tortoises may reduce exposure to the environment and predators by quickly finding food, eating their fill, and returning to shelter. If several good-to-average years occur in succession, a juvenile tortoise may reach adult size sooner and with minimal risk. Tortoises hatching near the beginning of such favorable interludes, which may occur years or decades apart, may be more likely to survive to maturity.

An integral component of the desert tortoise life-history strategy is that adults have a long reproductive life. Tortoises cannot predict climate, of course, so a mature female's chances of having young survive to adulthood may be low unless she has a few young almost every year, even in dry years if her energy reserves allow it, because good times may begin unpredictably. Environmental conditions may change markedly during the long incubation period, through which developing embryos are buffered from the external environment by the nest and hard-shelled amniotic egg. A female that lives a long time and reproduces most years is likely to have young hatch near the beginning of one of the sporadic favorable periods at least once or twice during her life. She may enjoy greater lifetime reproductive success than those using other reproductive strategies, such as producing more eggs per year over a shorter life (high reproductive effort is usually costly to health and survival) or strongly concentrating reproduction in good years and forgoing reproduction in all other years.

Population biologists may assess an individual's value to the future of its population by its genetic contribution to future generations. This reproductive value can be measured as the number of offspring a female is expected to produce during the remainder of her life. Most hatchling female tortoises will not reproduce before dying, so a hatchling's reproductive value is low. Adult females have high annual survivorship, so they are likely to produce many young before dying and their reproductive value is high. A tortoise's reproductive value probably plateaus early in reproductive life and remains fairly high until death, because wild tortoises rarely reach senescence.

Because of their high reproductive value, an increase in adult mortality is likely to have negative consequences at the population level. The larger size (and therefore visibility) of adult tortoises leaves them more vulnerable to human impacts like collecting and shooting, and adults' tendency to move further and more frequently than juveniles may cause higher levels of road mortality. Tortoise populations are slow to rebound from declines. USFWS (1994) predicted that even under favorable conditions, populations may not be capable of doubling in less than 70 years. To maintain population viability, protection of adult tortoises is essential (although it is not safe to ignore juvenile mortality; Congdon et al. 1993). Resource managers must take measures to preclude mortality in excess of natural rates.

Development projects in tortoise habitat cause incidental mortality to tortoises, but, more importantly, habitat is lost permanently. Translocation of tortoises and other animals and plants has become a popular quick-fix to mitigate habitat loss caused by development projects. Permanent habitat loss necessarily reduces the number of tortoises that can be supported by the environment as a whole. In the Sonoran Desert of Arizona, most intact habitats host healthy populations of tortoises, so translocation from developing areas, even if it could be made to work (Berry 1986b), is not usually a useful conservation measure. Furthermore, disease, genetic, and behavioral concerns argue for cautious use of translocation (e.g., Jacobson et al. 1999). To allocate limited conservation funds effectively, resource managers must protect remaining habitat and the tortoises already living there. For habitat and tortoises on the periphery of developing areas, requiring mitigation and compensation for damage is an important tool in achieving broader conservation goals. Translocation of tortoises whose habitat is permanently lost to development is usually futile and wastes the limited resources available for tortoise conservation. Exceptions may include experimental introductions of tortoises into completely isolated, uninhabited areas to study the effectiveness of translocation and translocation into areas from which tortoises are extirpated (or reduced to inviable population size), but where the threats that caused the original extirpation are no longer present. For areas that host healthy populations, risks inherent to translocation will usually outweigh its potential, but uncertain, benefits. We recommend allowing natural processes to operate in recovery of local populations that have suffered declines. If threats are removed, natural population and metapopulation dynamics will determine local population size or density in the long term.

Where habitat is intact and protected, management should focus on protection of tortoises by reducing human-induced mortality. Where roads pass through Mohave tortoise habitat, population density may be depressed for 1.6 km or more on each side of the road (Nicholson 1978; Sazaki et al. 1995). Roads probably affect Sonoran tortoises similarly. Tortoise-proof fencing along roads reduces road mortality (Boarman et al. 1997; fig. 14.3), and tortoise-friendly culverts may make roads less of a barrier to movement, but

Figure 14.3. Desert tortoise barrier fencing constructed by the Bureau of Reclamation at Lake Pleasant County Park, northwest of Phoenix, to prevent mortality of tortoises on roads and to keep tortoises out of recreational areas. (Photograph by Jim Rorabaugh)

their effectiveness is not well known. It is crucial to mitigate and compensate for impacts to habitat and movement corridors. For projects causing temporary or light disturbance, effective mitigation measures include minimizing the size and intensity of disturbance, preproject clearances, biological monitors, short-distance translocation of tortoises during project implementation (AIDTT 1996b), and worker education (e.g., Olson et al. 1993; LaRue 1996). Compensation fees can be assessed for residual impacts (Desert Tortoise Compensation Team 1991). Finally, public education and protection of tortoises from collecting or killing can control the loss of tortoises from intact habitats.

Disease

Disease has not emerged as a major threat to Sonoran desert tortoises (Dickinson et al., ch. 10 of this volume), but federal listing of the Mohave population was hastened by extensive mortality in the western Mohave Desert, later attributed to upper respiratory tract disease (URTD) caused by *Mycoplasma agassizii* (Brown et al. 1994; Berry 1997). A five-year study of tortoise health at two sites in west-central Arizona found no clinical signs of URTD, although biochemical testing revealed limited occurrence of *M. agassizii* (Dickinson et al. 1996). Clinical signs of URTD have been minor and infrequent in examinations of about 1,000 different tortoises in thirty-five surveys at fourteen population monitoring sites across the Sonoran Desert of Arizona since 1990 (see Averill-Murray et al., ch. 6 of this volume).

The same surveys found high frequency and severity of shell disease, or cutaneous dyskeratosis, in most populations. The cause of this disease and its effects on populations or individuals remain unclear. Limited evidence suggests heavy-metal toxicity as a possible cause (Homer et al. 1996). Shell disease was common at Chuckwalla Bench, California, before a large-scale mortality event, but the cause of mortality could not be established (Berry 1997). Dyskeratosis has been conspicuous at several Sonoran Desert sites for at least ten years, with little or no consequence (Dickinson et al., ch. 10 of this volume, table 10.4).

Although we lack evidence that disease currently threatens Sonoran tortoise populations, we cannot discount its potential future importance. Several species of exotic tortoises and turtles are common as pets and have been found living free in Arizona, presumably after escaping or having been released. Such animals are known to carry a variety of diseases that may be transmissible to desert tortoises, with unknown consequences (James L. Jarchow, pers. comm. 1999). If disease does strike the Sonoran Desert population, it may be less devastating than in the Mohave population because the patchy and isolated distribution of Sonoran tortoises should slow or restrict the spread of disease.

Disease and Social and Genetic Disruption: Escape or Release of Captive Tortoises

Protection of wild tortoises came only after large, captive tortoise populations were established in many Southwestern cities. Captive desert tortoises breed readily, so the captive population is probably growing. Illegal importation of desert tortoises from California and Nevada is common, and Texas tortoises *(Gopherus berlandieri)* are occasionally but consistently found in Phoenix (Arizona Game and Fish Department, [AGFD], unpubl. data). Escape and release of captive desert tortoises (and other chelonians) are common, but not well documented (Bury et al. 1988; Howland 1994a). Escaped tortoises are frequently found in suburban areas (Arizona-Sonora Desert Museum and AGFD, unpubl. data), especially in Phoenix and Tucson, and tortoises are released into urban parks and natural areas (AGFD, unpubl. data).

Illegal release and accidental escape of captive tortoises may disrupt social systems (e.g., elicit agonistic responses, displace resident tortoises); cause inappropriate genetic mixing, even between species; or spread disease to wild populations (Jacobson 1993). The hypothesis that *M. agassizii* is an introduced pathogen (spread to wild tortoises via sick captives), at least in the Mohave population, has yet to be rejected. Released captives, especially unhealthy ones, suffer higher rates of mortality than wild tortoises (and much higher than captives), so release for humanitarian or health reasons is misguided. Pet tortoises should never be released and should be kept in escape-proof pens to prevent them from entering the wild.

Conservation Efforts

Agency Roles

Given the number of government agencies having responsibility for natural resource management, ranging from city parks departments to the federal Departments of Interior, Commerce, Defense, and Agriculture, it is not surprising that specific roles of each agency are often a mystery to the general public and sometimes to agency personnel as well. Various federal, state, and local laws establish the statutory authorities and responsibilities of each agency, but the lines separating them are not always clear.

WILDLIFE MANAGEMENT. Although some disagreement arises in specific cases, AGFD has primary authority for wildlife management in Arizona, including wildlife on private, state, and most federal (e.g., Bureau of Land Management [BLM], U.S. Forest Service, Department of Defense) lands. Exceptions include national parks and monuments (in part) and Indian reservations. The USFWS has primary authority for management of migratory birds and most federally listed species and some additional authority on national wildlife refuges, but AGFD usually participates in and often leads management. The National Park Service has substantial authority over wildlife on national parks and monuments, but again, AGFD is often an important co-operator. On Indian reservations, Native American nations manage wildlife, often through their own resource management agencies.

HABITAT MANAGEMENT. The BLM, U.S. Forest Service, National Park Service, USFWS, and Arizona State Land Department, the primary public land managers in Arizona, have statutory authority to manage wildlife habitat on most public lands (the BLM also manages extensive habitat on Department of Defense land). Through their livestock grazing and timber management programs, regulation of mining and other extractive uses, oversight of recreational use, and designation of wilderness areas, these agencies have substantial control over habitat condition on Arizona's public lands. The BLM and Arizona State Land Department manage the majority of Sonoran desert tortoise habitat in Arizona. On federal lands, a variety of laws and policies dictate that wildlife values receive consideration during land use planning. On state trust lands, the Arizona State Land Department is legally mandated to generate the maximum revenue possible for the state's schools. Management for wildlife habitat is secondary and cannot interfere with the primary mandate. Other significant Sonoran tortoise habitat managers include Native American nations (Gila River, San Carlos Apache, San Xavier, and Tohono O'odham Reservations), the Department of Defense, Maricopa and Pima County parks departments, the Arizona State Parks Department, and the City of Phoenix (municipal parks). Relatively little Sonoran tortoise habitat is privately owned (although in total this privately owned land exceeds the holdings of some agencies).

AGFD provides environmental review of projects proposed by public and private parties. If the project proponent or AGFD's Heritage Data Management System identifies the desert tortoise as a resident of the project area, then tortoises are considered during project review. Department personnel assess potential impacts and make recommendations for voluntary mitigation. AGFD's ability to protect wildlife is largely limited to regulation of take under authority of hunting, fishing, and other licenses they issue. Take (e.g., mortality and harassment) occurring incidental to construction or other activities is not within AGFD's purview, unless animals or parts thereof are taken into possession.

The Legal Protection of Desert Tortoises

Protection by the State of Arizona

Historically, desert tortoises were taken from the wild in large numbers as pets. The Mohave tortoise population was probably the primary source of tortoises for the pet trade, simply because tortoises were easier to collect there. However, collection of tortoises for personal use as pets was common in Arizona until the animals were protected in 1988. Interstate commerce in tortoises was formerly common, but we are unaware of substantial current trade.

From 1988 to the present, AGFD has considered the desert tortoise a candidate for threatened status (AGFD 1988) or a species of special concern (AGFD 1996). Desert tortoises are fully protected by the State of Arizona. There is no open season, so it is illegal to kill or capture a desert tortoise except under special permits issued by AGFD (plus federal permits for the listed Mohave population). Tortoises salvaged from urban or developing areas must be relinquished to AGFD or the Arizona-Sonora Desert Museum for adoption. Any tortoise in a person's possession must have been legally taken prior to season closure (which began 1 January 1988), be captive-bred progeny of such legally taken and owned animals, or be held under authority of a sanctioned adoption program (AGFD's Adobe Mountain Wildlife Center or the Arizona-Sonora Desert Museum, which is authorized by AGFD). The possession limit is one desert tortoise per person. Progeny of captive tortoises may be held in excess of the possession limit for up to two years after hatching, but the hatchlings must be disposed of by gift to another person or as directed by AGFD before the two years expire. Desert tortoises (and other native wildlife in Arizona) may not be sold, bought, traded, or otherwise used commercially, nor may they be exported from Arizona for commercial purposes.

Release of captive wildlife into the wild is prohibited in Arizona. Although widely unknown to the public, this prohibition is important for maintenance of healthy wildlife populations. All species of the genus *Gopherus*, including the desert tortoise, are on Arizona's restricted live wildlife list (Arizona Game and Fish Commission Rules, R12-4-406). Live *Gopherus* may be imported into Arizona only under special permit or under specific circumstances excepted under the Live Wildlife Rules (Arizona Game and Fish Commission Rules, Article 4). This restriction is intended to prevent captive tortoises imported from elsewhere from entering the wild in Arizona and threatening wild populations.

Federal Actions

In 1985, the USFWS was petitioned to list the desert tortoise as an endangered species in Arizona, Nevada, and California. The agency determined that the listing was warranted but precluded by higher-priority listing actions. In 1989, the petitioners presented new information, asking the USFWS to list the tortoise as endangered under emergency listing procedures. The new information included documentation of high mortality, which was later attributed to URTD, in large areas of the Mohave Desert. The agency listed the Mohave population, but, for many of the reasons outlined earlier, found that listing of the Sonoran Desert population was not warranted (USFWS 1989). In the final rule listing the Mohave population (USFWS 1990), the USFWS committed to evaluate status of the Sonoran Desert population. Barrett and Johnson (1990) prepared a status summary for the Sonoran population, and the USFWS (1991) determined that listing was not warranted.

After the 1991 ruling, the Sonoran Desert population became a category 2 candidate (a species of concern for which inadequate information exists to make a listing determination). The USFWS (1996) discontinued this category, so the Sonoran desert tortoise now has no status under the Endangered Species Act, but the agency informally considers it a species of concern.

International Treaties and Protection in Mexico

The desert tortoise is a threatened species in Mexico (Secretaría de Desarrollo Social 1994). It is listed in appendix 1 of the Convention on International Trade in Endangered Species of Wild Fauna and Flora. Permits are required to transport such species between member nations. Criteria for permit issuance are intended to protect wild populations and individual tortoises being traded. We are unaware of present or historical international trade in desert tortoises.

Conservation and Management Planning

The BLM Rangewide Management Strategy and Compensation Policy

In 1988 the BLM adopted a rangewide management strategy for desert tortoise habitat (Spang et al. 1988). The strategy focused on interagency coordination, inventory and monitoring, research, and habitat management and conservation. The BLM established three categories of tortoise habitat based on population densities and trends and manageability of the habitat for viable populations. Category 1 habitats support moderate- to high-density populations and can be managed for long-term viability without significant land use conflicts. Category 3 habitats generally support low-density populations and/or have significant land use conflicts, making it difficult to manage for viable tortoise populations. Category 2 habitats are intermediate. Implementation of the rangewide strategy is guided by the *Strategy for Desert Tortoise Habitat Management on Public Lands in Arizona* (BLM 1990).

In 1991 the BLM, the USFWS, and state wildlife agencies in the four states with desert tortoises adopted a policy of compensating for residual impacts of habitat-disturbing projects authorized or conducted by these agencies on BLM-managed lands (Desert Tortoise Compensation Team 1991). When required, compensation is provided either by purchase of privately owned desert tortoise habitat for transfer to a land management agency or by direct payments to fund the purchase of tortoise habitat or other tortoise management actions. Compensation ratios are linked to habitat categories and range from 1:1 (project proponent protects an area of habitat equal to that disturbed) for projects in category 3 habitat to 6:1 (six acres protected for every acre disturbed) in category 1.

The Arizona Interagency Desert Tortoise Team

The AIDTT, a team of state and federal agency biologists, was formed in 1985 to promote interagency coordination and discussion of research and management issues for Sonoran desert tortoises. An interagency memorandum of understanding formalized the AIDTT in 1995. The team has produced a management plan (AIDTT 1996a) and standard guidelines for project mitigation measures (AIDTT 1997) as well as handling of tortoises at development sites (AIDTT 1996b).

These documents were an important step in developing consistent and effective management of Sonoran desert tortoises and their habitats. The AIDTT plan calls for designation of Sonoran Desert management areas, which are similar to desert wildlife management areas designed for recovery of the Mohave population. The goal for the management areas is to maintain viable tortoise populations and conserve the ecosystems upon which they depend. The plan recommends maintaining current state protection of tortoises and provides recommended surface and forage management options to help land managers address management issues. The AIDTT also completed recommended standard mitigation measures for projects in Sonoran desert tortoise habitat (AIDTT 1997).

Research and Monitoring

The ecology of Sonoran desert tortoises was poorly known prior to the 1990s (Johnson et al. 1990), but recent research, much of it summarized in this volume, has greatly expanded our knowledge. Additional research needs were summarized by the AIDTT (1996a).

Population monitoring was identified by the AIDTT (1996a) as critical to desert tortoise conservation in Arizona. In 1990, after initial surveys in the 1980s (see Averill-Murray et al., ch. 6 of this volume), agencies began cooperative funding of monitoring across the Arizona range of the Sonoran desert tortoise (Howland 1994b; Averill-Murray et al., ch. 6 of this volume). Levels and sources of funding varied considerably from 1990 to 1999. Current budgetary trends, especially for monitoring and management of nonlisted species, make continuation of this important element of tortoise conservation uncertain. Population monitoring and basic research on desert tortoise ecology are essential to the development and maintenance of an effective conservation program and require that resource management agencies commit to long-term funding.

The Future of Desert Tortoise Conservation in Arizona

To conserve the Sonoran desert tortoise in Arizona, resource management agencies should implement the AIDTT plan (AIDTT 1996a). This may entail revision of land use plans to meet the AIDTT plan's habitat management recommendations. Projects carried out or permitted by agencies should conform with AIDTT mitigation recommendations (AIDTT 1997).

Where feasible and appropriate, tortoise-proof fencing and culverts or overpasses should be designed into road- and canal-building projects to reduce vehicle and drowning mortality and allow safe passage of tortoises. Where direct reduction of isolating barriers is not feasible (e.g., extensive urban or agricultural developments, large reservoirs), translocation of tortoises between neighboring but artificially isolated mountain ranges should be investigated as a management option. This would require careful deliberation, including research on genetic relationships among populations, institution of measures to ensure that only healthy tortoises are moved, and designing manipulations that mimic natural frequency and geographic patterns of movement. Only those age and sex classes that naturally move among populations and no diseased or otherwise unhealthy tortoises should be moved. Tortoises should be moved only to areas they could have reached on their own, in the absence of artificial barriers. If we find intact historical tortoise habitats with few or no tortoises, we should determine whether benefits of manipulations would justify costs. The effectiveness of translocation as a conservation tool could be tested by attempting to establish a breeding population in an artificially isolated area where tortoises have been extirpated.

ACKNOWLEDGMENTS

We thank Roy Averill-Murray and Terry Johnson for critical reviews of earlier drafts of this chapter. Terry Johnson contributed valuable insights into the history and politics of Sonoran desert tortoise conservation and management in Arizona. Discussion with Ted Cordery provided various views, including that from the perspective of a land management agency. Support was provided by the U.S. Fish and Wildlife Service and the Arizona Game and Fish Department, through its Nongame Checkoff and Heritage Funds.

LITERATURE CITED

Arizona Game and Fish Department. 1988. *Threatened native wildlife in Arizona.* Arizona Game and Fish Dept., Phoenix.

———. 1996. Wildlife of special concern in Arizona. Public Review Draft. Arizona Game and Fish Dept., Phoenix.

Arizona Interagency Desert Tortoise Team. 1996a. *Management plan for the Sonoran Desert population of the desert tortoise in Arizona.* R. C. Murray and V. Dickinson, eds. Arizona Game and Fish Dept. and U.S. Fish and Wildlife Service, Phoenix.

———. 1996b. *Guidelines for handling Sonoran desert tortoises encountered on development projects.* Arizona Game and Fish Dept. and U.S. Fish and Wildlife Service, Phoenix.

———. 1997. *Recommended standard mitigation measures for projects in Sonoran desert tortoise habitat.* Arizona Game and Fish Dept. and U.S. Fish and Wildlife Service, Phoenix.

Avery, H. W. 1994. Digestive physiology and nutritional ecology of the desert tortoise fed native versus non-native vegetation: Implications for tortoise conservation and land management. *Proceedings of the Desert Tortoise Council Symposium* 1994:143.

Barrett, S. L., and T. B. Johnson. 1990. *Status summary for the desert tortoise in the Sonoran Desert.* Report to the U.S. Fish and Wildlife Service, Albuquerque.

Berry, K. H. 1986a. Incidence of gunshot death in desert tortoises in California. *Wildlife Society Bulletin* 14:127–32.

———. 1986b. Desert tortoise *(Gopherus agassizii)* relocation: Implications of social behavior and movements. *Herpetologica* 42:113–25.

———. 1997. Demographic consequences of disease in two desert tortoise populations in California, USA. Pp. 91–99 *in* J. Van Abbema, ed. *Proceedings: Conservation, restoration, and management of tortoises and turtles—An international conference.* New York Turtle and Tortoise Society, New York.

Boarman, W. I. 1997. Predation on turtles and tortoises by a subsidized predator. Pp. 103–4 *in* J. Van Abbema, ed. *Proceedings: Conservation, restoration, and management of tortoises and turtles—An international conference.* New York Turtle and Tortoise Society, New York.

Boarman, W. I., M. Sazaki, and W. B. Jennings. 1997. The effect of roads, barrier fences, and culverts on desert tortoise populations in California, USA. Pp. 54–58 *in* J. Van Abbema, ed.

Proceedings: Conservation, restoration, and management of tortoises and turtles—An international conference. New York Turtle and Tortoise Society, New York.

Brown, M. B., I. M. Schumacher, P. A. Klein, K. Harris, T. Correll, and E. R. Jacobson. 1994. *Mycoplasma agassizii* causes upper respiratory tract disease in the desert tortoise. *Infection and Immunity* 62:4580–86.

Bureau of Land Management. 1990. *Strategy for desert tortoise habitat management on public lands in Arizona.* Bureau of Land Management, Arizona State Office, Phoenix.

Bury, R. B., T. C. Esque, and P. S. Corn. 1988. Conservation of desert tortoises *(Gopherus agassizii)*: Genetics and protection of isolated populations. *Proceedings of the Desert Tortoise Council Symposium* 1988:59–66.

Congdon, J. D., A. E. Dunham, and R. C. van Loben Sels. 1993. Delayed sexual maturity and demographics of Blanding's turtles *(Emydoidea blandingii)*: Implications for conservation and management of long-lived organisms. *Conservation Biology* 7:826–33.

Desert Tortoise Compensation Team. 1991. *Compensation for the desert tortoise.* Report to the Desert Tortoise Management Oversight Group, Phoenix.

Dickinson, V. M., J. L. Jarchow, and M. Trueblood. 1996. *Health studies of free-ranging Mohave desert tortoises in Utah and Arizona.* Technical Report no. 21, Arizona Game and Fish Dept., Phoenix.

Ehrlich, P. R. 1988. The loss of biodiversity: Causes and consequences. Pp. 21–27 *in* E. O. Wilson, ed. *Biodiversity.* National Academy Press, Washington, D.C.

Frankel, O. H., and M. E. Soule. 1981. *Conservation and evolution.* Cambridge University Press, Cambridge, U.K.

Germano, D. J. 1992. Longevity and age-size relationship of populations of desert tortoises. *Copeia* 1992:367–74.

———. 1994. Comparative life histories of North American tortoises. Pp. 175–86 *in* R. B. Bury and D.J. Germano, eds. *Biology of North American tortoises.* Fish and Wildlife Research no. 13, U.S. Dept. of the Interior National Biological Survey, Washington, D.C.

Germano, D. J., R. B. Bury, T. C. Esque, T. H. Fritts, and P. A. Medica. 1994. Range and habitats of the desert tortoise. Pp. 73–84 *in* R. B. Bury and D. J. Germano, eds. *Biology of North American tortoises.* Fish and Wildlife Research no. 13, U.S. Dept. of the Interior National Biological Survey, Washington, D.C.

Henen, B. T., C. C. Peterson, I. R. Wallis, K. H. Berry, and K. A. Nagy. 1998. Effects of climatic variation on field metabolism and water relations of desert tortoises. *Oecologia* 117:365–73.

Homer, B. L., K. H. Berry, and E. R. Jacobson. 1996. *Necropsies of eighteen desert tortoises from the Mojave and Colorado Deserts of California, 1994–1995.* Report to National Biological Service, Washington, D.C.

Howland, J. M. 1994a. Observations and activities of the naturalist for the Desert Tortoise Natural Area, Kern County, California: 12 March–12 July, 1989. *Proceedings of the Desert Tortoise Council Symposia* 1987–1991:228–44.

———. 1994b. *Sonoran desert tortoise population monitoring.* Technical Report no. 38, Nongame and Endangered Wildlife Program, Arizona Game and Fish Dept., Phoenix.

Howland, J. M., and C. M. Klug. 1996. Results of five consecutive years of population monitoring at three Sonoran desert tortoise plots. *Proceedings of the Desert Tortoise Council Symposium* 1995:74–87.

Jacobson, E. R. 1993. Implications of infectious disease for captive propagation and introduction programs of threatened/endangered reptiles. *Journal of Zoo Wildlife Medicine* 24:245–55.

Jacobson, E. R., J. L. Behler, and J. L. Jarchow. 1999. Health Assessment of Chelonians and release into the wild. Pp. 232–42 *in* M. A. Fowler and R. E. Miller, eds. *Zoo and wild animal medicine: Current therapy.* Vol. 4. W. B. Saunders, Philadelphia, Pa.

Johnson, T. B., N. M. Ladehoff, C. R. Schwalbe, and B. K. Palmer. 1990. *Summary of literature on the Sonoran Desert population of the desert tortoise.* Report to U.S. Fish and Wildlife Service, Albuquerque.

LaRue, E. L. 1996. *Federal biological opinion analysis for the Eagle Mountain Landfill Project.* Report to CH2M HILL, Santa Ana, Calif.

Luckenbach, R. A. 1982. Ecology and management of the desert tortoise *(Gopherus agassizii)* in California. Pp. 1–37 *in* R. B. Bury, ed. *North American tortoises: Conservation and ecology.* Wildlife Research Report no. 12, U.S. Fish and Wildlife Service, Washington, D.C.

Martin, P. P. 1996. *Songs my mother sang to me.* University of Arizona Press, Tucson.

McArthur, E. D., S. C. Sanderson, and B. L. Webb. 1994. *Nutritive quality and mineral content of potential desert tortoise food plants.* Research Paper no. INT-473, U.S. Dept. of Agriculture Forest Service, Intermountain Research Station, Ogden, Utah.

Nagy, K. A., B. T. Henen, and D. B. Vyas. 1998. Nutritional quality of native and introduced food plants of wild desert tortoises. *Journal of Herpetology* 32:260–67.

Nicholson, L. 1978. The effects of roads on desert tortoise populations. *Proceedings of the Desert Tortoise Council Symposium* 1978:127–29.

Olson, T. E., K. Jones, D. McCullough, and M. Tuegel. 1993. Effectiveness of mitigation for reducing impacts to desert tortoise along an interstate pipeline route. *Proceedings of the Desert Tortoise Council Symposium* 1992:209–19.

Peterson, C. C. 1994. Different rates and causes of high mortality in two populations of the threatened desert tortoise *Gopherus agassizii. Biological Conservation* 70:101–8.

Ralls, K., and J. Ballou. 1983. Extinction: Lessons from zoos. Pp. 164–84 *in* C. M. Schonewald-Cox, S. M. Chambers, B. McBryde, and L. Thomas, eds. *Genetics and conservation.* Benjamin/Cummings, Boston.

Sazaki, M., W. I. Boarman, G. Goodlett, and T. Okamoto. 1995. Risk associated with long-distance movements by desert tortoises. *Proceedings of the Desert Tortoise Council Symposium* 1994:33–48.

Secretaría de Desarrollo Social. 1994. Poder ejecutivo diario oficial de la Federación. Tomo 488 no. 10, Mexico, D.F., lunes 16 de mayo de 1994.

Shields, T., S. Hart, J. Howland, T. Johnson, N. Ladehoff, K. Kime, D. Noel, B. Palmer, D. Roddy, and C. Staab. 1990. *Desert tortoise population studies at four plots in the Sonoran Desert, Arizona.* Report to Arizona Game and Fish Dept., Phoenix.

Spang, E. F., G. W. Lamb, F. Rowley, W. H. Radtkey, R. R. Olendorff, E. A. Dahlem, and S. Slone. 1988. *Desert tortoise habitat management on the public lands: A rangewide plan.* Report to U.S. Bureau of Land Management, Washington, D.C.

Turner, F. B., K. H. Berry, D. C. Randall, and G. C. White. 1987. *Population ecology of the desert tortoise at Goffs, California, in 1985.* Report to Southern California Edison Co., Rosemead.

U.S. Fish and Wildlife Service. 1989. Endangered and threatened wildlife and plants: Emergency determination of endangered status for the Mojave population of the desert tortoise; emergency rule. *Federal Register* 54:32326–31.

———. 1990. Endangered and threatened wildlife and plants: Determination of threatened status for the Mojave population of the desert tortoise. *Federal Register* 55:12178–91.

———. 1991. Endangered and threatened wildlife and plants: Finding on a petition to list the Sonoran population of the desert tortoise as threatened or endangered. *Federal Register* 56:29453–55.

———. 1994. *Desert tortoise (Mojave population) recovery plan.* U.S. Fish and Wildlife Service, Portland, Ore.

———. 1996. Endangered and threatened wildlife and plants: Notice of final decision on identification of candidates for listing as endangered or threatened. *Federal Register* 61:64481–85.

Wilson, E. O. 1992. *The diversity of life.* Harvard University Press, Cambridge, Mass.

Woodbury, A. M., and R. Hardy. 1948. Studies of the desert tortoise, *Gopherus agassizii. Ecological Monographs* 18:145–200.

Woodman, P., S. Hart, P. Frank, S. Boland, G. Goodlett, D. Silverman, D. Taylor, M. Vaughn, and M. Walker. 1995. Desert tortoise population surveys at four sites in the Sonoran Desert of Arizona, 1994. Unpubl. report to Arizona Game and Fish Dept. and U.S. Bureau of Land Management, Phoenix.

CHAPTER 15

When Desert Tortoises Talk, Indians Listen

Traditional Ecological Knowledge of a Sonoran Desert Reptile

GARY PAUL NABHAN

S-am wo wo'ikud g Komik'ceḍ
Tortoise would be lying in bed,
c wo ne'id. S pi hekid s-amiced g Ban
singing. Coyote just never understood it
mas ha'icu a:gc ñe'e i:da s-pad-makam.
when that lazy one was singing.
T hab wo kaidam ne'icudad g ha:sañi:
He would be heard singing for saguaro cacti:
"Mant hemu ba'i yia ke:k,
"I've now ripened, standing here before you,
Mu'i u'uhig mu'i kuhu c ia ni-i'ajid.
so many birds are cawing, swarming,
Mu'i na:nko kaij.
saying/the seeds of various things,
Mu'i u'uhig mu'i kuhu c ia ni-i'ajid."
so many birds are cawing, swarming."

—O'odham story recorded by Juan Dolores, included in Saxton and Saxton (1973), retranscribed in Tohono O'odham orthography and retranslated

Recently, indigenous hunters, farmers, and foragers have been recognized by conservation biologists as contributors to and potential allies in the protection of endangered species and other constituents of biodiversity (Williams and Baines 1993; Orlove and Brush 1996; Tuxill and Nabhan 1998). In the case of desert tortoises *(Gopherus agassizii)*, it is clear that indigenous peoples of the Sonoran Desert have had a long and intimate relationship with certain local populations (Felger et al. 1981; Schneider 1996).

Like other people within tortoises' range of the binational Southwest, indigenous communities within the Sonoran Desert used this creature for food and medicine and its carapace for ladles, dippers, bowls, and shovels (BioSystems, Inc. 1994). They also celebrate the spirit of the tortoise in songs, stories, petroglyphs, and ritual observances, as suggested by the above-quoted O'odham song and story fragment. It refers to the tortoise as the keeper and planter of giant saguaro cacti, a key cultural resource for these people, which the tortoise first kept near the Sea of Cortez and then let its seed be spread to other places (Saxton and Saxton 1973).

As they do for other species, indigenous paraecologists make fairly precise assessments of tortoises' local distributions and ecological interactions, as well as threats to their persistence (Nabhan 1990). However, some Western-trained scientists have remained skeptical that local anecdotal information about tortoises from "untrained individuals" is of much biological value. It should be no surprise that such information has seldom been systematically compiled, evaluated, or incorporated into threatened-species management plans.

Indigenous residents with long tenure in one desert landscape have often made observations and recognized patterns of tortoise behavior, habitat preferences, and forage-plant use that scientists—for lack of time or access to certain lands—have failed to notice. Such observations can generate hypotheses for testing and be followed by more rigorous assessments. In addition to their many hours of field observations of desert tortoises, indigenous residents may have oral histories that predate changes in environmental conditions that have dramatically affected tortoise populations. Their observations help orient biologists who are newcomers to an area and who have few other clues to when and why a tortoise population may have declined.

Of course, some cultures more than others are interested in particular species and their ecological interactions. These cultures tend to encode this traditional ecological knowledge in specialized vocabularies that may seem as foreign to the uninitiated as the technical jargon of scientists. There is a need to translate and decode this knowledge and to appraise its validity, which is not necessarily of equal quality for all individuals, cultural communities, or languages. It is time to define the natural sciences broadly, "not [as] the exclusive property of any one group . . . [for] every group has a 'science' based on observation, experiment and tradition . . . whether scientific knowledge is developed through 'experiment' or 'trial and error observation,' and whether knowledge is organized into a 'systematic body' or 'diffused throughout a range of cultural expression and behavior' " (Green 1981:205).

I include the folk science of several Sonoran Desert cultures in the following discussion. These indigenous groups include the people known to the outside world as the Seri (Comcáac) of Sonora; the Yaqui and Mayo (Yoemem and Yoremem) of Arizona, Sonora, and adjacent Sinaloa; the River Pima (Akimel O'odham) of Arizona; the Lowland Pima (O'odham) of Sonora; the Desert Papago (Tohono O'odham) of Arizona and Sonora; the Sand Papago (Hiá c-eḍ O'odham) of western Arizona and Sonora; the Cocopa (Kwapa) of Arizona, Sonora, and Baja California; and the Maricopa (Tipai) of Arizona. Throughout the rest of the chapter, I will use the groups' own names for themselves. I am grateful to many people who live and work in these indigenous communities for patiently and graciously sharing their knowledge with me and with others.

Indigenous Folk Taxonomies for Turtles and Tortoises

A desert tortoise was once way up in the sky.
It had been traveling very slowly on ground,
but in the sky it wanted to go faster.
—Francisco Barnet Astorga explaining with a story a Comcáac song sung by
Angelita Torres of Desemboque, Sonora

When we decide to learn from Native Americans' knowledge of local desert tortoise populations, it is critical that we become familiar with the specific names in indigenous languages by which local residents call this species as opposed to other turtles in the region. Table 15.1 summarizes the nomenclature used for desert tortoises in native languages of the Sonoran Desert. It also notes names for other turtles that live nearby, updating an earlier provisional linguistic summary by Felger et al. (1981).

One difficulty that nonnative speakers must contend with is that some terms are polysemous, that is, they are used in multiple ways, to refer either to a specific taxon or distributively to a set of related taxa for which the specific taxon is the most representative (type specimen). For instance, the Comcáac term *moosni* is polysemous, referring to green sea turtle *(Chelonia*

TABLE 15.1. Indigenous names for the desert tortoise and other turtles in the Sonoran Desert region.

Culture	Desert Tortoise	Other Turtles	References
Seri	*xtamoosni*[a]	*moosni*	Felger et al. 1981
(Comcáac)	*ziix catotim*	(sea turtle)	
	ziix hehet cöqiij	*xtamaiija*	
		(mud turtle)	
O'odham	*cecio komik'c-eḍ*	*vo'o komik'c-ed*	Pennington 1979
(Papago, Lowland Pima)	*comicturhu*	(mud turtle)	
Yoeme (Yaqui, Mayo)	*mochik*	*moosen*	Felger et al. 1981;
	caumaris	(sea turtle)	Thomas R.
		mochic	Van Devender,
		(mud turtle)	field notes
Tipai (Maricopa)	*kape't*		Spier 1970
Kwapa (Cocopa)	*xnyar*		Crawford 1989

[a] This name is archaic and is almost never used today.

mydas), to all sea turtles, and to turtles and tortoises collectively. An archaic name for the desert tortoise, *xtamoosni*, is almost never used today except when singing traditional songs or referring to forage plants associated with desert tortoises called *xtamoosni oohit* "desert tortoise what-it-eats" (table 15.1). Instead, most Comcáac interchangeably use the two descriptive nicknames for desert tortoises listed in table 15.1, *ziix hehet cöqiij* "thing that sits among plants" and *ziix catotim* "thing that slowly scoots along."

For other extant tribes in the Sonoran Desert, it appears that the term for desert tortoise is also used generically as the distributive term for all turtles and tortoises. For example, Tohono O'odham individuals will casually call a desert tortoise *komik'c-ed* "shell with [living thing] inside." However, when asked if it is the one that lives in *charcos (vo'o komik'c-eḍ* for *Kinosternon flavescens* or *K. sonoriense)*, they will add a descriptive modifier and say that no, it is the *do'ag komik'c-eḍ* "mountain turtle" or the *ce:cio komik'c-eḍ* "cave[-dwelling] turtle." If they were to see a mud turtle by itself, Tohono O'odham individuals may casually refer to it by the shorthand *ha'icu komik'c-eḍ* "some kind of turtle." Nevertheless, O'odham elders laughingly remind younger people that not knowing the differences among kinds of *komik'c-eḍ* can lead to trouble—or at least to distasteful meals. A Hiá c-eḍ O'odham elder told me of a woman who relished eating desert tortoises, but almost inadvertently ate a mud turtle. It was one that some mischievous boys brought her to try to fool her; only after several minutes of handling the turtle did she realize that it was the "wrong kind for eating."

Finally, the Comcáac use particular terms to describe the morphology and behavior of desert tortoises. For instance, *cpoin* "shutting-down time" or "closing-in time" refers to hibernation and estivation periods in caves or burrows, which are called *xtamoosni iime* "tortoise homes" or "resting places." *Moosni ipojc* refers to a carapace, whereas *moosni iti ihimoz* refers to a plastron. Finally, Comcáac can tell tortoise gender by carapace and plastron shapes, so that both men and women readily sex desert tortoises within seconds of picking them up.

Indigenous Knowledge of Desert Tortoise Behavior and Life History

Los rayos del sol
The rays of the sun
están pasando atrás
are passing behind
de la tortuga.
the tortoise.
Siempre anda
The tortoise always walks
la tortuga desde los cerros
from the hills
en una dirección
in a direction
opueste de la puesta del sol.
away from the setting sun.
—translation of a Comcáac song of desert tortoises sung by Jesús Rojo, elder of Punta Chueca, Sonora

Most Comcáac are familiar with both the times and place of desert tortoise hibernation and estivation, which they call *cpoin*. They suggest that in

Figure 15.1. A desert tortoise eating caliche in July 2000 in the Tucson Mountains, Pima County, Arizona. (Photograph by James L. Jarchow)

November the tortoises enter caves on rocky upland slopes, cavities in arroyo banks, and sometimes unsheltered depressions and emerge from them as early as March and as late as May. Tohono O'odham elders in southern Arizona claim that the time of emergence near their homes is March or April and that the tortoises are only seen actively moving around during the warmer seasons. According to Justina Rodríguez Valenzuela Guerrero, a Yoeme resident of Estación Llano near Benjamín Hill, Sonora, tortoises came out after the summer rains and were active in early mornings and late afternoons.

Once the tortoises emerge from hibernation, it is reported that they initially eat white rocks and talc-like silt or clay dust (fig. 15.1). The Yoremem claim that tortoises' urine turns white because they eat caliche-rich soil. Comcáac elders report that the caves and rock shelters in which desert tortoises hibernate will also harbor commensals such as Gila monster *(Heloderma suspectum)*, desert spiny lizard *(Sceloporus magister)*, coachwhip snake *(Masticophis flagellum)*, and white-throated packrat *(Neotoma albigula)*. There is one early report that the Comcáac believe that desert tortoises are gregarious and stay in family groups (Malkin 1962). Recently, Comcáac elders have reported to me that two to three tortoises may live together, especially pairs of a male and a female. They have also seen males fight one another, particularly when "one of their ladies is nearby." Others believe that although each tortoise has its own territory, it may cover considerable ground during its diurnal meanderings.

The Comcáac claim that they can hear desert tortoises mating from distance away, that the cries of a male while mounting a female in one canyon could be heard from the next canyon over in the same mountain range. From the 1950s, when first questioned by Malkin (1962), to the present, the Comcáac have consistently suggested that females lay eight to ten eggs per mating. The Yoemem claim that the tortoises they know lay a dozen eggs that are lightly buried in soil where the little tortoises are later seen to hatch. That clutch size is higher than the six to eight eggs per female documented among tortoises in Arizona (Averill-Murray et al., ch. 7 of this volume) and offers another potential topic for research—geographic variation in egg number per season.

Unspecified birds, snakes, and lizards as well as humans prey upon the eggs. These eggs were opportunistically gathered and eaten by the Comcáac in the recent past. Although contemporary Comcáac adults mention that tortoise eggs have been a traditional food during their lifetimes, the practice is declining for reasons mentioned in the final section of this report.

Indigenous Uses of Desert Tortoises

Una tortuga estaba caminando
A tortoise was walking on a hill
cuando se falseó en el cerro
when he hurt himself
donde trataba de comer algo.
where he tried to eat something.
Entonces, los cazadores
Then, hunters from afar
lejanos llegaron para matarla.
arrived to kill him.
Ya hay peligro
Now there is danger

Figure 15.2. A Comcáac preparing live tortoises for cooking over a wood fire on Isla Tiburón around 1948–1951. (Photograph by William Smith, courtesy of the University of Arizona Special Collections)

> —*tiene que esconderse.*
> —he has to hide.
> *Será muy peligroso que la*
> There will be danger if the
> *vean los cazadores*
> hunters see him
> —*tiene que esconderse.*
> —he has to hide.
> —Alfredo López Blanco, explaining a Comcáac song through a related story

In addition to eating tortoise eggs, indigenous dwellers of the Sonoran Desert seasonally ate tortoise meat, preparing it in a number of ways (fig. 15.2). The most detailed description I know of the culinary preparation of desert tortoises comes from the field notes of Ana Lilia Reina Guerrero, who interviewed Justina Rodríguez Valenzuela, a woman of Yoreme descent born in 1919, who told about her experiences from the 1930s through the 1950s.

Justina said that to clean a tortoise, the shell was split on one side with an ax and the insides extracted with a knife. The heart was taken out first so the animal would die and the limbs would relax. Except for the stomach, intestines, and head, all of the muscles and glands were eaten. The limbs were singed in a fire to remove the skin. The toes were cut off. Justina said that the meat was like chicken but better. The meat was used for flavor in soups with noodles or rice, vegetables, or chile. Some people ate the eggs either found in the soil or extracted from the female, although Justina's family did not. Meat was not cooked in the shells, which were discarded.

Patricia Preciado Martin (1992:13) documented the traditions followed by Livia León Montiel, a Mexican-American woman born in Rillito, Arizona, in 1914:

> A wonderful memory is the day we went with my dad to get *tortugas* [desert tortoises]. . . . My father liked turtle soup very much. The tortugas do not have much edible meat; just the *patitas* [legs] are good for eating. The patitas were cut off and cooked with bones and all, like neck bones. They boiled them until the thick skin and nails loosened. Then they peeled off the hide — nails and all — and the meat remained. Mother cooked the meat with onion, green chile, and tomato, and made *caldo* [soup] that was like *cazuela* [a soup with *machaca*, dried shredded meat]. Very little meat but with a lot of substance and flavor.

Not all members of a community took to tortoise meat. In one rather maudlin reminiscence, a Hiá c-eḍ O'odham elder recalled an incident from his childhood that made him shy away from ever eating desert tortoises. When he was a boy, an old O'odham man living along the lower Gila River was known for hunting and eating desert tortoises. When several boys caught one, the old man cracked its plastron open with a hammer, turned the tortoise on its shell, and then put hot rocks from his campfire on it with additional coals heaped around the wounded animal. The boys went away to play when the old man said it would take thirty minutes to cook; when they returned, the tortoise was gone. At first, they assumed that the old man had removed it and eaten it without sharing the meat with them; then, they realized that there was a track of ashes leading from the campfire. The tortoise had righted itself and escaped. They caught up with it, but when the boys saw that it was still walking around with a big hole in its ventral side, they didn't want to eat it. The O'odham teller of this tale implied that if the animal had wanted to live that badly, he himself could refrain from recatching and eating it or any others of its kind.

Tortoises were put to other uses as well. During the past century, their shells have been made into hinged jewelry boxes (Martin 1992) and into musical instruments and baby rattles (Felger et al. 1981; Russell 1983). Comcáac girls made doll clothing out of tortoise bladders and adorned their clay figurines with these ash-dried, softened bladder fragments (Felger et al. 1981). There was also a medicinal use among the Comcáac: Poor eyesight could be cured by heating a stone in the body cavity of a desert tortoise, putting one's face over the steam that rises from it, and opening one's eyes to the steam.

Indigenous Knowledge of Desert Tortoise Diets and Habitat Use

Wo'i': "Hitisa empo bwaka sikili iktena mochik?"
Coyote: "What did you eat that your mouth is red, desert tortoise?"
Mochik: "Yoremta hiwa iani intok enci hiwa"
Desert Tortoise (in his own talk): "I have eaten . . . [garbled: A person like you/ a prickly pear fruit/ a Coyote/ a Christian] and now I will eat you."
Yoeme: "Yoemtane bwa'aka ian intokoenchine bwa'ane."
True Yaqui talk: "I ate a person and now I will eat you."
—Yoeme story fragment collected between 1930 and 1932 by Ralph Beals (1945) kindly retranslated by Felipe Molina, Yoem Pueblo, Marana, Arizona, 1997

In a Yoeme story recorded in Sonora in the 1930s and again in Arizona in the 1970s, it is clear that the Yoemem were so familiar with desert tortoise consumption of prickly pear (*Opuntia* spp.) that they could base word play around this ecological interaction (Beals 1945; Valenzuela K. 1977). Beals (1945:224) translated version is as follows:

> Coyote speaks Yaqui correctly, Turtle does not. One time as a Yaqui sometimes wanders the woods hungry for food, so Coyote wandered. He saw Turtle below a prickly pear *(nopal)* which was

dropping its fruit. Coyote saw Turtle's mouth was red and thought Turtle would give him meat. He ran toward him, and said, "*Hitisa empo bwaka sikili iktema mocik?* (What hast thou eaten that thy mouth is red, Turtle?)"

Turtle replied, "*Yoremta hiwa iani intok enci hiwa.* (I have eaten a Christian and now I will eat you."

Coyote was frightened and ran off. But this was not correct Yaqui. It should have been said, "*Yoemtani bwaka ianintoko encine bwana.*" Thus, those who deny their own language or do not speak it correctly are called turtles.

A more recently transcribed version was told by Yoeme elder Carmen García (1901–1971) to Mini Valenzuela K. (1977:49):

One time, in the month of August when it is very, very hot, at the time when the *nabo* [prickly pear plant in Yoeme] is ready for picking, a turtle was walking beneath the branches of a prickly pear cactus. She was eating the *tunas* [prickly pear fruits in Spanish] which were ripe and which had fallen to the ground. She traveled along with her mouth all red with the juice of the *nabo*.

While she was walking along, she encountered a coyote dying from hunger. This one greeted her with much respect, and then asked, "*Jitasa eh buaka, tenek?*" which is to say, "What have you eaten that your mouth is all red?"

"I have just finished eating a coyote. And if you bother me, I am going to eat you too," replied the turtle, opening her mouth very wide and showing all her teeth.

Well, the coyote got scared, and thought no more of eating her.

The association between prickly pear fruit and desert tortoises is not surprising to any naturalist, but indigenous people of the Sonoran Desert also associate this reptile with other, less obvious forage plants. These include six plants known to the Comcáac as *xtamoosn-oohit* "desert tortoise what-it-eats": brittle spineflower *(Chorizanthe brevicornu)*, *Fagonia californica*, *F. pachyacantha*, horse purslane *(Trianthema portulacastrum)*, pebbly pincushion *(Chaenactis carphoclinia)*, and windmills *(Allionia incarnata*; Felger and Moser 1985). The horse purslane, known by the Comcáac as *jomete hant kopete*, and pebbly pincushion have not been previously documented in desert tortoise diets (Van Devender et al., ch. 8 of this volume). The brittle spineflower, a rather inconspicuous, short-lived spring wildflower, has only recently been recognized as a significant component of desert tortoise diets. Further attention to the culturally identified forage species not yet on scientists' foraging observation lists is warranted, because areas with abundant forage reserves may better serve as protected areas for viable populations of desert tortoises than will forage-limited areas.

During springs with abundant winter wildflowers, Hiá c-eḍ O'odham elders claim that desert tortoises will rapidly gain weight after emergence, grazing "just like cattle in a green pasture." We do not know if this observation is true over the entire historic range of the Hiá c-eḍ O'odham or valid only for the Mohave Desert populations. Yoremem people preferred to collect tortoises after the rainy season began, and they judged from the stomach contents of captures that the tortoises not only ate caliche-rich clays but grasses as well.

It is clear that indigenous people have developed a gestalt sense of where their encounters with desert tortoises have been more frequent and, potentially, where desert tortoise population densities may be higher. For instance, Yoremem families who lived in Estación Llano from the 1930s through the 1950s told Ana Lilia Reina Guerrero that tortoises were easy to find on certain hills or along arroyos where sheltersites were abundant. Although factors such as forage availability, shelter, and nesting site densities are cited as reasons why these areas harbor more tortoises, other, less obvious factors may also be imbedded in their reasoning: human history, current land uses, surface water availability, or predator density. As of spring 2001, indigenous paraecologists among the Comcáac are now carrying geographic positioning systems to help field biologists map such favorable habitats for desert tortoises. The Comcáac, at least, appear willing at this time to assist in such surveys, particularly where they feel that poachers from the outside are beginning to have a negative impact on this traditional resource.

Indigenous Knowledge of Desert Tortoise Predators and Threats

La tortuga está muy contenta
The tortoise is very pleased

aunque no puede entender
even though he cannot understand
al mexicano o al americano.
the Mexican or the American.
Por eso, se mete
Because of this, he hides
abajo de unas ramas
under a branch
y se encuentra
and there he finds
un montón
a bunch
de cilantro—
of coriander—
como los quelites del campo,
like wild greens,
y sigue comiendo muy contenta.
and he continues eating happily.
—translation of a Comcáac tortoise song by Lydia Félix de Blanco, Punta Chueca, Sonora

There are a few native animals that the Comcáac have observed as predators on desert tortoises in their homelands: coyotes, bobcats, mountain lions, and hawks. According to Malkin (1962), the Comcáac believed that mountain lions were the only effective predator on tortoises; however, one Comcáac community member explained to me the death of a desert tortoise at the scene of predation by inferring how a hawk had struck at it. Another one commented that coyotes are ineffective at predation because the tortoises can retreat into their shells, which coyotes are unable to penetrate either with their bite or with their claws.

Other than infrequent predation, the only other two factors affecting tortoise densities noted by the Comcaac paraecologists were the effects of drought and poaching to sell animals to Anglos visiting Sonora. One Hiá c-ed O'odham elder, who lives west of Ajo, Arizona, noted that the frequency with which he encountered desert tortoises varied greatly between "green" (wet) years and drought (dry) years, as well as with the appetites of his Mexican-American neighbors.

Indigenous Ritual Observances and Taboos Relating to Desert Tortoises

Esta prohibido tener una tortuga
It is prohibited to have a tortoise
en tu casa — nada va a crecer.
in your house — nothing can grow there.
La mala suerte caerá sobre la gente
Bad luck will fall on the people
si una familia encierra una tortuga.
if a family puts a tortoise into captivity.
La madre o padre sufrirá mucho
The mother or father will suffer greatly
porque sus niños no crecerán.
because their children will not grow.
Un nido es un lugar sagrado.
A nest is a sacred place.
No debes destruir
You should not destroy
nidos de tortugas, tecolotes,
nests of tortoises, owls,
o águilas del mar."
or ospreys.

—Alfredo López Blanco, Comcáac elder, Punta Chueca, Sonora

Although most Comcáac over fifty years of age claim to have hunted and eaten desert tortoise at one time or another, it is our impression that this use was culturally regulated by a variety of taboos and other practices constraining the who, when, and where of tortoise capture and consumption. Despite hundreds, perhaps thousands, of years of Comcáac use of tortoises on Isla Tiburón, demographic surveys there suggest that the island has one of the densest desert tortoise populations known from the Sonoran Desert (Reyes O. 1979; Reyes O. and Bury 1982). If Comcáac hunting pressure had been that great for centuries, we would not expect that desert tortoise populations would have rebounded so quickly after the Comcáac abandoned residency on Isla Tiburón in the 1940s. But as Felger et al. (1981:117) concluded, "Seri [Comcáac] hunting pressure, even in earlier times, does not seem to have been a major factor affecting tortoise populations, since there was at least as great Seri population density on the island as on the mainland." It may be worthwhile to entertain several hypotheses as to why Isla Tiburón's tortoises are so abundant, including the one outlined below.

For both the Comcáac and the O'odham, desert tortoises cause a sickness if not respected. Even though men and women may capture desert tortoises as food or as pets, they cannot do this at all times and in all places. For instance, a Comcáac community member out hunting mule deer *(Odocoileus hemionus)* on Isla Tiburón is not allowed to capture a desert tortoise during the hunt, for it will bring bad luck. Pregnant women are discouraged from hunting tortoises. If a young woman gives birth to only female offspring, her detractors claim that she had eaten the reproductive organs of a female tortoise; if her children are all male, they claim that she had earlier been hit in the small of the back with the reproductive organ of a male tortoise thrown at her by one of her girlfriends (Felger et al. 1981). However strange such beliefs sound to outsiders, they certainly put a psychological damper on a person's willingness to hunt tortoises.

In perhaps the most remarkable story about desert tortoises among the Comcáac, a legendary figure named *Ziix Taaj* is remembered for mysteriously arriving on the Sonoran mainland, where he incited an incident on a tall dune named *Comis*, south of Bahía Kino and Tastiota near San Nicolás. There, as Alfredo López Blanco tells the story, this man known for his supernatural powers was seen by other Comcáac up on the dune, playing a gambling game with someone, even though there was no sign that he had come there by boat or, once he had crossed over from the island, by foot.

When the people approached him, they saw that he was discussing the reed gambling game with a desert tortoise, who sat upright facing *Ziix Taaj*, with his carapace toward the newly arrived group of Comcáac. They watched amazed as *Ziix Taaj* and the tortoise gambled and talked. There are then two versions of the ending.

In one version, the tortoise won several consecutive rounds. This infuriated *Ziix Taaj*, who yelled, "When I lose, you better get out of here!" He threw a towel and a knife at the tortoise, hitting him in the chest and knocking him over onto his carapace. The people saw this terrible act — *Ziix Taaj* striking the tortoise and knocking him over after he had yelled at him — and the tortoise was driven away, back to his family. In another version of this story, the tortoise won several rounds, before *Ziix Taaj* finally wins all the booty back, but the tortoise will not give him all of his winnings. That is why *Ziix Taaj* drives the tortoise away. Alfredo added that, according to his uncle, *Ziix Taaj* was the last and only person ever to speak with a desert tortoise and to fully understand this animal. Nevertheless, many Comcáac today hold tortoises in high regard, claiming that they can understand the Comcáac and that long ago they spoke to their ancestors.

The O'odham, too, believe that desert tortoises can inflict staying sickness upon anyone who is involved in wrongful actions, including thoughtless killing or impoliteness (Bahr et al. 1973). The maligned tortoise can cause chest pains, crippled legs, or sores on the body or feet, which are curable only by using a tortoise shell rattle and singing certain shamanistic songs. Although Russell (1983) presented some traditional O'odham songs about "turtles" (i.e., tortoise and mud turtle), none of the ones included appear to be for curing staying sickness (Culver Cassa, pers. comm. 1997). The net effect of such beliefs has been widely debated, but perhaps they reduce the probability of wanton overexploitation of desert tortoises and other psychologically dangerous creatures and objects. This does not mean that the Comcáac or O'odham necessarily lived in balance with or never harmed desert tortoises; they certainly hunted and consumed the animals for centuries. Nevertheless, it is my impression that desert tortoises were so important as a survival food during extended periods of drought that some of these cultural restraints served to slow down the harvesting of them during other periods.

Ethnobiological Education and Conservation

The black tortoise now approaches us,
wearing and shaking his belt of night,
the black tortoise now approaches us,
wearing and shaking his belt of night.

The harlot arose and ran about,
beating her breast and the air.
The harlot arose and ran about,
beating her breast and the air.

Understand, my younger brothers,
that it is the sun that gives me
the trance vision that I see.
The Sun gives me magic power.

—from José Luis Brennan's translation of a desert tortoise song of the Akimel
O'odham (Pima), presented by Frank Russell (1983)

Whatever the aboriginal patterns of desert tortoise hunting and consumption were historically, it is clear that fewer indigenous people in the Sonoran Desert regularly eat the animals today. Rapid cultural and dietary change has made it difficult to reconstruct the past patterns of human-tortoise interactions. As local people become less dependent upon native animals as symbolic harbingers and as utilitarian resources, they become less familiar with their ecology and behavior as well. They no longer remember the tortoises once spoke to their people about another way of living in this world. This "cycle of disaffection" is what naturalist Robert Pyle (1993) claims leads to "the extinction of experience" and the decline in informal stewardship of local animal populations.

Concerned that this may have already happened in many indigenous communities, the Arizona-Sonora Desert Museum has been involved in surveying grade school and junior high students in Indian schools to see the degree with which they recognize, honor, and participate in the same traditions that their O'odham, Yoremem, and Comcáac grandparents participated in. Our preliminary findings suggested that animal lore transmitted exclusively in the native languages is less common than in the past in O'odham, Yoremem, and Yoemem communities in the United States (Nabhan and St. Antoine 1993). A preliminary survey in Comcáac communities revealed that Comcáac children still speak and sing animal songs and tell stories in their indigenous language, but know only a third of the animal songs that their elders know (Rosenberg 1997). Whether this means that songs are being lost from the younger generation or simply that the entire repertoire is not accumulated until a Comcáac individual reaches adulthood, we do not know. Nevertheless, it is clear that at least some of the tortoise lore is being passed on. The Comcáac children recently interviewed by Rosenberg (1997) selected desert tortoises as one of the three native animals they liked the most, and most knew their names in *cmique iitom* "Seri language."

I recently followed up on these two preliminary surveys with a set of interviews with fifty-one Comcáac individuals; we focussed not merely on taxonomic knowledge, but on their active participation in Comcáac traditions that engaged them in direct contact with desert tortoises. The results demonstrated that many (77.8 percent) of the Comcáac had collected live tortoises, although somewhat less in people born after 1974 (75 percent versus 90.5 percent) and slightly less in women (70 percent). There was a substantial decrease in women born after 1974 who had eaten tortoise eggs (20 percent versus 42 percent) but not in men (41 percent versus 38 percent). In both cases, I believe that nearly all the adults born prior to 1975, when the Mexican government formally opened its office in Comcáac territory, first participated in these traditions when they were ten years old or younger. In essence, desert tortoises are still locally abundant and the children still have access to them, but there is less of an economic or nutritional need and less lore being taught about desert tortoises today than a quarter century ago.

The issue here is not whether Comcáac youth should be encouraged to hunt tortoises or to eat their eggs; it is whether the natural-history lessons once taught by Comcáac elders during such subsistence activities can still be transmitted. To reinforce and celebrate the retention of such traditional knowledge, Northern Arizona University and University of Arizona researchers have worked with Comcáac school teachers on a number of means to teach traditional knowledge about desert and marine reptiles.

We first produced a bilingual booklet, *Los Animalitos del Desierto y del Mar*, which is now being used in all primary school classrooms (Rosenberg et al. 1997). It uses Comcáac riddles, stories, and songs to teach about desert reptiles, including tortoises, and has been well received by both adults and children. Edited by Janice Rosenberg, with guidance from me and Comcáac teachers Pedro Romero and Rodrigo Méndez, it draws upon knowledge from Comcáac elders and from biologists such as Howard Lawler. Next, Laurie Monti produced a booklet on desert foods and diabetes prevention for the Comcáac that also uses reptiles to discuss healthy diets for desert dwellers. With support from the Amazon Conservation Team, Jack Loeffler, Laurie Monti, Tom Vennum, and I then made archival-quality recordings of Comcáac songs about reptiles for a cassette to be shared with all Comcáac families, and in another edition, with local conservation organizations and agencies. In collaboration with the Columbus Zoo, we presented the Comcáac tribal governor a turtle and tortoise learning kit of puppets, anatomical models, maps, coloring books, and posters to be used in Comcáac schools. The Desert Tortoise Council has also trained Comcáac adults in tortoise surveying, monitoring, and public education. Similarly, O'odham paraecologists have put out a flier on the Tohono O'odham reservation urging their people to protect desert tortoises that wander out onto roadways and to refrain from taking them as pets.

Our hope is that Comcáac, O'odham, Yoremem, and Yoemem paraecologists can serve as a bridge between their own communities' rich knowledge of tortoises and that of Western scientists. Such cultural scientific exchanges may become increasingly more important as protected areas become comanaged with indigenous peoples (Orlove and Brush 1996) or additional animal populations become at risk on or near tribal lands. For centuries, the indigenous peoples of the Sonoran Desert have listened to desert tortoises whenever they spoke to them; they have now become spokespersons on behalf of the tortoises as well.

ACKNOWLEDGMENTS

I am grateful to Felipe Molina, Culver Cassa, Pedro Romero, Ernesto Molina, Nacho Barnett, Jesús Rojo, Angelita Torres, Adolfo Burgos, Amalia Astorga, Alfredo López Blanco, Victoria Astorga, José Luis Blanco, the late Justina Rodríguez Valenzuela, Socorro Guerrero Rodríguez, Gabriel Vega, Vicente Tajia, Frank Jim, Fillman Bell, Delores Lewis, and Francisco Suni for cross-cultural exchanges. Thanks also to Mary Beck Moser, Amadeo Rea, Mercy Vaughn, Richard Felger, Betsy Wirt, Craig Ivanyi, and especially Ana Lilia Reina Guerrero for additional information from her field notes, translation, and interpretation. This work was partially funded by the DeGrazia Foundation, the Pew Scholars on Conservation and Environment Program, the Amazon Conservation Team, Agnese Haury, Evelyne and David Lennette, and the Desert Tortoise Council.

LITERATURE CITED

Bahr, D., J. Gregorio, D. López, and A. Alvarez. 1973. *Pima shamanism and staying sickness* (ka:cim mumkidag). University of Arizona Press, Tucson.

Beals, R. L. 1945. The contemporary culture of the Cahita Indians. *Smithsonian Institution Bureau of American Ethnology Bulletin* 142:1–244.

BioSystems, Inc. 1994. *Life on the edge: A guide to California's endangered natural resources— Wildlife.* C. G. Thelander, ed. Heyday Books, Berkeley, Calif.

Crawford, J. M. 1989. *Cocopa dictionary.* University of California Press, Berkeley.

Felger, R. S., and M. B. Moser. 1985. *People of the desert and sea.* University of Arizona Press, Tucson.

Felger, R. S., M. B. Moser, and E. W. Moser. 1981. The desert tortoise in Seri Indian culture. *Proceedings of the Desert Tortoise Council Symposium* 1981:113–19.

Green, R. 1981. Culturally-based science: The potential for traditional people, science and folklore. Pp. 204–12 *in* M. Newall, ed. *Folklore in the twentieth century.* Rowman and Littlefield, London.

Malkin, B. 1962. Seri ethnozoology. *Occasional Papers of the Idaho State Museum* 7:1–68.

Martin, P. P. 1992. *Songs my mother sang to me: An oral history of Mexican-American women.* University of Arizona Press, Tucson.

Nabhan, G. P. 1990. El papel de la etnobotánica en la conservación de recursos fitogenéticos en reservas de la biósfera. *Bio Tam* 1:1–4.

Nabhan, G. P., and S. St. Antoine. 1993. The loss of floral and faunal story: The extinction of experience. Pp. 219–30 *in* S. Kellert and E. O. Wilson, eds. *The biophilia hypothesis.* Island Press, Covelo, Calif.

Orlove, B. S., and S. B. Brush. 1996. Anthropology and the conservation of biodiversity. *Annual Reviews of Anthropology* 25:329–52.

Pennington, C. W. 1979. *The Pima Bajo of central Sonora, Mexico.* Vol. II. University of Utah Press, Salt Lake City.

Pyle, R. M. 1993. *The thunder tree.* Houghton-Mifflin, New York.

Reyes O., S. 1979. Aspectos biológicos de la tortuga del desierto *(Gopherus agassizii)* en la Isla Tiburón, Sonora. Pp. 142–53 *in Memorias del IV Simposio Sobre el Medio Ambiente del Golfo de California.* Publicación Especial no. 17, Instituto Nacional de Investigaciones Forestales, Hermosillo, Sonora.

Reyes O., S., and R. B. Bury. 1982. Ecology and status of the desert tortoise *(Gopherus agassizii)* on Tiburón Island, Sonora. Pp. 39–49 *in* R. B. Bury, ed. *North American tortoises: Conservation and ecology.* Wildlife Research Report no. 12, U.S. Fish and Wildlife Service, Washington, D.C.

Rosenberg, J. 1997. Curriculum development for the Seri ethnozoology environmental education project. M.S. thesis, University of Arizona, Tucson.

Rosenberg, J., P. Romero, G. P. Nabhan, and H. E. Lawler. 1997. *Los Animalitos del Desierto y del Mar.* Arizona-Sonora Desert Museum, Tucson.

Russell, F. 1983. *The Pima Indians.* 1908. Reprint, University of Arizona Press, Tucson.

Saxton, D., and L. Saxton. 1973. *O'othham Hoho'ok A'agitha: Legends and lore of Papago and Pima Indians.* University of Arizona Press, Tucson.

Schneider, J. 1996. *The desert tortoise and early peoples of the western deserts.* Special Report, Desert Tortoise Preserve Committee, Riverside, Calif.

Spier, L. 1970. *Yuman tribes of the Gila River.* Cooper Square, New York.

Tuxill, J., and G. P. Nabhan. 1998. Plants and protected areas: A guide to *in situ* management. *In World Wildlife Fund people and plants manual.* Stanley Thornes, London.

Valenzuela K., M. 1977. *Yoeme: Lore of the Arizona Yaqui people*. Sun Tracks Series. University of Arizona Press, Tucson.

Williams, N. M., and G. Baines. 1993. *Traditional ecological knowledge: Wisdom for sustainable development*. Centre for Resource and Environmental Studies, Australian National University, Canberra.

Contributors

Pamela J. Anning, Saguaro National Park, Tucson, Arizona

Roy C. Averill-Murray, Non-Game Branch, Arizona Game and Fish Department, Phoenix

Scott Jay Bailey, Natural Resources Department, Tohono O'odham Nation, Sells, Arizona

Sheryl L. Barrett, U.S. Fish and Wildlife Service, Tucson, Arizona

Alberto Búrquez M., Instituto de Ecología, Universidad Nacional Autónoma de México, Hermosillo, Sonora

R. Bruce Bury, Forest and Rangeland Ecosystem Science Center, U.S. Geological Survey, Corvallis, Oregon

Michael J. Demlong, Non-Game Branch, Arizona Game and Fish Department, Phoenix

James C. deVos, Research Branch, Arizona Game and Fish Department, Phoenix

Vanessa M. Dickinson, Research Branch, Arizona Game and Fish Department, Phoenix

Todd C. Esque, Western Ecological Research Center, Las Vegas Field Station, U.S. Geological Survey, Las Vegas, Nevada

David J. Germano, Department of Biology, California State University, Bakersfield

Peter A. Holm, Luke Air Force Base, Luke, Arizona

Jeffrey M. Howland, Santa Ana National Wildlife Refuge, Alamo, Texas

Craig S. Ivanyi, Arizona-Sonora Desert Museum, Tucson

James L. Jarchow, Sonora Animal Hospital, Tucson, Arizona

Trip Lamb, Department of Biology, East Carolina University, Greenville, North Carolina

Howard E. Lawler, El Tigre Journeys, Iquitos, Peru

Brent E. Martin, Department of Ecology and Evolutionary Biology, University of Arizona, Tucson

Robert D. McCord, Mesa Southwest Museum, Mesa, Arizona

Ann M. McLuckie, Utah Division of Wildlife Resources, St. George

David J. Morafka, Department of Biology, California State University–Dominguez Hills, Carson

Gary Paul Nabhan, Center for Sustainable Environments, Northern Arizona University, Flagstaff

Michelle J. Nijhuis, *High Country News*, Paonia, Colorado.

Olav T. Oftedal, Department of Conservation Biology, National Zoological Park, Smithsonian Institution, Washington, D.C.

F. Harvey Pough, Department of Life Sciences, Arizona State University West, Phoenix

James C. Rorabaugh, Arizona Ecological Services Field Office, U.S. Fish and Wildlife Service, Phoenix

Cecil R. Schwalbe, Western Ecological Research Center, Sonoran Desert Field Station, U.S. Geological Survey, University of Arizona, Tucson

Ellen M. Smith, Department of Life Sciences, Arizona State University West, Phoenix

Mark H. Trueblood (deceased), Apollo Animal Hospital, Glendale, Arizona

Thomas R. Van Devender, Arizona-Sonora Desert Museum, Tucson

Elizabeth B. Wirt, Luke Air Force Base, Luke, Arizona

A. Peter Woodman, Kiva Biological Consulting, Inyokern, California

Index

Page numbers in *italics* indicate illustration on that page.

About the Editor

Tom Van Devender is the Senior Research Scientist at the Arizona-Sonora Desert Museum in Tucson, a position that he has held since 1983. Originally from Texas, he completed his Ph.D. at the University of Arizona in 1973 with a doctoral dissertation entitled *Late Pleistocene Plants and Animals of the Sonoran Desert: A Survey of Ancient Packrat Middens.* Over the next two decades, he used remarkably well-preserved fossils to reconstruct biotic communities and paleoenvironments for the last 50,000 years in the Sonoran and Chihuahuan Deserts. His lifelong interest in amphibians and reptiles led to studies of their fossils from middens and sediments from Arizona, Chihuahua, New Mexico, Sonora, and Texas as well as the diet of the Sonoran desert tortoise. Since 1990 Tom has studied local floras in eastern and southern Sonora and Baja California. He has authored more than one hundred publications in scientific journals and books. Tom coedited *Packrat Middens: The Last 40,000 Years of Biotic Change* (1990), *The Desert Grassland* (1995), and *Gentry's Río Mayo Plants: The Tropical Deciduous Forest and Environs of Northwest Mexico* (1998) and coauthored *Cactáceas de Sonora: Su Diversidad, Usos y Conservación* (2000) and *Mayo Ethnobotany: Land, History, and Traditional Knowledge in Northwest Mexico.*